CAMBRIDGE LIBRARY COLLECTION

Books of enduring scholarly value

Botany and Horticulture

Until the nineteenth century, the investigation of natural phenomena, plants and animals was considered either the preserve of elite scholars or a pastime for the leisured upper classes. As increasing academic rigour and systematisation was brought to the study of 'natural history', its subdisciplines were adopted into university curricula, and learned societies (such as the Royal Horticultural Society, founded in 1804) were established to support research in these areas. A related development was strong enthusiasm for exotic garden plants, which resulted in plant collecting expeditions to every corner of the globe, sometimes with tragic consequences. This series includes accounts of some of those expeditions, detailed reference works on the flora of different regions, and practical advice for amateur and professional gardeners.

Journal of a Tour in Marocco and the Great Atlas

This 1878 account of a scientific tour of Morocco and the Atlas mountains in 1871 was compiled from the journals of Sir Joseph Hooker (1817–1911) and his travelling companion, the geologist John Ball (1818–89). Their plan had been for Hooker to publish their findings soon after the journey, but his work as Director of Kew Gardens and President of the Royal Society, and Ball's frequent absences abroad, as well as his own writing commitments, caused delays. However, they argue that their information is unlikely to be out of date when, from a comparison with earlier accounts, 'no notable change is apparent during the last two centuries'. The botanical and geological interests of both men take centre stage in an engaging narrative which provides interesting details about the government, customs and daily life in an area which even in the late nineteenth century was little visited by Europeans.

Journal of a Tour in Marocco and the Great Atlas

With an Appendix
Including a Sketch of the Geology of Marocco

JOSEPH DALTON HOOKER
JOHN BALL
GEORGE MAW

CAMBRIDGE
UNIVERSITY PRESS

CAMBRIDGE
UNIVERSITY PRESS

University Printing House, Cambridge, CB2 8BS, United Kingdom

Cambridge University Press is part of the University of Cambridge.

It furthers the University's mission by disseminating knowledge in the pursuit of
education, learning and research at the highest international levels of excellence.

www.cambridge.org
Information on this title: www.cambridge.org/9781108077651

This edition first published 1878
This digitally printed version 2015

ISBN 978-1-108-07765-1 Paperback

The original edition of this book contains a number of oversize plates
which it has not been possible to reproduce to scale in this edition.
They can be found online at www.cambridge.org/9781108077651

MAROCCO

AND

THE GREAT ATLAS

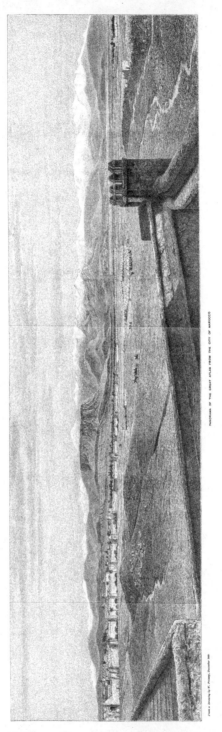

PANORAMA OF THE GREAT ATLAS FROM THE CITY OF MAROCCO.

From a Drawing by W. Prinsep, December 1825.

The material originally positioned here is too large for reproduction in this reissue. A PDF can be downloaded from the web address given on page iv of this book, by clicking on 'Resources Available'.

JOURNAL OF

A TOUR IN MAROCCO

AND

THE GREAT ATLAS

BY

JOSEPH DALTON HOOKER, K.C.S.I., C.B.

PRES. R. S.
DIRECTOR OF THE ROYAL GARDENS, KEW; ETC.

AND

JOHN BALL, F.R.S., M.R.I.A.

ETC.

WITH AN APPENDIX
including
A SKETCH of the GEOLOGY of MAROCCO, by GEORGE MAW, F.L.S., F.G.S.

CAPE SPARTEL

London

MACMILLAN AND CO.

1878

PREFACE.

THE EXPEDITION of which an account is given in the following pages was undertaken in the year 1871, and it was originally intended that a narrative of the proceedings should be given to the public soon after our return to England. Sir Joseph D. Hooker, who made careful notes throughout the journey, hoped to complete the work without much delay, and actually wrote the greater part of the first two chapters ; but the constant demands upon his time arising from his official duties at Kew, and the important botanical works to which he is a chief contributor, further increased by his election, in 1873, to the Presidency of the Royal Society, so far interfered with the completion of the original design as to compel him to request his fellow-traveller, Mr. Ball, to undertake the completion of the work. The latter was at the time engaged in preparing for publication a memoir on the Flora of Marocco, which has since appeared in the Journal of the Linnæan Society, wherein the botanical collections made during the journey are enumerated and described : and his performance of the task allotted to him has been further delayed by several prolonged absences from England.

As regards many countries visited by travellers a delay of several years in publication might seriously affect the accuracy of a narrative intended to represent the existing condition of the country and its inhabitants; but in the case of Marocco, where, from a comparison with the accounts of early travellers, no notable change is apparent during the last two centuries, the effect of a few years' interval may be considered insensible. Up to the date of our visit the Great Atlas was little better known to geographers than it was in the time of Strabo and Pliny; and it may be hoped that whatever interest belongs to our journey is as great now as it was at the moment of our return.

The narrative now published is mainly founded on the journals kept by Sir J. Hooker and Mr. Ball, supplemented in some particulars by that of our fellow-traveller, Mr. G. Maw. To the latter we owe a sketch of the Geology of Marocco, which appears in the Appendix. Along with this we have published some interesting contributions received from Mr. H. B. Brady and Mr. Freeman Rogers, as well as some papers upon various matters connected with the physical geography and the flora of Marocco.

It is impossible to present these pages to the public without repeating the expression of our obligations to some of those to whose assistance we largely owe whatever success we were able to attain. Foremost amongst these we must name H. E. Sir John Drummond Hay, K.C.B., British Minister Plenipotentiary in Marocco. From the moment when, in compliance with the request of Sir J. Hooker, Lord Granville, then Foreign Secretary, instructed our Minister to apply for the permission of the Sultan to visit the Great Atlas, Sir J. D. Hay, by his extensive knowledge of the country and the people, and by his great personal influence, afforded invaluable assistance to the expedition.

We were also much indebted for assistance and hospitality to the British Consular agents on the Marocco coast, and especially to the late Mr. Carstensen, then Vice-Consul at Mogador. We should not omit our acknowledgments of the courtesy and valuable information received from the late M. Beaumier, French Consul at the same port.

We trust that in the course of the following pages we have not omitted to express our thanks to other friends who have kindly contributed valuable information. The scope of this volume being mainly to give an account of our personal experience and observations, we have used, but sparingly, other materials, which might be in place if we had aimed at the production of a work of a more elaborate character.

CONTENTS.

CHAPTER I.

CHAPTER II.

CHAPTER III.

CHAPTER XII.

CHAPTER XIII.

CHAPTER XIV.

APPENDICES.

APPENDIX H.

APPENDIX I.

APPENDIX K.

APPENDIX L.

Errata.

P. 388, line 10, after 'Pharmacographia,' 502, insert Cosson, in Bull. Soc. Bot. Fr. xxi. 163.
,, 394,   ,,   4 from bottom, for 'Sus' insert *Sous.*
,, 395, lines 4 and 9 from top, for 'Sus' insert *Sous.*

ILLUSTRATIONS.

WOODCUTS IN TEXT.

MAP

JOURNAL

OF A

TOUR IN MAROCCO.

CHAPTER I.

On Saturday, April 1, 1871, our party, consisting of Sir
Joseph (then Dr.) Hooker, Mr. Maw, and Mr. Ball, with a
young gardener, named Crump, from the Royal Gardens
at Kew, left Southampton for Gibraltar, in the Peninsular
and Oriental Company's Steamship *Massilia*.

Even for the ordinary tourist it is a pleasant thing to
turn his face towards the South in the early part of the
year, and to feel that he is about to exchange six or eight
weeks of bitter easterly winds for the bright skies and
soft breezes of the Mediterranean region. Still more does
the botanist rejoice to quit the poverty of our slowly unfolding spring flora for the wealth of varied vegetation that
is spread around the shores of the Inland Sea. But for us,
the occasion was one of deeper and more special interest.
We were starting, under unusually favourable conditions,
to explore a country which, though close to Europe, is
among the least known regions of the earth. Although
the obstacles we were sure to encounter and the limited time
at our disposal, might not allow us to accomplish much,

we felt a confident hope that we should learn something of a great mountain chain all but absolutely unknown to geographers, and be able to fill up some missing pages in the records of our favourite science. The thrill of pleasurable anticipation at the prospect of setting foot within the boundaries of *terra incognita* was heightened by the fact that for each of us this land of Marocco had long been the object of especial interest and curiosity.

From an early period Hooker had conceived the desire to explore the range of the Great Atlas, to become acquainted with its vegetation, and to ascertain whether this supplies connecting links between that of the Mediterranean region and the peculiar flora of the Canary Islands. This desire was increased during a journey in Syria, in 1860, made in company with Admiral Washington, the late Hydrographer of the Navy, one of the very few Europeans who had reached the flanks of the Great Atlas chain, when, as a young naval officer, he accompanied the late Sir John Drummond Hay on his mission to the city of Marocco in 1829.

Maw had already made collections of living plants in the neighbourhood of Tangier, and had also visited Tetuan, where he had pushed his excursions farther than any but one preceding traveller.

Ball had landed at Tetuan in 1851 with the hope of attaining some of the higher summits of the neighbouring Riff Mountains; but the disturbed state of the country in that year made it impossible to advance beyond the immediate outskirts of the city.

From the moment when it seemed likely that the permission to visit the Great Atlas sought for by Hooker, through the intervention of our Foreign Office, would be accorded by the Sultan of Marocco, no time was lost in making the requisite preparations. Although everything was done within about a fortnight, our equipment was tolerably complete; and when, after the first excitement of departure had subsided, we thought it over on board ship, we found but one serious omission to deplore. Two mercurial barometers, provided by Hooker, had been entrusted to Crump, and were by him left behind at the last moment. Thus, in the important matter of determining heights, we were forced to rely upon aneroid barometers and boiling water observations. It was fortunate that Ball carried an excellent aneroid, by Secrétan of Paris, which has before and since been severely tested in the Alps with very satisfactory results, and whose indications during our journey agreed closely with those given by the thermometer in boiling water.

Among the various preparations made for our journey there was none more important for our purpose than a manuscript catalogue of all the plants hitherto known or believed to have been found in the Empire of Marocco, which we owed to the kindness of our excellent friend M. Cosson, the eminent French botanist. Up to that date the information to be found in books was extremely scanty, and scattered throughout various systematic works, and the whole when summed up would have given a most incomplete account of the two or three districts partially explored by botanists. M. Cosson, by his unequalled knowledge of the North African flora, and by careful study of all the collections made in Marocco, many of which are in his exclusive possession, was the only person who could have supplied the materials which were so serviceable throughout our journey.

In the agreeable society of old friends and new acquaintances, whom we met on board the rather crowded steamer, the voyage to Gibraltar did not appear too tedious, but we were well pleased when, on the afternoon of the 6th, the moment came for landing.

We were not destined to see much of the famous ' Rock ' or its native ' scorpions,' whether biped or hexapod. Scarcely had our voluminous baggage been transported to the hotel, when news reached us that an English steamer was about to sail within two hours for Tangier,

and we at once decided that not a moment's time should
be wasted.   Back again our heavy goods, in which botanical paper was a chief ingredient, were carried to the mole,
and after paying the innkeeper a pretty heavy ransom, on
account of rooms ordered but not used, and a hastily swal-
lowed dinner, we once more found ourselves afloat.   So
much haste was not necessary, for the steamer did not
start till some time after midnight; but the time was not
badly spent, for the steamer was one of those that ply
between London and the Canary Islands, touching at the
ports on the Atlantic coast of Marocco; and the skipper,
who was an old stager, and had formed his own opinions
about the country, had plenty of information, of a more
or less authentic, but mainly discouraging, character,
which he was most ready to impart.

The distance from Gibraltar to Tangier is not more
than thirty-five miles, and we came to anchor in the
open roadstead soon after daylight on April 7.   Unlike
the ports on the Atlantic coast, the shape of the land
here gives some protection from the prevailing westerly
seas and winds; but in other respects this is a bad one.
The ruined mole, round which sand has accumulated,
forms on one side a dangerous reef, and elsewhere the
shore shelves very slowly to a moderate depth.   Ships of
any burthen are forced to lie out far from shore, and the
landing from boats is usually effected on the backs of Jews,
inasmuch as no Moslem will degrade himself by perform-
ing such a service for a Christian.

On Good Friday the Jews were all engaged in the
ceremonies of the Passover; but, as the sea was unusually
calm, we were able to land on the ruins of the mole, and,
after floundering through slippery seaweed, we were not
long in reaching the sea gate of the city.

We had already perceived that, although no longer in
Europe, we were yet under the shadow of European manners and customs.   High above the city walls we espied,
as we neared the shore, several conspicuous inscriptions,

announcing the titles of various places of entertainment. In the centre the 'Hôtel de France' gave promise of culinary skill; but we preferred the 'Royal Victoria Hotel,' whose title, in quite gigantic letters, first attracted our notice, and which had been well recommended for cleanliness and comfort. Our subsequent experience justified the choice, and we had every reason to be satisfied with the attention we received from the intelligent and obliging coloured proprietor, Mr. Martin.

Tangier stands on the western side of a shallow bay, on rocky ground that rises steeply from the shore. Westward the hills gradually rise in swelling undulations towards the Djebel Kebir, or Great Mountain, covered with dwarf oaks and flowering shrubs, that ends in the promontory of Cape Spartel. On the opposite, or eastern, side the shores of the bay are low and sandy, but are backed by the rugged range of the Angera Mountains, culminating in the Ape's Hill opposite Gibraltar.

As seen from the sea the town has a singular, though not an imposing, appearance. Cubical blocks of whitewashed masonry, with scarcely an opening to represent a window, rise one above another on the steep slope of a recess in the hills that faces the NE. A few slender square towers belong to as many mosques of paltry proportions. Numerous consular flagstaffs remind the European that he still enjoys the protection of his own government, and on the summit of the hill a massive gaunt castle of forbidding aspect shows where he might expect to lodge if that protection were removed, and he were to give offence to the native functionaries. Zigzag walls encompass the city on all sides, pierced by three gates, which are closed at nightfall.

The stranger, who knows that Tangier is one of the most important towns of Marocco, and the residence of the representatives of the chief civilised States, is apt to be shocked when he first sets foot within its walls. The main street is as rough and steep as the most neglected of

Alpine mule-tracks, and disfigured by heaps of filth—
importunate beggars of revolting aspect, led about by
young boys, assail him at every step—there is no bazaar,
as in eastern towns, and the miserable shops are mere
recesses, where, in an unglazed opening, little larger than
a berth in a ship's cabin, the dealer squats surrounded by
his paltry wares.

On longer acquaintance, he will somewhat modify
his first unfavourable impression. Unlike the towns of
Southern Europe, where the main thoroughfares are cared
for by the local authorities, while filth is allowed to accumulate in the byeways, the dirt and offal are here let to lie
under his nose in the most public places, while the steep
narrow lanes—reminding him of Genoa—that intersect
the masses of closely packed houses, are generally kept
clean and bright with frequent whitewash. The silent
dead walls that front the public thoroughfares conceal the
interiors of houses that are rarely opened to the eyes of
Europeans, but are not wanting in the signs of wealth and
of artistic taste. The dread of arbitrary exactions, that
elsewhere in Marocco drives the Moor as well as the Jew
to conceal the possession of property as carefully as men
elsewhere hide the evidence of guilt, is less keenly felt
here. For in and around Tangier, but nowhere else in
this country, it may be said that life and property are
tolerably secure, not only from outward violence, but from
the caprice and cupidity of men in authority. The presence of foreign diplomatic agents, and the constant communication with Europe, have brought the Moorish
authorities at this spot to some extent under the control
of civilised opinion, and the disastrous encounters with
France and Spain have convinced the Moor that, with all
his personal bravery, he cannot resist the regular forces of
his European neighbours, and must not provoke an unequal conflict.

Such historical recollections as are connected with

Tangier are not flattering to the self-love of the two nations of Europe that have had most to do with it.

In 1437 the Portuguese, who then held Ceuta, attacked the town, but their army was defeated under the walls, and they were forced to conclude an ignominious peace. The terms included the cession of Ceuta to the Moors, and the delivery as a hostage of Dom Fernando, the king's brother. The other stipulations not having been executed, the victors threw Dom Fernando into prison at Fez, and when he died in captivity hung up his body by the heels over the city walls.[1]

The fortune of war was changed in 1471 when the Portuguese took Tangier and several of the towns on the Atlantic coast, and the Moorish Sultan was forced to pay tribute to King Emanuel. Under less vigorous guidance, the Portuguese were unable to retain their ascendancy, but they kept possession of Tangier till, after nearly two centuries, it was, by a secret treaty, ceded to England as part of the dowry of Catharine of Braganza on her marriage with Charles II. When the brave Governor Dom Fernando de Menezes received the information, he entreated the Queen Regent to spare him the grief of seeing the city made over to the enemies of the Catholic faith. Her answer was the offer of a Marquisate if he obeyed, and dismissal from her service if he persisted in resisting her will. He chose the latter, threw up his command; and devoted the rest of his life to writing a history of the city. The English Court set great store by the new acquisition, believing, as the Earl of Sandwich said, that if it were walled and fortified with brass it would yet repay the cost. But English policy was then at its lowest ebb, and neither vigour nor intelligence directed any branch of our affairs. The English settlers sent out were an ill-conditioned rabble, ignorant of the country, its

[1] This episode forms the subject of Calderon's noble play—*El Prencipe Costante.*

language and manners, and the Governor and the garrison
were no better than the rest. After accomplishing one
useful work by constructing a mole that converted the
roadstead into a secure harbour, they were disappointed
in their expectation of an extensive trade with the interior, and, what was more galling, were worsted in every
encounter with the Moors, till, in 1685, the Government
in London decided to abandon Tangier. When this became known at Lisbon, the Portuguese strongly urged the
impolicy of abandoning such a position to pirates, and
requested that it should be restored to them on condition
that the English should have free use of the port. With
characteristic meanness and imbecility the Duke of York
—soon afterwards James II.—opposed the gift, and urged
that the honour of England required that the place
should be dismantled, and be left for occupation to whoever could hold it. His advice prevailed; and, on the
retirement of the English force, the mole was effectually
blown up, destroying the only good harbour for shipping
on the seaboard of Marocco—a distance of fully nine
hundred miles.

Nature, however, has made Tangier the port of North
Marocco, and, in spite of human perversity, it is a place of
some importance. Ready access to the fertile provinces
lying between the Straits of Gibraltar and Fez has made
it the centre of a considerable trade in hides and grain,
which go to France and England, to say nothing of cattle
and other supplies for the garrison of Gibraltar. Its
nearness to Europe has made it the residence of the
representatives of the principal civilised Powers, and its
admirable climate has attracted invalids from Gibraltar
and elsewhere, in spite of such drawbacks as dirt, bad
smells, and the utter absence of roads.

On our arrival, we were most kindly received by Sir
John Drummond Hay, to whose intimate knowledge of
the country and justly acquired influence with the Moorish
Court we are largely indebted for whatever success at-

tended our journey. We learned from him that the
Sultan had issued orders to the Governor of the Atlas
provinces to allow Hooker to visit the range of the Great
Atlas south of the city of Marocco, and to take every
precaution for his comfort and safety; but he added
that, although there was no reason to doubt the Sultan's
good faith, every artifice would be used to defeat the
object, and that it would not be prudent to start for the
south without an autograph letter from the Sultan himself, for which he had already made application. The
Court was at this time at Fez—several days' journey from
Tangier; and, as business moves at a slow pace in this
country, it was probable that we might have to wait some
time for the necessary document. We therefore at once
decided on devoting the interval to excursions in the
neighbourhood of Tangier and Tetuan. The latter city
lies at no great distance from the lofty peaks of the *Beni
Hassan*, probably the highest part of the north-western
range of the Lesser Atlas, best known as the Riff Mountains. There could be no doubt as to the botanical
interest attaching to a visit to that range, the higher
region of which is entirely unknown to naturalists, and
we were very desirous to make an attempt in that direction. After full consideration, however, Sir J. D. Hay
felt it necessary to object to our project, as involving
undue risk. The Riff mountaineers enjoy a virtual independence, merely paying tribute to the Sultan. They
are fierce and fanatical; and the presence of a Christian
on the highest mountain, which is rendered sacred by a
famous *marabout*—tomb of a Mohammedan saint—would
be regarded as a profanation. Meantime, we were led to
hope that we should be able to ascend the mountains
nearer to Tetuan, and there was no difficulty whatever
about excursions in the neighbourhood of Tangier.

Our first walk, in the afternoon of the 7th, was in the
agreeable society of Sir J. D. Hay, to Ravensrock, his
summer residence, on the wooded slope of the Djebel

Kebir, overlooking the straits. Near the city gate we
passed the cemetery, where turbaned tombstones almost
disappear amidst the copious growth of prickly pear
(*Opuntia vulgaris*), and then went some way through
dusty lanes between lines of American aloe (*Agave ameri-
cana*), and quickset hedges surrounding gardens where
palms, acacias, and a few poplars were the prevailing trees.
As we cleared the enclosures, and got into irregular, open
ground, where steep slopes of uncultivated land alternate
with patches of tillage, our eyes were gladdened by the
sight of many a bright southern flower, already blossom-
ing abundantly, in spite of the weather which, till lately,
had been unusually cold. Trefoils, Medicagos, vetches,
and other leguminous plants were here the predominant
forms, as they are everywhere in the spring flora of the
Mediterranean region. As we began to ascend the flanks
of the Djebel Kebir, the character of the vegetation
changed. Where the ground has not been cleared to
make a garden for some of the European residents, whose
little villas are scattered over the slope, the ground is
covered with masses of luxuriant shrubs, and climbing
herbaceous plants, among which some familiar forms of
the North are mingled with many exotic species. Thus
we saw roses, brambles, bryony, honeysuckle, and white
convolvulus holding their ground amidst masses of lentisk,
myrtle, Phillyrea, Alaternus, dwarf prickly oak (*Quercus
coccifera*), gum cistus, and the golden profusion of five or
six species of the *Cytisus* tribe that replace our native
broom and gorse. After ascending several hundred feet
by the roughest of paths, carried along a shaded gully, we
entered through a gate the terraced garden whereon stands
the house.

Nothing of its kind can surpass the beauty of the view.
The steep slope below is planted with oranges and pome-
granates—the first laden with golden fruit, the second with
crimson flowers—broken here and there by palms, figs,
olives, and carob trees, standing against a background of

deep blue water, dancing in the gentle westerly breeze. On our left the steep slope of the mountain, rising over against the blue outline of Cape Trafalgar, forms the portal through which the Atlantic pours its current into the Mediterranean. Along the opposite shore of Spain every undulation, from the coast to the distant purple sierra, is plainly seen. The little town of Conil and the very houses of Tarifa are discernible with the naked eye, and visitors are enabled through a glass to watch the people as they come and go, and that extraordinary phenomenon for Southern Spain, the diligence, that of late years has plied between Algeciras and Cadiz. Turning to the right, the eye reaches the entrance to the Mediterranean, between the rock of Gibraltar and the loftier summit of Ape's Hill; and in clear weather the range of the Serrania de Ronda, stretching towards Malaga, is seen on one side, while on the other the snowy peak of the Beni Hassan, south of Tetuan, closes the view. To give variety, if that were wanting, there is the ceaseless passage of shipping through this greatest of maritime highways, in a double stream of vessels, of every size and every nation, from the great Peninsular and Oriental steamer to the Moorish felucca. It is an example of the readiness with which sound travels over an unbroken surface, that the morning and evening gun at Gibraltar, nearly forty miles distant, are usually heard at this spot.

In the course of several delightful evenings passed in the agreeable society of Sir J. D. Hay and his family, we obtained much curious and valuable information respecting the country and its inhabitants, most of which was confirmed by our own subsequent observation and experience. We already knew that Marocco is the China of the West, and that while other Mohammedan States have been drawn, though at a tardy and halting pace, into following the general movement of European progress, this has remained more isolated and more impenetrable than even the Celestial Empire itself. But we were scarcely pre-

pared to find that the utmost excesses of barbarism are
matters of daily occurrence in a country so close at hand;
and though we had read startling statements in the books
of preceding travellers, and heard confirmatory tales during
our stay in North Marocco, we were inclined to think that,
at the worst, these referred to solitary acts of cruelty,
probably magnified by the proverbial tendency to exaggerate all that is strange and horrible.   It was not until we
had spent some time in the southern provinces, beyond
the reach of European prying observation, that we could
persuade ourselves that these terrible stories of cruelty and
wrong merely give a true representation of the ordinary
condition of the country.   Sir J. D. Hay, who probably
knows it better than any other European, was not slow to
testify to the good qualities of the rural population of
Marocco, and the general absence of crime.   We were
afterwards led to believe that if life and property may be
said to be tolerably secure throughout the portion of the
empire really subject to the Sultan's authority, this is due
rather to the fact that temptation is rare, and the danger
of swift and bloody retribution imminent, than to the
existence of any high moral standard among the people.
It is a strange inversion of all notions of government, that
crime should come from above rather than below, and that
the dread that men feel for the safety of their persons and
goods is directed rather to the constituted guardians of
order than to the outcasts from society.   The first feeling
of one unused to a barbarous government is surprise that
it should be allowed even to exist, much more that it
should possess considerable stability, and be handed on
from one generation to the next, without a general outburst
of resistance.   Observation tends to explain this seeming
enigma.   Bad as it may be, the oppression exercised by
the few strikes only those who are in some way conspicuous.
The common mass, who offer no special temptation to
extortion, escape comparatively unhurt, and feel little
sympathy for the victim.   Accordingly it is only when a

Sultan or a Governor indulges in mere gratuitous acts of cruelty against his humbler subjects, that we hear of a general revolt. Oppression is, after all, less intolerable than anarchy; and at that very time most men would have chosen to live in Marocco rather than in Sicily.

Among other objects of interest Sir J. D. Hay showed us a coloured view of the Great Atlas range, as seen from the neighbourhood of the city of Marocco, executed at the time of his father's mission to that city in 1829, and this naturally engaged our special attention.[1] The most singular point in the structure of the mountains was a very long range of what were represented as precipitous rocks of seemingly uniform height and structure, that appeared to rise abruptly from the plain, and to form an almost continuous outer wall or rampart on the north side of the chain. We were also shown a copy of Hollar's[2] rare engraving, representing Tangier at the period of the English occupation, with the soldiers of Charles II., in their cumbrous uniforms, strutting on the mole.

Those who have read his interesting and lively little work, ' Morocco and the Moors,' will not be surprised that so keen a sportsman and close an observer of the habits of wild animals as our host should have many curious anecdotes to tell; but we were not prepared to hear that less than twenty-five years before a lion had been killed close to the spot where his beautiful villa now stands. At the

[1] A copy of this view is given in the frontispiece.

[2] Wentzel Hollar (or Hollard), a native of Prague, was sent to Tangier in 1669 by the king to take views of the town and its fortifications, which he afterwards engraved. Being one of the most distinguished engravers of the time, he settled in England, and executed some 2,400 prints, chiefly etchings, which are remarkable for their spirit, freedom, lightness, and finish. Hollar was one of the most conscientious of men; he worked for the booksellers at the rate of 4d. an hour, and always with an hour-glass on his table, which he invariably laid on its side, to prevent the sand from running, when not actually at work with his pencil or graving tool, and even when conversing on his business with his employers. He is said to have died in great poverty, with an execution in his house and a prison in prospect.

present time no animal of prey larger than a jackal is seen
in this part of the country, but the wild boar is as abundant there as it is everywhere throughout Marocco. No
doubt the religious scruples that forbid the use of the
flesh have gone far to prevent the natives from reducing
the numbers of these mischievous brutes. One anecdote
in favour of an animal whose moral character stands in
low repute may here be permitted.

Sir J. D. Hay had brought up a young leopard in his
house until the animal had reached his full size and
strength, and it seemed a scarcely safe companion for the
younger members of his family. He therefore resolved to
present it to the Zoological Gardens in London, where it
was duly installed. Some two years later, when on a visit
to England, its former master bethought him of the
leopard, and, going to the gardens, recognised the animal
and spoke to him in Arabic. The once familiar sounds
immediately awoke the animal's memory, and it at once
displayed the appearance of unbounded, but joyous,
excitement. On explaining the circumstances the cage
was opened, and the animal showed the utmost delight at
the approach of its early friend and master.

On the night of Easter Sunday, while enjoying the
cool air and the view from the roof of the British Resi-
dency, we beheld that grand display of the Aurora Borealis,
which was visible at the same time throughout Western
Europe. As in the equally brilliant auroras of the preceding autumn, which the popular imagination in many
different parts of Europe had attributed to the burning of
Paris, the characteristic feature of this display was the
pale flickering crimson tinge that rose from the northern
and western horizon towards the zenith. Brilliant auroral
phenomena are rarely seen in so low a latitude as Tangier;
but thirty-two years earlier Hooker had beheld them from
a still more southern station, during the visit of the
Antarctic Expedition to Madeira in 1839, as described by
Sir James Ross in the narrative of that voyage.

We were much impressed by the accounts we received of the remarkable salubrity of the climate of North Marocco, and we gathered abundant evidence to the same effect in regard to other parts of the territory. Nothing is more rare than to find a country where neither the natives nor foreign visitors have any complaint to make against the climate, and in that respect Marocco is almost unique. As regards the season of our visit, however, our case was that of nearly all travellers in whatever country they may find themselves. We had arrived in an exceptional season! How often is this fact gravely stated as something remarkable and unusual in the experience of the narrator, whereas, if he would but reflect, it merely represents the common experience of mankind in most countries of the earth! Excepting some portions of the equatorial zone, where the seasons recur with tolerable constancy, our notions of the climate of a place are got at by taking an average among a great many successive seasons. It is true that our own islands afford an extreme instance of variability; but elsewhere in the temperate zones of both hemispheres, the difference between corresponding seasons in successive years is often very great. Any one who watches the meteorological notices published in our newspapers, must be aware that if any particular day, week, or month be compared with the general average for the same period during a long term of years, he will find it to be either considerably hotter, or colder, or drier, or moister than the corresponding average day, week, or month; and when registers shall have been kept for a sufficient time in other countries, the same result will be seen to hold good, though in a somewhat lesser degree. Travellers will then be prepared to find that they should expect to enjoy or suffer from an exceptional season, and will think it more remarkable when they happen to alight on a season that approaches near to the average. That preceding our visit had been unusually severe; snow had been seen at Tangier, and had lain for some hours on the

rock of Gibraltar, and, as a consequence affecting the
object of our journey, the spring vegetation in North
Marocco was unusually retarded. At the same time, so
far as our sensations went, nothing could be more agreeable than the climate of this season, the thermometer
in the shade during the day varying from 60° to 66° Fahr.,
and the air being delightfully clear and bracing.

On April 8 we started for a short excursion to the
headland of Cape Spartel. In the immediate neighbourhood of Tangier Europeans may safely walk or ride unattended; but, as we were going a little beyond the ordinary
limits, it was considered prudent to give us the escort of
two soldiers, and to these we added a baggage mule and
a native guide. In a botanical sense we were about to
travel over beaten ground—the only spot in all Marocco
where a naturalist can without difficulty wander at will
over rocky hills that retain their natural vegetation. The
little that was then known of the flora of the empire
would have dwindled to a scanty list if we had struck out
the rich collections that successive botanists during the
last 100 years have brought from the Djebel Kebir and
the adjoining hilly district west of Tangier. Although
there was little prospect of new discovery, the expedition
could not fail to offer a veritable feast to a botanist, and
especially to one not already familiar with the vegetation
of the opposite coast and the adjoining region of southern
Portugal.

After standing the fire of some harmless 'chaff' from
the Jew and Moorish boys that loitered about the city
gate, we soon got clear of the enclosures near the town, and
descended through cultivated land into a little grassy valley that lies below the hilly range of the Djebel Kebir.
Bright spring annuals—blue and yellow lupen, crimson
Adonis, a deep orange marigold (*Calendula suffruticosa*),
blue pimpernel, and other less conspicuous flowers—enlivened the tillage ground; but the northern botanist is
more struck by the perennial species that hold their ground

on the large portion of the soil which the plough has not touched. Predominant among these, as elsewhere throughout a large part of the Mediterranean region, is the palmetto, or dwarf palm (*Chamœrops humilis*). Where unmolested by animals, and protected from the periodic fires that the native herdsmen renew for the sake of getting herbage for their cattle, it forms a thick trunk, ten or twelve feet in height, which probably takes a long time to attain its full size; but in the open places it is commonly stemless, and covers the ground with its radiating tufts of stiff fan-shaped leaves. Many plants of the lily tribe abound ; but in this mild climate most of them had flowered in winter, and few now showed more than their tufts of large root-leaves. Most conspicuous is the large maritime squill (*Scilla maritima* of Linnæus). The flowers are not large or showy, and do not correspond with the size of the bulb which often equals that of a man's head. Another species of the same genus (*Scilla hemisphærica*) is more ornamental, as are the two common asphodels. The slender iris (*I. Sisyrhynchium* of Linnæus), whose delicate flower lasts only a few hours—opening one at a time on successive days, appearing about mid-day and withering in the afternoon—is very abundant.

On reaching the hollow ground, where a slender stream runs through damp meadows, we were charmed by the delicate tint of a pale blue daisy that enamels the green turf. It is merely a slight variety of the little annual daisy (*Bellis annua*), so common in many parts of Southern Europe ; but the blue tint does not seem to have been noticed elsewhere. The larger blue daisy, afterwards seen as one of the ornaments of the mountain region of the Great Atlas, was at first supposed to belong to the same species; but, besides that this is perennial, it shows other less obvious differences.

It was on the slopes of the Djebel Kebir, where the stony ground is almost exclusively occupied by a dense mass of small shrubs, few of them rising more than three

or four feet from the ground, but nearly all covered with
brilliant flowers, that we first began to seize the really
characteristic features of the North Marocco flora.   A
great variety and abundance of flowering perennials of
shrubby habit is, indeed, a distinguishing feature of the
whole Mediterranean region ; but very little observation
was needed to show that we were here in that well marked
division that includes Southern Portugal, South-western
Spain, and the opposite corner of Africa.   This may be
called for distinction the Cistus and Heath region ; for
though most of the same kinds of Cistus and Helianthemum
extend as far as the south of France, and many species of
heath inhabit the Atlantic coasts of Europe as far north
as Connemara, it is only here that both these tribes
flourish together, and give a prevailing character to the
vegetation.   Most conspicuous of all is the gum-cistus
(*C. ladaniferus*), which in the Sierra Morena and the
adjoining parts of Spain and Portugal obtains such predominance that for twenty miles together one may ride
through a continuous thicket where the peculiar scent of
the gum that covers the leaves and young branches is
never absent.   About Tangier the rich purple spot that
usually adorns the base of the large petals is wanting, and
the flowers show unmixed snowy white.   Of the same
tribe, besides several true *Cisti*, there are many species of
Helianthemum.   Of heaths, along with the commoner
kinds (*Erica arborea* and *E. scoparia*), we saw in abundance the rarer and more characteristic forms, *E. australis*
and *E. umbellata*. *E. ciliata*, one of our English rarities,
is here very scarce, though it grows on the opposite side
of the Strait.   Our common heather (*Calluna vulgaris*)
still holds its ground, but in a poor and stunted condition.
The rhododendron of the East (*Rh. ponticum*), that is at
home in the mountain region of Asia Minor and Syria,
and which strangely reappears here and there among the
low hills between Tarifa and Algeciras, on the north side
of the Straits, has not been found on the African shore ; but

until the coast between Tangier and Ceuta has become more accessible, it will not be safe to assume that it is wanting. Among the many shrubby leguminous plants whose flowers give the prevailing golden tint to the hill sides, two of the Broom tribe (*Genista triacanthos* and *Cytisus tridentatus*), plants of very peculiar aspect and characteristic of this region, attracted our attention. It is impossible to omit another ornament of the hills—a plant rather widely diffused but nowhere common (*Lithospermum fruticosum*), whose azure blue flowers formed a charming contrast with the surrounding masses of golden colour.

The botanical district to which the northern corner of Marocco belongs has been already called that of the Cistus and Heath, but no single species of those tribes exactly conforms to the limits above pointed out. There are, however, several less conspicuous plants whose distribution more closely agrees with those limits. The most singular of these is the *Drosophyllum lusitanicum*, a plant of the sun-dew tribe, whose branched stem bears several large yellow flowers. The numerous slender strap-shaped root-leaves, nearly a foot in length, that are gradually contracted to the thickness of whipcord, are beset with pellucid ruby-tipped glands, and present a peculiarity that appears to be unique in the vegetable kingdom. Any one who has remarked the growth of ferns must have seen that in the young state the leaves are rolled or curled inwards, so that in the process of unfolding the face or upper side of the leaf, which was at first concealed, is gradually opened and turned to the light. A similar process occurs in many other plants ; but in *Drosophyllum* alone, so far as we know, the young leaf is rolled or curled the reverse way, so that the upper side of the leaf is that turned outwards. It appears to grow in many parts of Southern Portugal ; reappears on the north side of the Straits of Gibraltar near Tarifa and Algeciras, and on the southern side about Cape Spartel and on the hills above Tetuan, where it commands a view of the opening of the Mediterranean, but extends no farther eastward. Very similar is the distribution in Europe of two ferns whose natural home seems to be in the Canary Islands—the graceful *Davallia canariensis*, and the *Asplenium Hemionitis* of Linnæus. Both occur here and there in shady spots, from the rock of Lisbon to Algeciras and Tangier, but are unable to travel eastward beyond the Pillars of Hercules.

The scarcity of trees in this country is mainly due to the mischievous interference of man. The same ignorant greed of the herdsman, who to procure a little meagre herbage for goats sets fire to wide tracts of brushwood, that has reduced whole provinces of Spain to a nearly desert condition, has been equally busy and equally effectual in Marocco. The evergreen oak, which might produce much valuable timber, is the chief indigenous tree of this country; but, except on the rocky western declivity of the hill above Cape Spartel, few here arrive at a moderate growth, and the same is true of the Portuguese oak (*Quercus lusitanica*). The latter, indeed, never attains a considerable stature; but, where preserved from damage, it forms thickets some twenty or thirty feet in height, and, if duly protected, would help to preserve the hilly districts of this region from being annually parched by the summer sun. One of the shrubby evergreen oaks of this country (*Quercus coccifera*, L.), whose dark green spiny leaves are more like those of a holly than of an ordinary oak, might perhaps be successfully introduced in the south-western parts of the British islands. Its very dense foliage would make it valuable as a screen, and it produces a good effect when mixed with other shrubs.

Although the distance did not exceed ten or twelve miles, we had so much to do in filling our tin boxes and portfolios that the sun was sinking in the Atlantic as we reached the lighthouse at Cape Spartel. It is impossible not to feel some interest in this structure that for so many a mariner marks the limit of the great continent, more than three times the area of Europe, that remains, in spite of all the efforts of modern enterprise, the chief home of all that is strange and mysterious and unknown in the world. It represents, too, the only concession that the Moor has made to the demands of modern civilisation; for the building has been raised at the cost of the Sultan of Marocco, though the expense of its maintenance is shared between the four Powers, England, France, Italy, and Spain. The representatives of these States at Tangier form a board of management, and each in turn undertakes the actual control and inspection of the building. It was by an especial favour, and on the ground of our scientific pursuits, that we received permission from the Spanish Consul-General, then Acting Commissioner, with the concurrence of his colleagues, to lodge for the night within the building. It stands on a rocky platform some 250 feet above the sea. The massive tower, or pharos, that bears the lantern, is about eighty feet in height, and, with the annexed building, is enclosed by a strong wall, forming an outer court. The interior of the building is singularly picturesque. An inner octagonal court, surrounded by pillared arcades, supported on round, slightly stilted arches, with a fountain of cool spring water in the middle, gives access to the rooms, small and bare but perfectly clean, of which three were given for our accommodation. Some fowls and eggs supplied by the lighthouse-keeper, eked out by the provisions we had carried from Tangier, produced an excellent supper, and the evening was fully employed till a late hour in arranging and laying out the spoils of our first day's work in Marocco. It was near midnight when, before turning in for the night, each in turn paused in the court to enjoy the exquisite beauty of the scene. The full southern moon poured a flood of silver light through the arched spaces, converting the pattering spray-drops of the fountain into pearls and diamonds. The shadows of the slender columns lay like bars of ebony on the white flags; while, for a roof, the Great Bear, every star twinkling its brightest, stretched upward towards the zenith. The great tower rose in dark shadow, for the lantern was turned away from us; but we could discern, streaming out to seaward, in spite of the apparent clearness of the air, two faintly marked cones of yellow light that were soon quenched in the moonlight. The air was still, the sea was quiet, and at first the silence seemed unbroken; but as the listener stood, the pulses of the great ocean, though they smote but gently the cavernous rocks below, beat distinctly on the ear, and marked the passing minutes.

We rose betimes next morning, finding fresh enjoyment in each breath that we drew of the delicious air, and after breakfast set out for a walk southward along the coast. For the first two or three miles the rocky ground sloped downward towards our right, and finally fell steeply to the beach. It was apparent that the season was not quite advanced enough to enjoy the full beauty of the flora, but we found, besides the *Drosophyllum* already mentioned, many interesting forms. Orchids were not so abundant as they usually are at this season in the warmer part of the Mediterranean region. *Plantanthera diphylla*, growing in shady spots, was the only uncommon species.

An indentation of the coast marks the spot where a slender stream descends to the sea through a stretch of white sand; and beyond this the rocky coast rises but slightly above the sea level. Our steps were directed towards the so-called Cave of Hercules. This was originally a mere hollow in the face of the sea cliff; but from a remote period of antiquity it has been quarried for the purpose of extracting the hand-mills universally used in this part of Marocco. These, which are quite the same as the Scotch *querns*, are cut out in the rudest way by hammer and chisel, leaving the surface of the rock marked by a series of circular indentations about eighteen inches in diameter. In this way the original dimensions of the cave have been greatly enlarged, and, as it is still worked for the same purpose, the process is sure to be continued. In connection with the question raised of late years as to change of relative level of land and sea within the historic period, we observed some very ancient markings that showed the works to have been carried somewhat below the present level of high tide; but we could trace none that appeared to reach so low as that of the ebb tide.

So far as the evidence at this point goes, it seems to prove a slight amount of submergence during the period for which the rock has here been quarried. This period may probably be reckoned at 2,000 years, and possibly much exceeds that limit. Taken in connection with still existing remains in Greece, Asia Minor, the Phœnician coast of Syria, and Egypt, it tends to show that the changes in the general level of the Mediterranean coasts, indicated by many geologists, must have proceeded very slowly during the historic period, and that the more considerable oscillations, that have undoubtedly occurred near Naples and on the east coast of Sicily, have been mainly due to the local influence of volcanic action.

The soil near the cave was much mixed with sand carried by the wind, and the plants seen were chiefly widely diffused species that find tolerably uniform conditions of life on the sandy shores of the west coast of Europe. The rocks near the cave produce samphire and the sea fern (*Asplenium marinum*), just as they do in Cornwall; while *Diotis maritima* and *Lotus Salzmanni*, a local variety of the widely spread *Lotus creticus* of Linnæus, were frequent on the sands. The chief ornament was *Statice sinuata*, whose delicate azure flowers were already in blossom, long before most of the species of that late-flowering genus.

Our course now lay inland; but, instead of following the direct way back to Tangier, we were led by a false report (our first experience of blundering interpretation of English by the help of Moorish Arabic) to bear to the left, and recross the Djebel Kebir, so as to take Sir J. D.

Hay's villa of Ravensrock on our way back to the town. Near the track we passed close to a native village, or *douar*, the first which we had seen. When we had heard that the native population is broadly distinguished into two classes by the fact that some retain their original nomadic habits so far as to live permanently in tents, moving from one spot to another during the course of the year, while the others live in houses, and have become rooted to the soil, it never occurred to us that there could be any difficulty in distinguishing between one class and the other with the help of such obvious characteristic marks. But we soon found that the difference is but slight, and not very apparent. The black camel's hair tent is often, both in seeming and in fact, a more durable dwelling than the miserable huts, composed chiefly of slender branches to which the dried leaves still adhere, covered sometimes with brown straw, and oftener with some tattered fragments of cloth, the remains of worn-out garments. Only the mountain tribes, the descendants of the ancient Bereber stock, whose southern descendants we were to become acquainted with in the valleys of the Great Atlas, have preserved the familiar use of stone masonry in this part of Africa. Laden with plants, and with appetites sharpened by our climb over the hill, we returned to our comfortable quarters at the Victoria Hotel. We did not pass over the very highest point of the Djebel Kebir; but an observation taken some sixty or eighty feet lower indicated an elevation of about 800 feet above the sea level.

CHAPTER II.

Start for Tetuan—Vegetation of the low country—Serpent charmers
—Twilight in the forest—The Fondak—Stormy night on the roof—
Breakfast on the hill—Riff Mountains—A Governor in chains—Fate
of high officials in Marocco—Valley of Tetuan—Jew quarter of the
city—Ascent of the Beni Hosmar—Vegetation of the Mountain—
A quiet day—Jewish population—Ride to Ceuta—Spanish campaign in Marocco—Fortifications of Ceuta—Return to European
civilisation—Spanish convict stations in Africa.

On April 10 we started, rather late, for Tetuan, leaving our tents and heavy baggage at Tangier. Our pompous
interpreter, Hadj Bel Mohammed by name, whose huge
blue spectacles seem to be permanent appendages of the
Victoria Hotel, we found forward and intrusive in manner,
and indolent and inefficient in action, and altogether of no
account as a companion to travellers. Of the two soldiers
who formed the escort—one recognised by his taciturnity
the inferiority of his position ; but the other by his quaint
appearance and jocular disposition afforded us much
amusement, if not much reliable information. This little
fellow is properly called Hadj Mohammed, but he seems
to be familiarly known among the English visitors to
Tangier by the name of Bulbo. There was nothing military about him, except a very long gun which, throughout
our journey, remained carefully covered up in an intricate
red cloth case. If by any chance his aid had really been
required, and such an unlikely suggestion were admitted
as that Bulbo would have done anything else than put
spurs to his horse and run away, he would have been
driven to beg the attacking party to give him a quarter
of an hour's delay to get ready for action.

The distance from Tangier to Tetuan is only about forty miles; but we decided on stopping for the night at the Fondak,[1] a solitary Moorish caravanserai, about thirty miles distant from Tangier. Hurrying past the accumulations of offal and filth that are shot over the seaward face of the city wall, and indulging in a ten minutes' gallop over the sandy beach, we left the seashore; and, after riding some way through deep sandy lanes, before long reached a stretch of low cultivated land that extends westward from Tangier to the hills that divide this from the neighbouring provinces of Laraish and Tetuan.

The season was not sufficiently advanced for the flowering of many seaside plants; but there was quite enough to rejoice the eyes of botanists who had escaped from the ghastly spring season of the North when the days grow longer, but only more dreary, and the bitter east wind parches and blasts the young leaves and blossoms that are tempted to their destruction by the mildness of our winter weather. As everywhere on the seaboard of Marocco, the great yellow chrysanthemum (*C. coronarium*), with florets varying in hue from orange to pale lemon colour, is conspicuous on sea banks, with several fine species of Heron's-bill (*Erodium*). In the sands a large purple-flowered *Malcolmia* (*M. littorea*) and many *Leguminosæ* already diversified the aspect of the vegetation; while robust *Umbelliferæ*, mingled with the familiar eryngo of our own shores, had as yet merely developed their showy leaves.[2] But the characteristic form which chiefly interests the stranger to this region is a grey leafless bush, with long pendulous whipcord-like branches waving in the breeze, that is common among the sandhills, and recurs elsewhere

[1] From this Moorish word the Spaniards have taken *Fonda*, the common designation for an inn of the better class; while it is more accurately preserved in the Venetian *Fondaco*—e.g. Fondaco dei Turchi, &c.

[2] These sandhills were revisited by one of our party in the month of June, and then supplied many interesting plants not seen during our first stay at Tangier.

in dry exposed situations. There is something sad in the meagre and drooping aspect of the plant that brings to mind those dismal mourning trinkets, wherein a lock of hair is made to form the effigy of a weeping willow. This is the R'tam of the Moors, whence botanists have formed the name *Retama* for a small group of brooms, containing a few nearly allied species, that are widely spread throughout the region extending from Spain to the Canary Islands. In the early spring our Tangier plant (*Retama monosperma* of Boissier) is covered with clusters of small white odoriferous flowers. These had nearly all disappeared, and were succeeded by little hard one-seeded pods, which in some of the varieties ultimately become thick and fleshy, and are much sought after by birds. Not uncommonly the slender branches are laden with clusters of a small species of Helix that at some distance might be taken for fruit.

Without halting, except at one spot to secure some specimens of the great onion (*Allium nigrum* of Linnæus), we rode pretty fast through the belt of cultivated land that lies between the shore and the hills. The agriculture of this country has probably undergone little change since the earliest historic period. The plough in daily use is the same that is figured on the monuments of ancient Egypt, and with two exceptions the crops are the same—barley, wheat, lentils, vetches, flax, and pumpkins. America has supplied two valuable articles of food—maize and potatoes—and two exotic plants that have become so common as to modify the appearance of many localities—the *Agave*, or American aloe of British greenhouses, and the Indian fig (*Opuntia vulgaris*)—both extensively used for hedges, and multiplying freely on waste ground. The last-named plant contributes to the scanty dietary of the natives; but the fruit, when eaten in any quantity, is said to be indigestible, and a potent ally to diarrhœa and dysentery. On reaching the hills, of which we merely crossed some low spurs, the aspect of the vegetation became more varied. The dominant plants were still those we had seen in similar situations about Tangier— the palmetto (*Chamærops humilis*), the great branched asphodel (*Asphodelus cerasiferus*), and some spiny species of the Cytisus tribe; but the slopes were covered with a brilliant and varied vegetation, presenting a marked contrast to the comparative monotony of the tillage region. Most of the common orchids were seen, and we admired the many climbing plants that cover the bushes, and even reach the tops of tall trees. The beautiful *Clematis cirrhosa* is, indeed, less common here than it is in Algeria; but the two forms of *Smilax*, the spiny and the smooth-stemmed (*S. aspera* and *S. mauritanica*), were abundant; and a wild vine is common here, as it is in similar positions on the northern skirts of the Great Atlas, where it is not known to have ever been cultivated for the production of wine. Our chief botanical prize in this part of the day's ride was a beautiful Cytisus, with silvery white leaves and numerous dense heads of bright yellow flowers (*Genista clavata* of Poiret).

Throughout all this part of Marocco we were struck by the abundance of a dwarf plant of the artichoke tribe (*Cynara humilis*), which plays an important part in the domestic economy of the natives. It is almost stemless, and produces (at a later season) a large blue head of flower from the midst of a great tuft of much divided and very spiny leaves. Though not cultivated, it grows in great abundance in waste spots and the margins of fields on clay soil. Great piles of it were exposed for sale about the land gate of Tangier; and every morning whole processions of men, women, children, and donkeys, all laden with the same substance, were to be seen taking the same direction. It was painful to watch the women, half veiled, but not so as to disguise their age and ugliness, staggering onward, with huge bare legs and feet, under balloon-like loads of this spiny burden, tied up in a large coarse cloth. At this season the foliage serves as fodder for animals;

somewhat later, when the heads are approaching the flowering state, they are extensively consumed as food for the human population, the end of the stem and the receptacle being eaten raw, as artichokes are in many parts of Southern Europe.

Though, to judge from the extent of tillage, the population cannot be very small in this part of the country, we saw but few habitations, and those of the most miserable description—chiefly low mud hovels in small groups, seemingly built with a view to avoid observation in out-of-the-way spots, and never near to the main track. In this region the natives are of mixed race, partly Moors and partly of Bereber stock, descendants of Riff people, who have come down from their mountains to settle in the low country.

We made our mid-day halt in a rich green level tract that lies between the first and lower hills, and a second and more considerable range which connects the Angera Mountains on the north with the higher mass of the Riff Mountains south of Tetuan. The drainage of this broad valley seems to flow southward till it falls into a considerable stream, descending from the high peaks of Beni Hassan and its neighbouring summits, that reaches the sea on the west coast some eighteen miles south of Cape Spartel. Our eyes were here gratified by the sight of comparatively fine trees, everywhere so scarce in Marocco. Of these the most conspicuous is a southern species of ash, very like the common tree. It is the *Fraxinus oxyphylla* of Bieberstein, which extends from Southern Russia and the Levant to Spain and Marocco. The leaves and fruit are smaller, but in this district the tree rivals in stature our native British species. Poplars are common beside the streams, which are fringed by tall oleanders and willows, and in drier spots the fig, carob, and olive grow to a large size. The almond tree is also common, but does not appear to have naturalised itself.

Animal life does not seem to be abundant; but some

of the birds were new enough to our eyes to diversify the
way. The commonest is the stork, which appears, from a
sense of entire security, to have assumed a tone of complete intimacy with his human neighbours. He may be
seen about the houses, familiar with the little brown-faced,
black-eyed boys, or striding majestically through the
crops, or wheeling slowly in wide circles through the air,
till he suddenly stops, drops his long legs that had been
stretched out behind him during flight, and, poising himself on them like an acrobat on loose stilts, comes to rest.
A blue headed bee-eater, apparently the same species that
is extremely common in South Marocco, was also seen
during our ride.

As we began to ascend the main range of hills that
still separated us from Tetuan we overtook a couple of
wild-looking fellows, one carrying a tambourine, the other
a cylindrical basket, who soon showed that they wanted
to attract our attention. Our stately interpreter, riding
along with his nose in the air, purblinded by his blue
goggles, took no notice of them till one sat down and
began tom-toming on the tambourine; and Bulbo, ever
ready for amusement, soon enticed us to see the snake
charmers. These have been so often described, that it is
enough to make a few notes on the natural history of the
exhibition. The object of the tom-toming—at first gentle
and lastly furious—with which the performance commences, is clearly to aid the charmer in his endeavours to
addle his brains, and deaden his nervous susceptibility, so
that he may better encounter the pain, which, though not
intense, must be considerable. His own share commenced
by frenzied dancing and bodily contortions, and above all
rolling his head violently from side to side. This accomplished, the basket was opened, and after a good deal of
hustling two magnificent snakes unwillingly glided out,
raised their beautiful heads, looking as proud as swans,
glanced scornfully about, and very naturally tried to get
back. This the charmer prevented, and still keeping up

his abnormal nervous condition by rolling his head and
eyes, bullied one of them into biting his arm, and then
his hand between the thumb and forefinger, and drawing
blood. He next vainly tried to make a snake strike at
his forehead, and then prevailed on it to seize on his nose,
and lastly on his protruded tongue, where it held on,
probably attracted by the moisture, for some seconds,
leaving two bleeding wounds on the upper surface of the
organ, and as many on the under. With the snakes still
hanging about him, the hero concluded the performance
by laboriously thrusting a skewer through his cheek,
which had no doubt been previously perforated for the
purpose ; after this the serpents were allowed to retire
into the basket, which they were nothing loth to do.
In these performances, which have been seen by most
travellers in Egypt and India, there is little doubt that
the poison-fangs have been previously extracted. Whatever may be said of the effect produced by music on
serpents, there is no reason to suppose that it can modify
the poisonous effect of their bite, and the real object in
these cases is to act on the nervous system of the snake
charmer himself.   We were glad when the disgusting
exhibition was over, and we left the performers well
pleased with a gratuity of about eighteen pence—quite
as much as five shillings would be to a poor man in
England. When once the secret had been learned, many
an English bumpkin could be got to undergo the operation
for a pot of beer.

    As we began to ascend the rugged track that winds
up the hills the aspect of the country soon changed.
Amidst the brushwood that covered the slopes, old gnarled
trunks of wild olive, carob, and lentisk stood here and
there—survivors of the forests that must once have
covered the country—whose charred stems and maimed
branches told a tale of the way in which man's reckless
greed has marred the face of nature here, as in so many
other parts of the earth. Our last halt for botanising was

near a spring, where the green turf was decked with
many small orchids—all of them possibly forms of *Ophrys
lutea.* We were not then acquainted with the careful
observations of the late Mr. Treherne Moggridge, who
completely proved that the differences in the form and
colouring of the corolla which have been supposed to
separate several species of the genus *Ophrys* are variable,
even on the same plant; but our passing remarks entirely
tally with his conclusions. As we lingered, the sun sank
below the horizon; we unwillingly hearkened to the exhort-
ations of our followers, who seemed to grow uneasy at
the chance of being benighted, and pushed on towards.
our resting place.

The weird figures of the stunted and maimed monsters
of the forest drew closer together as we neared the crest
of the hill, and, in the fast growing gloom, assumed at
each moment a more wild and threatening aspect. Bare
branches standing against the sky, and eye-like holes in
the black hollow trunks, were transfigured by the fancy;
and to at least one of us the tale of Sintram, and Albert
Dürer's quaint old woodcut, supplied additional elements
to the mental picture; until, as we emerged from the
wood, the note of the cuckoo, bringing a whole train of
home associations, suddenly broke the spell. We rode
onward, and soon stood before El Fondak, the most stately
place of shelter for travellers in the Marocco Empire.

From without this shows a rather imposing aspect,
resembling that of a hill fort. A strong wall, some
eighteen or twenty feet in height, without window or
opening of any kind, except a central gate, surrounds a
large court-yard. We had been warned that the accommodation within was not good, and we were not long in
coming to the same conclusion.

The large quadrangle formed a sort of stable-yard,
wherein were littered camels, horses, mules, and donkeys.
The surrounding enclosure, covered with a flat stone roof,
was walled in on two sides, and on the others formed a

range of open sheds wherein the camel drivers piled their
burdens, or the keeper of the caravanserai sheltered his
cattle. On the other sides a series of doors gave admission
to as many small cellars, or dungeons, with no other
opening than the door for admitting light or air, empty,
except for remnants of dirty straw and rubbish, but apparently tenanted by every imaginable variety of insect and
creeping thing. The keeper of the caravanserai, a repul-
sive-looking old man, threw open one of the doors, and
explained that the apartment had been reserved for our
use. No deliberation was necessary on this occasion, for
a unanimous declaration burst from our lips—nothing
would induce us to enter such a filthy den—and we at
once announced our intention to pass the night upon the
roof. Our luggage was accordingly conveyed up through
a narrow stone staircase, and we proceeded to prepare our
frugal supper, of which portable soup was the chief ingredient, and soon afterwards to make our arrangements
for the night.

Our so-called interpreter had become altogether ob-
noxious to us. During our mid-day halt he had coolly
appropriated the most comfortable spot in the shade,
devoured most of our oranges, and plainly showed that he
had no notion of taking the slightest trouble about a set
of Frankish lunatics, who spent their time in grubbing up
little weeds by the roots, and looking at them through bits
of glass. He relieved us altogether of his presence this
evening ; and we felt a certain satisfaction in thinking that
his well-fed carcase would during the night supply whole-
some and abundant food for the legions of hungry insects
that tenanted the ground-floor of our hotel. Old Bulbo,
whether because he shared our preference for the clean
and airy quarters on the roof, or because he wished to
display his zeal for our protection, installed himself with
the long gun in the red case at a convenient distance,
while we, after slowly consuming the evening cigar, un-
rolled our cork mattresses, and prepared our bivouac. We

scarcely noticed at first the peculiar construction of the
roof. Round three sides of the building there was a low
parapet wall, but none whatever towards the front, where
the flags sloped slightly outwards, and ended abruptly at
the edge of the outer wall of the building. The stars
shone brightly in the sky, and a pleasant breeze from the
east fanned our faces as we lay down to rest on the front
part of the roof, congratulating ourselves on the excellence
of our quarters, when compared to the misery we had
escaped below. Before long the breeze freshened, the
night grew cooler (55° Fahr.), and we were glad to lace
the oilcloth covers of our mattresses so as to keep out
the keen air. Before doing so, Hooker judiciously laid
an empty box on the windward side, and steadied it by
placing within it two or three bottles of wine, and a few
other luxuries for our consumption, his watch, and such
other miscellaneous articles as lay at hand. Snugly ensconced in our coverings, oblivion soon crept over us, and
we slept, it is hard to say how long. A horrid crash, and
the fall of a heavy body between the adjoining sleepers,
startled two of them into sudden consciousness. It was
something like what happens in the saloon of a steamer,
when a heavy sea strikes the ship, and, amidst a smash of
broken glass and crockery, one is suddenly roused from
one's sofa by the unexpected visit of one's neighbour's travelling bag and hat-box. The cause of the phenomenon
was the same, though the position was very different. The
wind had risen to something more than half a gale, and
seemed much inclined to sweep clear away from the stone
roof everything that was not firmly fixed in its place. As
we lay tightly laced in our oilcloth covers, like the
chrysalis in its case, it cost some struggling and wriggling
to get ourselves free, and rush to the rescue of our property, which was careering along the roof before each gust
of wind that struck the building. Several articles had
already been carried away over the edge; but the moon,
shining brightly from amidst the light scudding clouds,

helped us to recover everything of importance. The watch and note-book were safe; but the contents of a broken bottle of claret had somehow run under the cover of Hooker's mattress, and, placed as we were, the attempt to rearrange it was something like the classical difficulty of ' swopping horses in the middle of a stream.' Cautiously creeping about to see what had befallen our companions, we found the faithful Bulbo (with more practical meteorological instinct than we had displayed) safely ensconced on the lee side of the low parapet. The shapeless heap, rolled up in the multitudinous folds of a white haïk, could not have been recognised, but for the inevitable long gun in the red case that lay beside it.

Little sleep was to be expected for the remainder of the night, and with the first light we began to move. Though the wind was falling, we could not attempt to avail ourselves of Maw's cooking apparatus, and we agreed to postpone breakfast till we should reach some more sheltered spot. The vegetation here was little advanced, and we saw but few plants in flower, save a little yellow Lithospermum (*L. apulum*), on our way to the top of the pass, which was covered with low brushwood and shrubs of the same species that we had seen near Tangier.

We halted in a hollow place near the highest point, where we strangely omitted to take observations for altitude; and after a slight repast hurried down the slope in a SSE direction, towards the valley of the Tetuan river. We here enjoyed a fine view of the snow-streaked mass of the Riff Mountains, which we may call, from their best known peak, the Beni Hassan Group.

The mountain ranges of the Riff—extending for about 180 miles from Tetuan to the mouth of the Oued Moulouya, which lies very near the French frontier—undoubtedly form a part of the system of the Lesser Atlas of Algeria; but, if we may trust the maps and such scanty

reports as can be picked up, they constitute a separate
group, not continuous with the coast range of Western
Algeria. The true relations between the main range of
the Lesser Atlas of Algeria and the diverging ranges of
the Great Atlas that extend over the region S. and SE. of
Fez must remain unknown so long as the latter region
remains inaccessible to European travellers. The river
Moulouya and its eastern branch, the Oued Za, mark the
existence of two considerable valleys, and it is probable
that the very sinuous course laid down for both those
streams in the French map may be founded on native
reports approximately correct; while it is quite certain
that the adjoining mountain ranges as shown on that map
differ very widely from the truth. A traveller going from
Fez to the mouth of the Oued Moulouya, in a direction
slightly north of due east, traverses a broad valley, with
the Riff Mountains on his left, lying between him and
the Mediterranean coast, and the northern branches of the
Great Atlas on his right. Somewhere near Theza he
reaches the watershed between the region that is drained
towards the Atlantic through the Oued Sebou and the
basin of the Oued Moulouya, but seemingly without having to make any considerable ascent. He descends to the
Moulouya—or rather he would do so if the powerful Halaf
tribe, who hold that region, allowed strangers to pass—
where that river, after cutting its way through the unknown region between the Great and Lesser Atlas, enters
a wide plain, some forty or fifty miles in extent each way.
Before reaching the sea, the valley is again narrowed. On
one side is the eastern extremity of the Riff Mountains,
and on the other a range of lofty hills that may be considered as spurs of the Lesser Atlas of Algeria.

Before quitting this dry subject, it is necessary to remark that, even as regards the relatively well-known
district near Tangier and Tetuan, the best maps are far
from complete accuracy. In the French War Office Map
—undoubtedly the best map of Marocco—the hill shading

gives far too much importance to the comparatively low hills running from WSW to ENE. on the south side of Tangier, and not enough to the range which we crossed between El Fondak and Tetuan. This extends from the main mass of the Beni Hassan to Ape's Hill opposite Gibraltar, and divides the waters running to the Atlantic from those of the Tetuan river. Over against this (which we had just crossed) rose a parallel and more lofty range, terminating in the bold craggy mass of the Beni Hosmar (B. Aouzmar of the French map), rising steeply from the valley opposite Tetuan, and to ascend this was the main object of our present excursion.

Soon after we entered the main valley, and were riding along a broad track parallel to the Tetuan river, we came upon a group that for the first time brought home to us an illustration of the true condition of society in this country. A body of armed horsemen, many of them true Negroes or mulattoes, were resting beside the way, broken up into lively groups, laughing and chattering together. Amongst them was a solitary man, poorly clothed, and, as we observed, laden with heavy chains. He kept his back turned towards the track, and seemed to take advantage of the halt to dip his feet into the brook that ran along beside it. So numerous an escort in charge of a single prisoner suggested something unusual, and we were led to make inquiry. According to the story retailed to us, the chained captive was lately the powerful governor of a distant province, who had offered a stout resistance when summoned to the capital to give an account of his administration. It is well understood in Marocco that such summons, whether framed as a peremptory order or a flattering invitation, has but one meaning—that the time has come when it seems to the Sultan or his counsellors that the wretched governor should be 'squeezed,' or, in other words, be forced by torture to surrender whatever wealth he may have hoarded. As the appointment of a new governor generally means that the province will be subjected to fresh impositions and extortions, the people are apt to side with the old governor, and sometimes, in a country where the central power is so feeble, a man, by a judicious combination of force and bribery, may long keep the government at bay, and escape the miserable fate that usually awaits him. Our prisoner, apparently, was too formidable a man to be safely kept at Fez or Marocco, and was therefore sent to Tetuan, the extreme limit of the territory, there to undergo such torture as might be necessary to extort confession of the hiding place of his treasure, unless, through ill-judged obstinacy, he should die in torments before disgorging as much as might be expected. No better illustration of the system can be found than the fact that strangers are informed, as of something extraordinary and unexampled, that one old man now lives at Tetuan who long held a high and confidential post in the government, and yet was allowed to retire without being 'squeezed!' The truth is, that he had gained the good-will and confidence of the representatives of the European Powers, and that it was urged upon the late Sultan that the credit of his government would suffer, if, after a long course of faithful service, the minister were to undergo the common fate of his colleagues.

Some twenty years before, when one of our party visited Tetuan, the whole province was thrown into confusion by one of these customary acts of the then reigning Sultan. Hash Hash, a man of unusual capacity and energy, had governed the province of Tetuan for many years with extraordinary success. He kept the turbulent Riff mountaineers in order, and, so it was said, Jew and Christian, under his rule, enjoyed the same security as the Moor. At length he received messengers from the Court with the gift of a white horse richly caparisoned, and an autograph letter from his sovereign full of commendation and winding up with an invitation to the capital, then fixed at Fez. He started on the fatal journey, but arrived only to be flung

TETUAN

J. B. delt.

into a dungeon and subjected to daily torture. Soldiers were sent to Tetuan, where his house was pillaged, his wives and children led to prison, while the absence of all control led to a rapid growth of crime in the district, and life and property were no longer thought safe in the surrounding country.

The approach to Tetuan presented the most picturesque scene that we anywhere beheld in Marocco. Begirt with a lofty wall, set at short intervals with massive square towers, the city shows from a distance only a few mosques and a heavy, frowning heap of masonry that forms the castle or citadel. It stands on the slope of a limestone hill, some two hundred feet above the river, which flows through a broad valley, rich with the most brilliant vegetation. After riding for hours over the thirsty hills, it was a delight to rest the eyes on the patches of emerald meadow, and on the darker green of the luxuriant orchards, where the best oranges in the world grow along with figs, almonds, peaches, and all our common tree fruit. Amidst all this wealth of greenery many a little white house—a mere cube of chalk—gleamed brightly. Most of these seem to belong to peasant owners, but some are kiosks to which the wealthier inhabitants repair to escape from the heat and bad air of the town.

We were not yet familiar with the squalor and neglect that seem the inevitable characteristics of a Moorish town, and it was a disappointment to find the interior of Tetuan correspond so ill to the picturesqueness of its outward aspect. After riding between high walls, apparently forming an inner defence to the town, we went through some streets of mean aspect, and, traversing one wide open space, passed under an interior gate guarded by a sentry, and found ourselves in a labyrinth of narrow alleys decidedly cleaner than the remainder of the city. This is the Jewish quarter, where, as in the Jewry or *Ghetto* of mediæval Europe, the children of Israel are required to live apart, within a wall and gates that are locked at night,

and where they seem to manage their own affairs with
little interference from the Moorish authorities. We soon
established ourselves in very fair quarters at the house of
Isaac Nahum, who acts as clerk and interpreter at the
single consulate which of late years has watched over the
safety of all Europeans who happen to reach Tetuan
whether by land or sea. Since the war in which Tetuan
was taken by the Spanish troops—their solitary achieve-
ment during the last sixty years—the Government of
Spain has desired to maintain its influence in this part of
the country by the presence of a consul ; and the other
European States have willingly taken advantage of his
presence. The duties cannot be heavy, for few strangers
now visit Tetuan, although up to the year 1770 it was the
residence of all the European consuls. The beauty of its
site, the excellence of its oranges and other fruit, and the
reported superiority in refinement of its inhabitants, both
Moorish and Jew, do not compensate for the difficulty of
access by sea, since none but the smallest class of coasting
vessels can cross the bar at the mouth of the river. This
is guarded (or was so up to the time of the Spanish war
in 1859) by a massive square tower, without door or other
apparent opening. A Christian boat from Gibraltar, in
which one of us had formerly arrived, was hailed from the
summit of the tower. After a preliminary parley, a rope
ladder was let down from the top, some seventy or eighty
feet, and a black soldier scrambled down with great activity,
the final result of the parley being that the strangers, after
payment of some trifling harbour dues, were sent to the
town, a distance of five or six miles, under the escort of a
soldier.

Whether because there really is some slight diminution
in the feeling that has so long excluded strangers, and
especially Christians, from the interior of Marocco, or that
previous travellers had happened to make the attempt at
unfavourable conjunctures, we found that the letter to the
Governor given to us by Sir J. D. Hay was scarcely re-

quired, and no difficulty was raised about the requisite
official permission to ascend the Beni Hosmar, as the
mountain mass is called, which forms the end of the chain
extending northward from the Beni Hassan.

One of our party had already succeeded in ascending
about half the height of the mountain ; but the only
European known to have reached the upper ridge was the
late Mr. Barker Webb, the author of the 'Phytographia
Canariensis,' and other important botanical works. He
effected his object by liberal expediture, having begun by
a present of 40l. to the Governor, besides handsome rewards to those who were sent with him.

We had no occasion to follow this example. The protection of the British Government, and the interest shown
in our journey by the British Minister, were quite sufficient arguments on our behalf, and with the courteous
assistance of the Spanish consul the arrangements for our
excursion were soon settled. The requisite orders were
issued by the Kaïd, and two soldiers were appointed, along
with our Tangier men, to escort us on the following
morning.

In spite of the usual delays, we started in good time on
the morning of the 11th, and, descending over successive
ledges of tufa, forming terraces for gardens and orchards,
soon reached the level of the river, which was easily forded.
The air was cool (55° Fahr. at 6 A.M.), the sky bright, and
the hedges gay with the evergreen rose (*R. sempervirens*),
and the large-flowered form of the hedge convolvulus (*C.
sylvatica*), which in the South replaces our more modest
Northern form, *C. sepium* of Linnæus. A short ascent
among trees and high hedges took us clear of the cultivated land, and the aspect of the country at once changed.
The upper part of the mountain is disposed in tiers of
limestone crags, irregularly disposed, and therefore offering
no difficulty for the ascent ; but round the base are rather
steep and very arid slopes, formed, in great part, of old accumulations of débris fallen from the upper crags. The

most conspicuous shrubs are lentisk, oak scrub, *Juniperus phœnicea*, and several *Cisti*; but the palmetto successfully contends against its rivals, and in some places quite covers the soil. It disappears, however, before one reaches the middle height of the mountain, and the limit of its free growth, not taking account of a few scattered and stunted specimens, was found to be 1,227 feet (374 mètres) above the sea. The prevailing species, however, were small shrubby *Leguminosæ*. Of these the most trying to the temper of the botanist is *Calycotome villosa*. This and the allied species (*C. spinosa*) are very common in the warmer parts of the Mediterranean region, and the stiff spiny points of the numerous branches are most effective in tearing the clothing and the skin of anyone who approaches them.

We followed a tolerably good cattle track which wound upwards to the right, in a southerly direction, towards the upper part of the mountain. Before reaching its middle height, on some crags facing towards Tetuan, we found a peculiar saxifrage (*S. Maweana*), first collected by Mr. Webb more than forty years before, but which, with several others, remained unknown and undescribed in his Herbarium. Maw refound the plant in 1869, and has successfully cultivated it, along with many other Marocco rarities, in his garden in Shropshire. On the same rocks, besides numerous interesting plants not yet in flower, we gathered a curious crucifer (*Succowia balearica*) which must flower very early as the fruit was already approaching maturity.

As we really desired nothing more than to be let to wander about on the mountain according to our own fancy, we were rather pleased than otherwise when our escort of four soldiers with the guide, seeming to think that they had done enough of mountaineering after an ascent of some two thousand feet, proceeded to instal themselves, with the horses, who enjoyed a day of rest, in a pleasant spot, and showed no sign of pushing the enterprise farther.

A steep slope now led us up to the rocky ridge of the mountain commanding a wide view, and overlooking a deep glen on the seaward side of the mountain. Here, in spite of the early season, we found several plants in flower that excited in us a lively interest. A little polygala, with rich purple red flowers, reminds one much of the red variety of *P. chamœbuxus* that is often seen in the Eastern Alps, but appears to be quite distinct. A chrysanthemum, differing little from an Algerian species, was our first acquaintance amongst a group of forms that is especially characteristic of the flora of the Great Atlas. But we were, perhaps, still more pleased to find on these heights, far removed from the nearest known station, some descendants of a suffering race that must, at some remote period, have been widely spread throughout Europe, the bright-flowered *Ranunculus gramineus*. Although it is still found at several places in France, in a few spots in the Alps, and in Spain, it appears to have disappeared from the Apennines within the last two centuries, and to be everywhere losing ground. When the rapacity of collectors shall have reduced it elsewhere to the condition of a vegetable Dodo, future travellers may rejoice that it has found a refuge in this corner of Africa. The distribution of the genus *Ranunculus*, in nearly every known country, supplies many topics for thought and inquiry. There are very few regions where the unbotanical traveller fails to recognise the familiar buttercup of his youth ; yet, if he examines the plants, he will find well-marked differences in the leaves, the fruit, the stem, or the root, though the flowers may be scarcely distinguishable. Since our first landing in Marocco, buttercups had met us in all directions ; but they nearly all belonged to one variable species, *R. chœrophyllos*, widely spread round the warmer shores of the Mediterranean. In shady places we had a few times gathered another North African species, *R. macrophyllus*, and on this mountain we found a few specimens, already past flower, of *R. spicatus* ; but of all the common species of Britain and Middle Europe, not one had been seen, unless we count the ubiquitous white-flowered species of our ditches, *R. aquatilis.*

From the time we first got a clear view of our mountain we had fixed on a range of beetling crags, not far below the summit, which promised to afford an excellent habitat for rare plants. The promise was kept, for we had scarcely approached their base when with joyful cries we saluted one of the chief prizes of our excursion. From clefts on the face of the rock hung great leafy tufts, quite a yard in diameter, supported on stems as thick as a man's arm. The flowering branches produced an abundance of yellow flowers, then just expanding and only partly opened. We should have set it down as a new and very luxuriant species of wild cabbage, but that we happened to know that the fruit is entirely different, so much so as to constitute a very distinct genus of *Cruciferæ.* Mr. Webb, who probably gathered the plant at this very spot, described and figured it, in the ' Annales des Sciences Naturelles,' under the name *Hemicrambe fruticulosa ;* but the original specimen seems to have been lost or mislaid, and no one had since laid eyes upon the living plant. The same rocks produced abundantly the beautiful *Iberis gibraltarica,* besides many fine plants not yet in flower, amongst which we recognised the rare Spanish centaurea, *C. Clementei.*

As seen from Tetuan, the ridge above the rocks appeared to lead very directly to the not distant summit of the mountain ; but when, after a short scramble, we had set foot upon it, we clearly saw our mistake. At about a mile and a half from where we stood, and separated from us by a rather profound depression, was another ridge, some three or four hundred feet higher, which might or might not be surpassed by more distant prominences in the same range. It would have been easy to reach the farther summits, but we thought our time better spent in carefully examining the part of the mountain within our reach.

Various indications, such as the disappearance of several species that are abundant lower down, and the much more backward state of the vegetation, went to prove that the climate of the upper plateau is sensibly different from that of its middle region ; but there was little to show that we had reached the limit of a true mountain, much less that of a subalpine flora. We had, indeed, already found a variety of the large-flowered *Senecio Doronicum*, which in the Alps and Pyrenees ascends even to the Alpine region ; and near our highest point Ball found a form of *Erodium petræum*, which in the Pyrenees and Northern Spain usually attains the subalpine zone. The season was still too little advanced ; and the naturalist who will follow our footsteps about the beginning of June may expect a much richer harvest.

Having taken observations for altitude, which give height of about 3,040 feet above the sea for our station, we halted a few minutes to enjoy the noble panorama that was spread out below us. On the western side successive undulations of the ground—range beyond range of low hills—melted away into the horizon, but as the eye turned northward it rested on a more varied picture. To the right of the Angera Mountains and Ape's Hill a small dark islet seemed to stand out from the Spanish coast. In this we scarcely recognised Gibraltar, for the shadow of a cloud happened to rest on its grey limestone cliffs. To the right extended a long reach of coast line, foreshortened from the promontory of Ceuta to the mouth of the river below Tetuan, with the much more distant outline of the Serrania de Ronda in the background. Then as we turned eastward, though the view was partly interrupted by projecting spurs of the mountain, we followed the long outline of the coast range of North Marocco, the secure refuge of the unconquered Riff tribes, whose fastnesses have never been profaned by the presence of an alien master. Some patches of dark shade evidently indicated forests, and these may probably consist wholly or in part of the Atlantic cedar, although that tree is not positively known to grow in Marocco.[1]

In order to cover as much ground as possible during the descent, we here agreed to take different directions, and lost sight of each other for some time. Hooker came upon a small mountain village, or hamlet, where several Bereber or Riffian families were crowded together in hovels built of mud mixed with stone, and rather better fitted to resist the weather than the sheds we had seen in the plain. Conversation was not practicable, but there was no indication of ill will on the part of these people. The only attempt at intercourse was on the part of one sturdy man who apparently requested a pinch of snuff, but declined the offer of a cigar. The use of tobacco for smoking appears to be unknown in Marocco, while *kief*—prepared from the chopped leaves of common hemp—is almost universally employed for that purpose both by Moors and Berebers; but snuff is in general request, and is imported in considerable quantities, both by regular traders and by smugglers who profit largely by the heavy duty.

In descending the mountain we observed large patches

[1] In the Herbarium of the late Mr. Webb, now in the Museum at Florence, the plants gathered by him during his short expedition to Marocco are preserved as a separate collection. Amongst these are some fragments of the Atlantic cedar, which would appear from the accompanying label to have been obtained by him at Tetuan from some native of the Riff Mountains. It is probable that the same tree may be widely spread throughout the unexplored mountain districts of North-eastern Marocco. Gerhard Rohlfs, the only European who is known to have traversed the high mountain region S of Fez, describing the fine valleys inhabited by the powerful tribe of the Beni M'ghill, says that the prevailing trees were larches of greater dimensions than he had ever seen elsewhere. He declares that he measured several stems from three to four mètres in girth, and that such were not uncommon. It is in the highest degree improbable that the larch, which in Europe finds its southern limit in the Pyrenees, should extend to Marocco; and, as Mr. Rohlfs has no knowledge of botany, it is most likely that the tree which called forth his admiration is the Atlantic cedar.

of a species of furze, smaller and stiffer in habit than our
common gorse—the *Ulex bœticus* of Boissier—one of a
group of nearly allied forms that replace our British
species in the south of Spain and Portugal, and the neighbouring shores of Marocco.

On rejoining our so-called escort, we agreed that the
track was too steep to make riding pleasant; and thus we
all descended on foot till near the foot of the mountain,
when a proper care for their dignity compelled the soldiers
and the guide to remount.

We returned to our quarters in the town before the
sun had set, and closed a very enjoyable day by reviewing
our botanical prizes as we laid them into paper to undergo
the first step in the process of their preservation. As
usual the evening cigar accompanied our discussion as to
future proceedings, and to its soothing influence we
doubtless owe the fact that these debates always led to a
satisfactory conclusion. On this occasion we agreed to
divide our small party into two sections and separate
for a few days. Maw was anxious to return at once to
Tangier, with a view to visit some swamps that lie about
ten miles south-west of the town, while Hooker and Ball
were desirous of examining the coast between Tetuan and
Ceuta. As it appeared that a small stock of Spanish
would serve all necessary purposes in the excursion to
Ceuta, Maw volunteered to take our disagreeable interpreter and one unnecessary soldier back to Tangier, while
Bulbo was willing to risk a visit to the infidels at Ceuta.

. On the morning of the 12th Maw departed, but
Hooker was unwell. It was decided that a quiet day
and the judicious exhibition of moderate doses of cognac,
which we owed to the kindness of the Spanish consul,
would be the most appropriate treatment; and the result
was quite satisfactory.

Ball spent the day in botanising over the hills near
the town, and was well satisfied with the result. The
rarest plant found was, perhaps, a curious and very dis-

tinct fumitory (*Fumaria africana* of Lamarck), which he
had gathered nearly at the same spot twenty years before.
The red-flowered *Polygala* of Beni Hosmar (*P. Webbiana*
of Cosson) was seen in a few spots near the town along
with *Arabis pubescens* ; and that singular plant, the
Drosophyllum, hitherto seen in Marocco only on the hills
west of Tangier, was here found within sight of the
Mediterranean, growing along with *Helianthemum um-
bellatum* and several other less rare species of the Cistus
tribe.

During our stay here we had a good opportunity of
seeing something of the life of the Marocco Jews, who
form a distinct and important element .in the population
of the empire. Tetuan has long been one of the head-
quarters of the Hebrew race. When most of the chief
Moorish families took refuge here after their expulsion
from Spain—and some are said still to preserve the keys
of their own houses in Granada—many Jews, flying from
the faggots of the Inquisition, preferred the comparative
toleration of Moslem rule, to the oppression and social
disabilities that awaited them in Christian Europe. It
was more tolerable to submit to occasional injustice and
cruelty which was shared by all classes of society around
them than to be daily reminded that they formed a class
apart—the proper objects of general contempt and aversion. It is true that until late years the Marocco Jew
was exposed to some vexatious regulations. He was
required to put off his sandals on passing the outside of a
mosque, to wear a peculiar dress, and is still confined to a
separate quarter in each town. But in ordinary intercourse between man and man the Jews of the coast towns
seemed to us to have attained a footing of almost complete
equality, due as well to their superior intelligence and
commercial instinct, as to the tolerance which affinity of
race and creed has developed among the people of Arab
stock. In truth, the Moor feels that the Jew is indispensable to him. In despite of his aversion to intercourse

with the Christian, trade, in which the Jew serves as intermediary, has become a practical necessity, and it has procured for him foreign luxuries which he is now little inclined to forego.

In point of fact, Tetuan boasts of being the cradle of more wealthy Jewish families than any other town in the world; and among the practical concessions enjoyed by them, there now appears to be no difficulty in the way of Jews leaving the empire and returning to it, and frequent intercourse is carried on between the city and Europe by the way of Ceuta. The ceremonial observances of the Mosaic law are strictly adhered to. The first question put to us on our arrival was to know whether we had with us leavened bread, as such could not be admitted to the house during the feast of the Passover; and during our stay we were given cakes, some of plain flour, others prepared with orange juice.

The houses are quite on the same plan as those of the Moors, or in other words they merely differ in architectural detail from the ancient type that is preserved for us in the smaller houses of Pompeii. A single court (atrium) has several small rooms or closets used for kitchen, offices, and sleeping place for servants, and one large apartment, the chief living room of the family, filling one side. This remains open to the court by day, but is closed at night by a curtain. On the upper floor a gallery surrounds the court, and into this open upper rooms of moderate size. In Nahum's house a second floor above the first had been added, but this appeared to be an unusual arrangement. On our arrival we had been struck by the superior neatness and cleanliness of the Jewish, as compared to the Moorish, quarter, and the same remark applied to their persons.

No European traders appear to have settled at Tetuan, and such trade as it possesses is in the hands of the Jews. Oranges, and a sort of brandy, called Mahaya, distilled from the grape, are the chief exports. The coarse pottery made here is much the same as that produced

E

in Algeria and throughout Western Marocco. Rude geometrical patterns in ill-defined blue and green tints are
usually enriched by round spots of bright red, laid on
with something like sealing-wax over the glazing, and
easily removed with spirit. The only thing deserving
notice as representing art-manufacture is the gold embroidery, usually worked on silk or velvet. This is used
for curtains or hangings by some wealthy Moors, and for
personal wear by the Jewish women and children. At
this festival season the younger children frequently
appeared with caps or diadems richly embroidered; but
the women more often wear a light silk handkerchief,
with the fringe hanging freely, but kept in its place by a
fillet of black ·or red velvet worked in gold, and forming
a very ornamental head-dress.

Travellers have indulged in enthusiastic descriptions
of the beauty of the Marocco Jewesses. Those who
have visited Tetuan will have seen a fair specimen in
the person of our host's sister, a tall comely girl, free
from the tendency to corpulence which is too common,
and whose regular features are set off by a pair of fine
dark eyes. But those for whom expression is an essential
element of beauty in the human countenance will usually
find something wanting to complete the attractions of the
undeniably handsome women of this country.

It so happened that the occasion was especially favourable for seeing something of the life of the Israelite society
of the city. This was the last of the festival days of the
Passover, and towards evening there was a large gathering
of neighbours in the ground-floor apartments of our house.
The women were richly dressed in loose garments of light
silk and a profusion of gold embroidery. It was almost
impossible to recognise our host's mother, a corpulent
woman, who had hitherto appeared in a shabby costume
of the scantiest proportions in which the developments of
her ample person were unpleasantly apparent. Arrayed in
festival splendour, she now assumed a regal attitude, and

her figure appeared to be modelled on that of the nearest
Christian potentate, the unregretted Queen Isabella. The
men wore long blue coats of the dressing-gown pattern,
with white cotton stockings and slippers, and, if not
picturesque in appearance, showed to advantage beside
our host who, mindful of his dignity as interpreter to
the Consulate, appeared in European black frock coat
and trousers. The children were especially gorgeous in
head-dresses of crimson or purple velvet richly embroidered
in gold. During the evening there was an attempt at
dancing to the music of an accordion; but the space was
too limited, and this was speedily given up. The party
continued, however, till a late hour, and midnight passed
before the sound of lively talk and laughter ceased in the
lower chambers of our house.

On the morning of April 13 we started for Ceuta, about
thirty miles distant from Tetuan. The track for several
miles lies at some distance from the coast, which on the
north side of the mouth of the river forms a projecting headland, called by the Spaniards Cabo Negro. After riding
through green lanes, we mounted gradually by a broad
path that winds amidst bushy hills for a couple of hours,
and then descended towards the sandy shore; and for the
remainder of the way kept close to the beach. After fording
one or two smaller streams issuing from the marshy pools
that lay between us and the hills on our left, we had a little
trouble in crossing a more considerable torrent that seems
to bring down most of the drainage of the Angera
Mountains lying behind Ape's Hill. The horses' feet sank
deeply in the yielding sand of the bed, though we were
able to wade across without difficulty. It was an anxious
moment for us as we watched the baggage mules struggling
and floundering, until the water rose very nearly to the
precious packages of paper that contained the fruits of
our work since we left Tangier. Several villages were
seen on the slopes of the hills to our left, but during the
entire day we passed only three or four small houses.

Our day's ride lay over the scene of the Spanish campaign in Marocco in the winter of 1859-60—a military
event so completely eclipsed by the great wars that have
since desolated many parts of Europe, as to be now almost
forgotten. An intelligent and animated account of it was
published by the late Mr. Hardman, who accompanied
the Spanish army as correspondent of the *Times* newspaper. The advance of O'Donnell, the Spanish com-
mander-in-chief, was slow and cautious; but considering
the natural difficulties, and his complete ignorance of the
resources and designs of the enemy, any other course would
have been chargeable with rashness. The Moors, although
at the last they showed the utmost personal intrepidity,
failed to display the slightest military capacity—even
such as has been found among many savage tribes—failing
to take advantage of natural difficulties, and exposing
themselves in fruitless and desultory attacks when the
Spanish force occupied strong positions. The most serious
difficulty for the Spanish general arose from the necessity
for moving his army along the narrow strip of shore,
where for several miles the ground between this and the
stony hills of the interior is partly covered by shallow
lagoons, and the soft soil is intersected by streams. An
active enemy knowing the ground might have inflicted
heavy loss on the advancing force; but, contrary to all
expectation, the Moors scarcely showed themselves at the
critical moment, and the Spaniards had none but the
natural obstacles to contend with. After crossing the pass
over which the ordinary track runs to Tetuan, the Spaniards
marched to the left, and established themselves in an
entrenched camp near the mouth of the Tetuan river,
where they received by sea reinforcements in men, heavy
guns, and provisions. After some delay, a brilliant action,
terminated by the storming of their camp near Tetuan,
cowed the Moorish leaders, and the Spanish occupied the
city, but only after it had been sacked by the irregular
forces of the retreating army. The Moors then sued for

peace; but whether the negotiations were merely opened
to gain time, or that the terms demanded by Spain, including the permanent cession of Tetuan, were deemed
exorbitant, hostilities were resumed in March, and the
Spanish army commenced to move towards Tangier. A
final effort was made by the Moors; and in the battle
which ensued, on the slopes of the hills by which we descended a few days ago into the valley of Tetuan, their
men, though fighting against nearly 25,000 regular troops,
well provided with artillery, seemed for a moment likely
to win the day by sheer desperate valour. The victory
cost the Spaniards some 1,300 men in killed and wounded,
but achieved the object of the campaign. Guided by wiser
counsels, the Spanish Government ceased to insist on the
permanent occupation of Tetuan, and the city was restored
to the Moors, on the payment of a war indemnity of about
4,000,000l. sterling. In the judgment of impartial foreign
critics, the Spanish troops behaved extremely well throughout this campaign : when well led they showed no lack of
fighting qualities, and to their patience under hardship,
their temperance, and general good conduct, all observers
bore testimony.

One result of the war was to increase the customs'
duties throughout Marocco, and to cause more strenuous
efforts to keep down contraband trade than had ever been
used before. The indemnity was partly provided by a
five per cent. loan, raised in London; and the customs
duties supply the means for paying the interest, with in-
stalments of the principal. These have been so punctually
discharged, that the stock usually stands at par. On the
Atlantic sea-board the points accessible to sea-going ships
are so few that little smuggling can exist. The long strip
of Mediterranean coast between Tetuan and the French
frontier is nearly all held by the semi-independent tribes
of the Riff mountaineers, and it may be presumed that
these pay no duties on the few articles of foreign produce
that they consume; but the southern shore of the Straits

of Gibraltar and the coast between Ceuta and Tetuan are easy of access in fine weather, and here the Moorish authorities are obliged to maintain a force of coast-guards. We met several wild-looking fellows, who became more frequent as we approached the Spanish lines before Ceuta, each scantily clothed and armed with a long gun. They must suffer much in cold and rainy weather, as they have no other protection than a slight screen of branches, interlaced with straw or reeds.

Ceuta stands upon a narrow promontory that forms the eastern extremity of a spur projecting from the high range of Ape's Hill. As this promontory is only the last in a series of conical summits that gradually diminish in height as they approach the Mediterranean, the fortress is completely commanded on the land side. But the Spaniards have erected small forts on the nearer heights, and with moderate watchfulness are secure enough from any assault that could be made by the Moors. As we rode over the neck of land connecting the fortress with the adjoining hills, and finally approached the only entrance, which is reached by a succession of gates and drawbridges, we had leisure to admire the elaborate character of the defences, in which every known resource of military engineering, as understood at the beginning of the last century, seems to have been accumulated. The soul of Uncle Toby would have delighted in the multiplication of ditches, curtains, ravelins, demi-lunes, hornworks, and palisades that have been expended here for the purpose of astonishing the untutored mind of the ignorant Moor.

The little town that forms the kernel of these vast fortifications far surpassed our expectations. Say what we will, there is a vast gap between the condition of the least advanced countries of Europe and the barbarism from which no Mohammedan State has yet contrived to raise itself. Ceuta, however, is a very favourable sample of a Spanish town, and is far superior in aspect to most places of equal importance in the mother-country. The well-

built houses in the main street, all dazzling with fresh
whitewash, were gay with bright flowers that stood in pots
and boxes on the balconies behind ironwork of elaborately
ornate character, and the inhabitants had an air of activity
and animation not common in Spain, anywhere out of
Catalonia. We drew our bridles at the Fonda Italiana,
the best looking of several inns, where we learned that all
the bedrooms were occupied, and were sent for sleeping
quarters to a neighbouring house. We got a large room
with two good beds, and found everything both there and
at the inn, where we were well fed, scrupulously clean.
Our remark, which probably would not have been approved
in Downing Street, was, 'What a pity, when they were
about it, that the Spaniards did not annex the whole of
North Marocco!' The course of events in Spain during
the last six or eight years has gone far to justify Downing
Street, and to show that European anarchy may be even
worse than Moorish misgovernment.

As, in accordance with our daily custom, we reviewed
the produce of our day's botanising, before committing
our plants to paper, it seemed to fall rather short of our
expectations. The season was not yet advanced enough
for many seaside species, and, besides, as every naturalist
knows, one's power of observation on horseback is comparatively limited. When the eye is carried forward by
an external agency, and its motion is not altogether regu-
lated by the will, many minute objects are too imperfectly
seen to convey a definite image; and however often one
may dismount, many slight suggestions that would be
tested by one on foot are allowed to pass without verifi-
cation. Along with most of the shrubs that we had seen
about Tangier, we passed many small trees of *Tamarix
africana* and stout bushes of *Juniperus phœnicea*. The
most ornamental plant that we gathered was *Phaca bœtica*,
with fine purplish blue flowers, very unlike any of the
forms of the same genus with which we were familiar in
the Alps. The most interesting plant, in a scientific

sense, that we found this day was so minute as to be altogether overlooked at the time; and it was only some time
after our return to England that two minute specimens
(less than an inch in height) were found engaged in a tuft
of some stouter plant. They belong to a little crucifer,
called *Malcolmia nana*. It has been found in a few spots
scattered at wide intervals throughout the Mediterranean
region, and as far eastward as the shores of the Caspian
Sea.

At Ceuta we had the spectacle—always a painful one
—of gangs of convicts chained together, and working
under the charge of soldiers, which meets the eye in so
many parts of Southern Europe. Difficult as is the
subject of penal discipline for criminals, it may safely be
said that this is one of the worst—if not the very worst—
system that has ever been devised. The punishment,
however hard, loses through familiarity most of its
deterrent effect; while, far from reforming, it seems to
be the most efficient method known for finally corrupting
the less hardened offender. The objections are somewhat
lessened when the convict station is removed from the
general gaze, and where the prisoners have little hope and
even little temptation to escape.

These conditions are satisfied in the three fortified
posts which, besides Ceuta, the Spaniards hold on the
coast of Marocco. The most considerable of these is
Melilla, on a promontory a few miles south of Cape Tres
Forcas, said to be a strong fort, but grievously damaged
by an earthquake in 1848. It must be little better than
a prison for the garrison as well as for the convicts, if
it be true, as we were told, that it is considered unsafe
to venture beyond musket-shot from the walls, and the
Riff mountaineers amuse themselves from time to time by
taking pot-shots at the sentries on the ramparts. The
other posts are on rocky islets near the shore. El Peñon
de Velez, also called Velez de Gomera, is about half-way
between Ceuta and Melilla, and only about eight miles

from the site of the Carthaginian city of Bedis—Belis of the Arabs—whence some etymologists derive the Spanish Velez. From the rank of an episcopal city in early Christian times, Bedis fell into bad repute as a pirate port, until it was taken and destroyed by the Spaniards The third Spanish post is on the larger of the Zaffarine Islands, that rise from the Mediterranean nearly opposite the mouth of the Oued Moulouya, not far from the French frontier. To judge from a small packet of plants collected there by Mr. Webb, the only scientific traveller known to have visited them, these are mere barren rocks, affording no shelter to any but the common seaside species.

Of late years the Riff people have kept to their mountain fastnesses, and piracy is no longer an habitual occupation; but it would not be safe to suppose that it has been completely extinguished. The coast has many inlets and creeks that shelter fishing boats, which may easily be used for cutting out unarmed merchantmen when becalmed near the coast. As late as 1855 two or three cases of that nature were reported to the home authorities by the Governor of Gibraltar: and as pursuit was out of the question, and the Moorish Government owns no control over the Riff population, no redress was obtainable. The increasing use of steam has probably made the occupation tedious and unprofitable.

CHAPTER III.

Sail to Algeciras—Vegetation of the neighbouring hills—Comparison
between the opposite sides of the Strait of Gibraltar—Return to
Tangier—Troubles of a botanist—Fez pottery—Voyage in French
steamer—Rabat and Sallee—Land at Casa Blanca—Vegetation of
the neighbourhood—Humidity of the coast climate—Mazagan—
View of Saffi.

WITH the previous permission of the Commandant, we
sailed from Ceuta in the Government felucca on the morning of the 15th, and had a pleasant run before a south-
west breeze, which took us before noon to Algeciras. Our
intention had been to return the same day to Tangier, but
we found that the ordinary steamer had been taken up to
carry sight-seers to a bull-fight at Seville. Resigning
ourselves to the delay, we found fair accommodation in an
inn upon the quay, and started for a walk over the wooded
hills behind the town, not sorry to have an opportunity of
comparing the vegetation of the opposite shores at this
point where Europe and Africa so nearly meet.

The general aspect of the floras is nearly identical,
but there is enough of difference to show that for a long
period a barrier has existed sufficient to limit the diffusion
of many characteristic species. Of these we found three
on the hill near Algeciras—*Rhododendron ponticum,
Sibthorpia europœa,* and *Helianthemum lasianthum,* a
fine species with large yellow flowers, approaching a Cistus
in stature and habit. A much longer list of European
plants that have not passed into Africa might be made
if all the known species found between Gibraltar and Trafalgar were taken into account; but it might with some
reason be objected, that our knowledge of the African side

of the Strait is too incomplete to speak confidently on this point. On the other hand, however, we may with some certainty assert that comparatively many well-marked species found on the southern side of the Strait are limited to the African shore, and have not been able to spread into Europe. From the accessible materials we find at least thirty-eight species belonging to this category, of which the large majority are species spread over a wide area in Northern Africa.

In attempting to draw inferences from these facts, it is necessary to bear in mind that the region where they occur—the southern part of the Iberian peninsula, and the opposite corner of Marocco—is remarkable for the variety of its flora, and for the large number of distinct species, each inhabiting a very restricted area. To those who suppose that the presence of numerous plants in two neighbouring districts, which are limited to one or the other, but are not common to both, is to be regarded as evidence for the existence of a physical barrier between them, an objector might reply that we have no more right to affirm that it is the prolonged existence of the Strait between Europe and Africa that has prevented the extension of so many species from one continent to the other, than we have to maintain that two neighbouring mountain groups, such as the Sierra Nevada of Granada and the Serrania de Ronda, each possessing a number of peculiar species, must have been formerly isolated by the sea, as otherwise the species would have been intermixed.

In answer to this objection, it may, with some plausibility, be urged that a large majority of the species with restricted areas are mountain plants ; that there is much reason to believe that most of these peculiar species did originate within insulated areas, at a time when these were separated by the sea from neighbouring masses, where the conditions of life for each organism must have been somewhat different ; and that in a few instances local peculiarities of soil, either chemical or mechanical, may explain the fact that a particular species is limited to a very small district. These considerations do not, however, fully explain the known facts regarding some regions of the earth possessing an exceptional number of peculiar species confined to small areas, the most remarkable of which are Asia Minor, South Africa, South-western Australia, and that which we are now discussing; and in weighing the evidence afforded by the floras of the opposite coasts as bearing on the probable duration of such a barrier as the Strait of Gibraltar, it is best to leave out of account all species that are not known to be widely distributed. Here our very limited knowledge of the flora of North Marocco opposes a considerable difficulty. Subject to such light as future observation will throw upon the subject, it may be said that, so far as mere botanical evidence goes, we should infer that the barrier was not present at the time when the great majority of the existing plants spread into this region; but that it has been established long enough to oppose a limit to the further diffusion of many species that otherwise would, in all probability, be found on both sides of the Strait, thus indicating a period geologically recent, but very ancient as compared with the historic record.

On the following morning we crossed the bay to Gibraltar, and, still finding no means of conveyance to Tangier, endeavoured to console ourselves by botanising on the ' Rock.' Later in the day the impatience natural to the British traveller induced us to open negotiations for the hire of one of the numerous tug steamers that make handsome profits by helping becalmed ships through the Strait. The first demand of one hundred dollars helped to moderate our ardour; and, though the more reasonable sum of forty-five dollars was afterwards named by another merchant, we finally decided to remain a second night in Europe, and await the ordinary steamer on the following day.

It is well known that all the rules which prevent unauthorised persons from prying into the arcana of a fortress are strictly enforced at Gibraltar; and on this account a naturalist wishing to explore the rock should always apply for the previous permission of the Governor. Not intending to remain more than a few hours, we had declined the hospitable invitation of Sir W. F. Williams, and not thought of obtaining an order to authorise our unrestricted rambling over the rock. Towards evening Ball had started with his tin box to examine the steep eastern face that looks towards the Mediterranean. While scrambling about in search of plants, he became aware that his movements were watched by two Irish soldiers, both decidedly the worse for liquor, and as he returned towards the path the word 'spy' was emphatically pronounced more than once. Anticipating any further unpleasant remarks, he addressed them some ordinary question, with a fair infusion of that national accent that is unmistakable to the Hibernian ear. The effect was immediate: the men were delighted to recognise a countryman; question and answer rapidly succeeded, and the only difficulty was to resist their pressing invitation to adjourn to a neighbouring wine-shop, where the poor fellows' remaining intelligence would have been finally quenched in the compound of grape-juice and ardent spirits that is sold at Gibraltar as Spanish wine—not much worse, perhaps, than the mixture that is drunk at home by not a few persons boasting a refined taste under the name of pale sherry.

It seems natural to ask whether it is or is not true, as one is often assured, that correct plans of all the chief fortresses in Europe are to be found in the War Office of each of the chief States; for in such case the attempt to maintain secrecy as against the ignorant curiosity of travellers seems to be a puerile occupation for the military authorities in command.

The rock of Gibraltar and the sandy tract called the Neutral Ground produce many rare and interesting plants; but these are already well known to botanists, being separately described in Kelaart's *Flora Calpensis*, and further illustrated in a work of first-rate authority, Boissier's *Voyage Botanique en Espagne.* The only tree that seems to prosper thoroughly on this barren sun-baked headland is the Chinese *Phytolacca arborea*, which was planted some fifty or sixty years ago in the Alameda and elsewhere, many of which have attained a great thickness. They remind one of the stunted clustered columns of some mediæval churches, each of the very numerous branches developing a projecting cylinder of woody trunk covered with grey bark.

The so-called Club House, which ranks as the head inn, being already full, we put up at the Fonda Española, and had no cause for complaint, either as to food or accommodation. On the morning of the 17th we had notice that the steamer for Tangier was to start at noon; and, after laying in additional stores of drying paper, and enjoying a delightful morning stroll along the road to Europa Point, we were ready at the appointed time.

After more than the usual delay, we at length set our faces towards the African shore with a fresh SW. breeze in our faces. Few places in the world can show a greater variety of fine atmospheric effects than the Strait of Hercules. To-day the horizon behind us was clear, while the hills that bound the entrance from the Atlantic were veiled in thin haze; and, as the sun sank low, a strange purple hue suffused one-half of the sky. The skipper managed to arrive late in the roads at Tangier, and we found that, although a bribe to the official of the port might obtain admission within the walls, our baggage could not be landed until the following morning. We therefore decided to sleep on board the little steamer, and at length, on the morning of the 18th, we returned to breakfast at the Victoria Hotel.

Maw had made good use of his time. In a first excursion to the 'Lakes' he had failed to find a beautiful iris, which we had first admired on Sir J. D. Hay's dinner-table, and which we had taken to be the *Iris tingitana* of

Boissier and Reuter. Not easily foiled from his purpose, Maw returned two days later, and succeeded in his object. Subsequent examination has convinced us that the plant growing near the lakes is a luxuriant form of the *Iris filifolia* of Southern Spain, though intermediate between that and *I. tingitana*. The latter may perhaps be an extreme form of the same plant, but is yet little known, and had not, as far as we know, been brought into cultivation until carried to England by Maw. Our plant, which is one of the most beautiful of a beautiful group, is figured, under the name *Xyphion tingitanum*, in the 98th volume of the 'Botanical Magazine,' No. 5981. Nothing can surpass in the scale of rich sombre decoration the gradations of dark purple and brown velvet that enrich the petals.

One of the troubles that most try the patience of a botanical traveller here awaited us. As we had already assured ourselves, the spring climate of North Marocco is delightful to the human frame. The sky had been clear, the air warm, and only one or two slight showers of rain had fallen since we first landed on the coast; but the breezes, whether they travel eastward from the Atlantic, or westward from the Mediterranean, are laden with aqueous vapour nearly to the point of saturation, and nothing dries spontaneously by mere exposure to the air. Although our system of drying our plants by ventilating gratings makes it quite unnecessary to change the paper in such a climate as that of the Alps, or most parts of Europe, we now found that all the collections left at Tangier were suffering from damp, many specimens covered with mildew, and some hopelessly destroyed. Many hours on this and the following day were consumed in the endeavour to remedy the mischief. So far as structure is concerned, damp, when not too long continued, does not disorganise the tissues; but it finally removes the remaining freshness of colour which makes the beauty of a well-dried specimen.

In the course of the day we made some purchases of Fez pottery, of which a large store is kept by a Jew dealer.

This ware, which combines elegance and variety of form
with vigorous geometrical designs and rough execution, is
now well known to the devotees of the prevailing fancy for
ceramics, who pay in London or Paris many times over
the original price. Through the kindness of the British
Consul, Mr. White, we obtained some small specimens of
a very scarce variety of unglazed pottery, of which the
decoration consists merely in dots of black and red, forming various patterns. These were said to be the handy-
work of two potters of Fez, who both died during the last
cholera epidemic.

During our seven days' absence from Tangier, the vegetation had advanced very rapidly, and many plants had
come into flower during the interval; so that we found
abundant occupation, even in the immediate neighbourhood of the town. If we had wanted further evidence as
to the character of the climate, it was afforded by the fact
of our finding the British royal fern (*Osmunda regalis*),
on bare sandstone rocks, close to the sea. In our proverbially damp climate it requires boggy or marshy soil to
grow freely; but then, in spite of proverbs, we have fits
of dry weather during the spring, and every now and then
prolonged summer droughts, that forbid delicate ferns to
flourish in exposed situations.

Early on the morning of the 20th we were awakened
by the news that the long expected French steamer, *Vérité*,
of Marseilles, had arrived, and would depart in the afternoon on her voyage to the Atlantic ports of Marocco and
the Canary Islands. We were fully prepared to depart;
the expected autograph letter of the Sultan had been
delivered to Sir J. D. Hay, and by him to Hooker; our
heavy baggage had already been forwarded to Mogador,
and we lost no time in completing our preparations, and
bidding farewell to those whose kindness and hospitality
had made our stay at Tangier so agreeable. In quitting
Martin's Hotel, the solitary inconvenience that we could
call to mind was the swarms of flies that invade the rooms,

not more abundant, however, than in many valleys of
Switzerland and North Italy; and we carried away from
Tangier the impression that even on the Mediterranean
shores there are few spots that combine such advantages
of climate, natural beauty, and material comfort.

We found the *Vérité*, though boasting a French name,
to be a nearly new Clyde-built steamer, owned by a Marseilles Company and commanded by Captain Abeille of
that port, far better fitted up than most of those that ply
along this coast. The passengers were few, and, as these
disembarked at the intermediate ports, we at last became
the sole occupants of the state cabin. On a fine evening,
with the gentle heaving of the broad Atlantic billows to
tune all to harmony, we passed the headland of Cape
Spartel, and received the first rays of the great lanthorn
as they shot out seaward when lighted for the night.

At seven o'clock next morning the engines were
stopped, and going on deck we found ourselves lying some
way off the shore, opposite the mouth of the river Oued
Bouregrag, that divides Sallee from Rabat. The latter,
as seen from a distance, is a place of somewhat imposing
appearance. The chief mosque has a great square tower,
rivalling those of Seville and Marocco; and a pile of
modern masonry, on a scale unknown elsewhere in modern
days in this country, marks the large barrack where the
Sultan's body-guard is lodged when he pays his annual
visit to the coast. Carpets are made here, and also a
peculiar sort of unglazed pottery, coarse in texture, but
admirable in form, and singular in ornamentation.[1] Over
against Rabat, on the north side of the river, is Sallee,
once a famous place, the last outpost of Roman civilisation, and afterwards the home of pirates who were dreaded
throughout the Mediterranean and along the coasts of
France and England. Looking at the bare coast, and the
paltry groups of mud boxes that make up a Moorish town,

---

[1] Some fine specimens have been exhibited at the South Kensington
Museum, by our companion, Mr. Maw.

and knowing that the bar at the river's mouth allows, except at spring tide, the passage only of ships of small tonnage, it seemed scarcely credible that the European Powers should so long have allowed such a nest of hornets to flourish at their very gates. When one reads that up to the middle of the last century it was not a very rare thing for the 'Sallee rovers' to lie under Lundy Island, and cut out Bristol merchantmen, one asks what the British navy was about, that the malefactors and their ships were not swept from the sea, and Sallee itself utterly destroyed. The false humanity that caused in our time such bitter lamentations over the chastisement of Bornean pirates had not been yet invented.

We lay for the greater part of the day within some two or three miles of the shore, but the Atlantic rollers were too heavy to allow a nearer approach, or permit the landing of cargo. This happens too frequently to excite remark; and these great waves, originating in the passage of cyclones in the mid-Atlantic, often arrive so suddenly in the calmest weather as to create a serious danger for the seaman. At the least it is prudent to keep up a sufficient pressure of steam in the boiler to make it easy to gain the offing on the shortest notice; and we heard of several cases where the coast steamers had called in succession at all the Atlantic ports of Marocco without being able to communicate with any one of them, and cargo and passengers had been carried on to the Canary Islands with the uncertain prospect of being landed on the return voyage. Fogs offer another serious impediment to navigation on this coast. During the summer the low country for a distance of eight or ten miles from the shore is not rarely covered during the morning with a thick mist that clears away before mid-day. At such times ships dare not approach the sandy coasts, and, when the sky clears, the scarcity of landmarks makes it extremely difficult for the seaman to ascertain his exact position. As the same difficulty prevented us from touching Rabat on our return voyage, we can add nothing to what has

been told by preceding travellers. Counting Sallee as a
suburb of the larger town, the population is estimated at
40,000, or more than all the other Atlantic ports put
together. The inhabitants are said to suffer from three
scourges—prolonged droughts, the invasion of locusts, and,
worst of all, the annual visits of the Sultan, whose body-
guard of several thousand soldiers has to be fed at their cost.

To the naturalist a stay of some days at Rabat might
be of great interest if he were able to accomplish a visit
to the famous forest of Mamora, which fills a large part of
the space, some twenty miles in width, between the mouth
of the Bouregrag and the larger river Sebou that carries
to the sea the drainage of the high mountains near Fez.
The scene of most of the wonderful tales that circulate
among the people of North Marocco—adventures with
lions, robbers, and other wild animals—is laid in the forest
of Mamora; but excepting one solitary plant, brought
thence by the Abbé Durand—a very distinct species of
Celsia—nothing is known of the fauna and flora of the
forests of this part of Marocco. These appear to cover
a considerable tract parallel to the Atlantic coast, and
probably consist mainly of the cork oak, which in any
other country might become a considerable source of profit.
Eastward of the forest the country south of the Oued
Sebou is a marshy tract, breeding endemic fevers that are
said to extend to Sallee and Rabat.

In the afternoon the swell became more moderate, and
a boat came out with passengers, including the family of
Mr. Dupuis, the British Vice-Consul at Casa Blanca. It
was decided that it would not be safe to land cargo, so the
captain resolved to start without further delay and run for
Casa Blanca—the Dar el-Beïda of the Moors. The sun
had set, and night was closing in as we approached the low
shore, where a few white houses mark a station which has
risen to some little importance owing to the preference
shown for it by French merchants, who carry on a considerable trade with the interior.

We accepted a courteous invitation from Mr. and
Madame Dupuis, and, landing early on the morning of the
22nd, went to breakfast at their house. A less attractive
spot than Casa Blanca it is difficult to imagine. A fea-
tureless coast of low shelves of red sandstone rock overlaid
by stiff clay, stretches on either side in slight undulations,
nowhere rising more than a couple of hundred feet above
the sea. Not a tree gives variety to the outline or shelter
from the blazing sun. The attempts made by the few
residents to cultivate the orange and other useful trees
have met with little success; and the eye seeks in vain
the gay shrubs that adorn the southern shores of the
Mediterranean. The Cistuses, Genistas, heaths, Arbutus,
and myrtle, as well as the more sober prickly oak and
laurel, are all absent, and the arborescent vegetation is
almost limited to stunted bushes of lentisk some three or
four feet high.

As we strolled for several hours over the surrounding
country, we at once perceived the influence of new climatal conditions. It was not that many new species
marked the passage from one botanical province to another,
for to our disappointment we found very few that we had
not already gathered in North Marocco, and, excepting one
rare *Celsia*, none that were not already well known. As
elsewhere, *Leguminosæ* were predominant, and especially
trefoils and medicks; grasses were both numerous and
varied in species; and *Umbelliferæ* were represented by
many conspicuous plants, of which *Ferula communis*,
growing to a height of ten or more feet, is especially
notable. In the absence of more substantial materials,
the thick stems are used for fences. The contrast offered
by the vegetation of this coast with that of the Mediterranean shores is caused altogether by climatal conditions,
which allow one set of species to flourish while the rest
are more or less rigidly excluded.

The information received from our obliging hosts
respecting the country and the native population agreed

well enough with what we heard elsewhere. The prejudices of the natives are not so strong as to make them
indifferent to the advantages of trade with the intrusive
Christians who are settled on the coast; and the unfortunate issue of the last war with Spain has taught them the
prudence of avoiding wanton provocation. Whatever may
be the case with the tribes farther inland, the people of
the coast provinces are quite disposed for commercial intercourse; but the jealousy of the authorities makes enterprise of all kinds too unsafe to be risked by an ordinary
native of the country. Some of the provincial governors
who live near the coast carry on trade with European
merchants; but for the rest such business as exists is in the
hands of the Jews. The only interference of the Government, which is at least ostensibly dictated by a regard for
the welfare of the people, relates to the corn trade. In
favourable years Marocco produces much more grain than
the population can consume, but drought and locusts
often destroy the crops throughout large districts. The
permission to export corn is therefore given or withheld
by sovereign order according to the reports received at
head-quarters. It is needless to point out how much the
uncertainty thus produced must interfere with the profits
of cultivation.

At Casa Blanca our skipper took on board a considerable quantity of maize for the Canary Islands, and a
good many bales of hides and wool for Marseilles; and we
found the decks in some disorder when we returned on
board our steamer in the evening. All next day—the
23rd—we remained in the roads of Casa Blanca, uncertain
at what moment we should continue our voyage. The time
did not hang heavily on our hands, for we had as much
work as we could accomplish in getting our collections
into tolerably good order. We here had to deal with an
enemy that was new to all of us, excepting Hooker, and
which for the next week was to cause more trouble and
anxiety than any one not a naturalist can easily realise.

Nothing is more common with us at home than to grumble at the dampness of the climate; and, as far as the effects on the human animal are concerned, our complaints are perfectly just.   Air at 50° Fahr. cannot at the utmost carry more than about 4½ grains of aqueous vapour to the cubic foot; but at that temperature it produces, when nearly saturated, that feeling on the nerves of the skin, familiar to every inhabitant of these islands, which is the ordinary forerunner of colds, sore throats, rheumatism, and many another ailment.   But the botanist, to whom the condition of his drying paper is even more important than that of his own body, finds an easy remedy for the inconvenience.   By exposing his damp paper to a temperature of from 80° to 90° in the sunshine, or before a fire, he readily obtains a satisfactory degree of relative dryness, and in a very few days his specimens are in a state to put away, and with ordinary care need give him no further trouble.   But the case is very different where the ordinary temperature of the air in the shade is about 75°, as was the case here, not to speak of 85° which is the common limit in the tropics.   To the human body there is nothing unpleasant in the effects of such air when nearly saturated with vapour, and so long as the temperature remains habitually between 70° and 80° it is decidedly favourable to health, if not to vigorous exertion.   But a cubic foot of air at 77° contains nearly 10½ grains of vapour, and when at all near to the point of saturation it has no perceptible drying effect on surrounding objects, and a moderate increase of 10° or 12° Fahr. in temperature has but a slight effect in increasing its desiccating power. We were first struck by remarking the very long time required to dry the decks as compared with what is usual in the Mediterranean, and we had still more painful experience of the difficulty of drying our paper.   We were now the sole occupants of the saloon, and our captain left us free to use every part of the steamer; the deck was soon turned to account, cords were stretched across the rigging, even the neighbourhood of the boiler was invaded, but with indifferent success. Few readers may care to sympathise with the distress of a naturalist who looks on his specimens, not only as scientific documents bringing some additions to our knowledge of the structure and relations of the organised world, but as things of beauty giving delight to the senses of form and colour, when, after much pains and care, he finds the flowers change their hues and drop off, the leaves turn black, and when mould, the sure sign of decomposition, begins to encrust the stems and fruits.

At 1 A.M. on the morning of the 24th we were again under steam, and soon after daylight speed was slackened as we lay off Mazagan. The abruptness of the transition from deep blue water in the offing to a somewhat milky green where the ship gets into shallower water here attracted our notice. It is of common occurrence even on coasts where there is reason to believe that the bed of the sea shelves vary gradually away from the shore, and one might expect a gradual change of tint; but no satisfactory explanation occurred to us.[1] It was some time before the land came in sight, and we were able to make out the square tower of the Portuguese fort that marks the position of Mazagan. The town stands on a slightly projecting point of land facing northward, and therefore especially exposed to the north-east breeze that prevails throughout the spring and summer. We lay all day rolling heavily, and the surf, breaking in hills of foam upon the shore, was too high to allow of the landing of cargo; but in the afternoon a small boat put off with provisions. Amongst these was a large freshwater fish, a species of shad, that had been caught in the Oued Oum-er-bia which runs into

---

[1] Professor Tyndall has shown that the differences of tint in sea-water depend upon differences in the amount and dimensions of the particles of solid matter held in suspension; but the abruptness of the transition from one tint to another has, we believe, not been fully explained.

the sea some five miles east of Mazagan close to the site
of Azemour, a ruined town once of some importance. The
freshwater fish of the streams from the Atlas may probably
offer many objects of interest to the ichthyologist, but do
not seem likely to add much to the resources of the cook.
We were told that the fine-looking animal which was
displayed at table is considered a delicacy; but we found
the flesh insipid and cottony, and during our subsequent
journey we failed to find any fish worth eating.

Neither on this occasion nor on our return did we see
any trace of the ruins of Azemour or of the great river
Oum-er-bia. This is apparently the chief stream of Marocco. It drains the northern declivity of the chain of
the Great Atlas for a distance of 150 miles, and nearly
the entire of the extensive mountainous region, a still unknown network of high ridges and deep valleys, that
covers nearly half the space between the main chain and
the Atlantic seaboard. Like all the other rivers of this
country the volume of water varies to an extent unknown
in Europe. In dry seasons, when a large part of the waters
that descend from the mountains is diverted into irrigation
channels, and never reaches the sea, the main stream runs
over a shallow bed fordable in many places; but after
heavy rains the swollen waters have such a rapid current
that we were told of travellers being detained a week or
ten days waiting for the opportunity of crossing it. Lieut.,
afterwards Admiral, Washington [1] estimated the breadth of
the river where he crossed it, near Azemour, at 150 yards,
and found it much the same at about eighty miles from
the sea on the return journey from Marocco to the coast.

Mazagan, though a small and poor-looking place, bears
many traces of its European origin, as we remarked when
we landed here on our return voyage from Mogador. It
was built by the Portuguese in 1566, and held in spite of
frequent assaults by the Moors for more than two hundred
years, having been finally surrendered in 1770.

[1] 'Journal of the Royal Geographical Society,' vol. i., pp. 132–151.

J. B. delt.

SAFFI

We left the roads at 9 A.M. on the 25th, and were glad
to see for the first time the land rising in bold cliffs. The
headland seen a few miles south-west of Mazagan is Cape
Blanco; but this projects little from the general outline of
the coast, which shows a tolerably uniform direction, rising
gradually towards the south-west, till we reach Cape
Cantin, the chief headland of this part of the Atlantic
seaboard. The summit is apparently about three hundred
feet above the sea, and the calcareous strata nearly horizontal. Here the coast line, which from Cape Blanco had
kept the direction from north-east by east to south-west by
west, turns abruptly to the south. The cliffs recede a little
at first and form a slight curve, then rising to a second
headland some two hundred feet higher than Cape Cantin.
Beyond this the shore again recedes, and the land subsides,
where a slender stream has cut its way through the plateau
inland, and affords space for the little seaport town of
Saffi, or Asfi of the Moors. The coast line again rises on
the south side of Saffi, forming a steep escarpment some
three or four hundred feet in height, called the Jews' Rock,
about four miles from the town.

Saffi is by far the most picturesque spot on the west
coast of Marocco. The extensive fortifications of the Portuguese, high walls and square towers, spreading along the
shore and up the broken declivity on which the town is
built, with several steep islets, whose rocks have been
gnawed into uncouth shapes by the Atlantic waves, produce, as seen from the sea, a striking effect. Though
fully exposed to the west, this port is better protected
from the north-east winds than any other on the coast,
except Mogador. Behind it lies the fertile province of
Abda, famed for its excellent breed of horses, and it is the
nearest port to the city of Marocco—about one hundred
miles distant—but the want of secure anchorage for shipping neutralises these natural advantages.

Our stay on this occasion was short, and soon after
dark we were again in motion. We spent pleasantly

enough our last evening on board the *Vérité*. Though he took little pains to conceal his strong prejudices against the English nation, our captain was thoroughly good-natured and obliging towards the individual Englishmen with whom he was associated. No doubt our scientific pursuits recommended us to his good offices, for the slight smattering of scientific knowledge acquired by half-educated persons in most Continental countries has the effect of awakening some interest in such pursuits. It may, indeed, be doubted whether, at least in France, the teaching of physical science goes far enough to convey any accurate knowledge, even of an elementary kind ; but, at all events, the national temperament leads Frenchmen to expose their deficiencies more than other people readily do. An Englishman who knows that he is not well grounded in a subject holds his tongue, or if pressed by questions will probably exaggerate the extent of his own ignorance, where a Frenchman will gaily lay down the law and span over the gaps in his knowledge by startling bridges of conjecture. Our worthy skipper amused us not a little when, in conversation on the climate of this coast, reference being made to the rainless zone of the Peruvian coast, he explained that in that country the moisture of the air is absorbed by the gases that accompany earth-quakes, thus accounting to his own satisfaction for the meteorological phenomenon. But the full vehemence of his nature was reserved for matters of much more im-mediate interest. He had left Marseilles after the Com-munist rising in that city had been suppressed, but while the miserable tragic farce that was to end in the horrors of May, 1871, was being enacted in Paris. He could not allude to the subject without a degree of fury that to us seemed utterly unreasonable. But it is easy for people at a distance to treat such matters with calmness, and there were not many Englishmen on the spot who at the time were able to share the noble calmness of Lord Canning during the Indian Mutiny.

CHAPTER IV.

Arrival at Mogador—The Sultan's letter—Preparations for our journey
—The town of Mogador—The neighbouring country—Ravages of
locusts—Native races of South Marocco—Excursion to the island—
Climate of Mogador—Its influence on consumption—Dinner with
the Governor.

AT 5 A.M. on April 26 we at length reached the port of
Mogador. Before many minutes a boat was alongside,
and we were warmly welcomed by a gentleman who introduced himself as Mr. Carstensen, the British Vice-Consul,
brother-in-law of Sir J. D. Hay. He was, indeed, no
stranger; for, as a correspondent and active contributor
to the Royal Gardens at Kew, he had long been in friendly
relations with the chief of our party. To his energetic
good offices and hospitable attentions we owe deep obligations, and it was with sincere regret that we subsequently
heard of his premature death in 1873.

At an early hour we were comfortably established in
the British Consulate, where our host and hostess received
us as old friends, and we were soon engaged in discussion
as to the arrangements for the prosecution of our journey,
in all of which Mr. Carstensen's familiarity with the
country and perfect command of the language were of the
utmost value. Having received previous notice of our
arrival and of the objects of our journey, he had already
prepared the way, and thus very much abridged the delays
that are inevitable in such a country.

The first step necessary was to call on the Governor and
present to him the Sultan's letter. We were courteously
received by El Hadj Hamara, a well-looking man of

middle age, in a small plain room, whose only furniture consisted in cushions laid round the walls. After shaking hands in European fashion, we proceeded to seat ourselves, cross-legged—no doubt looking very uncomfortable during the experiment—while the Sultan's letter was produced. This was written on a small sheet of inferior paper, folded to the size of a note, and sealed with coarse sealing-wax. It was received by the Governor, the seal reverently applied to his forehead, and then broken. After reading aloud the few lines of writing, the Governor handed the letter to Mr. Carstensen, who proceeded to translate literally for our benefit. It ran thus: ' On receiving this, you will send the English hakeem and his companions to the care of my slave, El Graoui, to whom I have sent orders what he is to do.' It should be explained that El Graoui, spoken of as the Sultan's slave, was the Governor of the portion of the Great Atlas that is practically subject to the Imperial authority, and precisely the person whose favour and assistance it was essential for our objects to secure.

To strangers unused to the style of the Marocco Court, the Imperial letter did not seem a very promising document; but it was evident that, so far as the Governor of Mogador was concerned, it conveyed the impression that we were to be treated with respect and attention; and this was doubtless confirmed by the arrival of a courier from Marocco, bearing a letter from the Sultan's eldest son, then acting as viceroy in the southern provinces of the empire, with orders to take every care for our safety and comfort during the journey to the capital.

We soon had a specimen of the shape in which official protection displays itself in this country. On a representation from Mr. Carstensen that we should require numerous baggage animals, besides horses and mules to ride, the order had gone forth a week before our arrival that no horses or mules should be sold or hired in the town of Mogador until we had selected such as we required. This accordingly was one of our first cares, and the embargo was raised in the course of the day. We followed local advice, confirmed by our own previous experience in warm countries, in choosing mules in preference to horses. On a long journey they are far less liable to be laid up, and, to a scientific traveller who has frequent occasion to dismount, they give less trouble. Their obstinate temper is, however, often annoying, and, though surefooted, they sometimes have a very unpleasant trick of tripping or stumbling over stony ground.

A precaution which we took this day is much to be recommended to travellers. This was to make a trial of pitching our tents on a piece of rough open ground. People readily suppose that a tent that is easily set up in an English lawn must answer their expectations on a march, and have little notion of the amount of discomfort caused by trifling defects. We speedily found that the pegs supplied in England are not nearly hard enough to pierce the stiff-baked clay or stony paste that forms the prevailing soil in this country ; and it was fortunate for our comfort that we took from Mogador an ample supply of rough pegs, made from the wood of the argan tree. We were each provided with a tent which satisfied our individual wants, but scarcely corresponded with the native ideas of what befits personages of distinction. We were well aware that in this country *prestige* was an essential element in success, and therefore willingly accepted the liberal offer of a large handsome native tent made by the local agent of Messrs. Perry & Co. of Liverpool. This was available only for the journey across the plains between Mogador and Marocco, as it was very heavy, forming a load for two camels, and therefore not suitable for a hilly country. It supplied a comparatively spacious saloon, wherein we passed our evenings very pleasantly, before retiring to our separate quarters for the night.

The next matter requiring attention was our costume. It was foreseen that during some part of our journey, at least, it might be expedient to adopt the native dress, or

such an approximation to it as would prevent our attract-
ing notice from afar as strange and outlandish creatures.
After due deliberation, the *haïk* was finally rejected. This
is the ordinary outer garment of natives of the upper
class. An ample robe of fine white woollen stuff is a
graceful and picturesque garment, especially on those who
know how to group its folds about the person; but it is
absolutely incompatible with the free use of the limbs, and
more especially for botanists, whose pursuit brings them
into frequent contact with the numberless spiny plants of
this region. The unsightly *jellabia*, a blouse of rough
white woollen stuff, with the addition of a hood that may
be drawn over the head, was adopted, and was not found
very inconvenient.

Anticipating unavoidable exposure to a nearly vertical
sun, we had provided ourselves with the grey pith ventilating helmets so commonly used by Englishmen in the
tropics. It was found that by winding round one of these
a moderate strip of the usual material for turbans, it
might be made to pass muster at a distance. But for
head-gear on important occasions the turban was indispensable. The material, a broad band of light muslin,
about thirteen or fourteen yards in length, is supplied from
England, but the art of winding it round the head requires
long practice, and we always resorted to the aid of one
of our attendants. It certainly gives protection against a
hot sun; but it is never quite convenient to a European
of active habits, who finds it hard to acquire the orthodox
gravity of Oriental demeanour, and is sadly apt to disturb
the folds of the turban by some abrupt movement.

There was one article of dress as to which no compro-
mise was possible. The slippers down at heel that are
commonly used by all classes of natives, and even the red
or yellow loose boots that are sometimes worn on a journey,
were equally unsuited to our habits and pursuits, and we
held fast to our accustomed foot-covering.

Mr. Carstensen had kindly made excellent arrange-

ments for our convenience during our journey by selecting
such native attendants as we should require. One was
told off to each of us as a personal servant, expected to
be always in readiness to render any required assistance;
and Hooker's English attendant, Crump, was included
in this arrangement. This may appear superfluous, and
so it might be to ordinary travellers; but for a party of
naturalists anxious to make the best use of their time,
it was almost indispensable. Several other men were
attached to the camp in various capacities, one of the
most useful being a saddler, daily in requisition to repair
damage done to leather work; but by far the most important member of our suite was the interpreter to the
British Consulate, whose services were spared for fully
five weeks. Even with Mr. Carstensen's thorough knowledge of the language, this must have been felt as a serious
inconvenience, for Abraham proved himself active and
intelligent; and the duties of a consular agent on the
Marocco coast being by no means of a hum-drum character, the need of a man familiar with the country and
the people in the capacity of secretary and assistant is
daily felt. Being a Marocco Jew, born in a position of
relative inferiority to his Mohammedan neighbours,
Abraham no doubt felt a keen satisfaction in the sense of
security which he derived from his position in the British
service. To be able to converse in a tone approaching to
equality with powerful officials; to emancipate oneself
from restrictions trifling, yet galling, in matters of dress
and demeanour; to share in some measure in the vague
sense of power vested in the representatives of the great
European States—must be the climax of ambition to a
member of a despised nationality in a land where neither
intelligence nor wealth nor good reputation give a man
security or social recognition.

It had been arranged that our escort was to consist of
four soldiers, under the command of a *kaïd*, nearly equivalent, as we were told, to a captain in European army

rank. This was more than was requisite for security, as, with all its barbarism, the Marocco Government is efficient enough within the parts of the territory where the Sultan's authority is recognised and feared. Within those limits it is enough to let it be known that a traveller enjoys the Imperial protection; no one will ever think of daring to molest him.

After devoting a good part of the day to indispensable preparations for our future journey, we were free to look about us in the singular little town which, as the chief port of South Marocco, is the last outpost of civilisation on the African coast at this side of the French settlements of Senegal. Like many other places in Marocco, this owes its existence to the caprice of a Sultan. It was founded in 1760 by Sidi Mohammed, the most energetic of recent Moorish sovereigns, and became a considerable place when, a few years later, the same ruler destroyed Agadir, and ordered the merchants established there to remove to Mogador. Jackson tells us that it received its European name from the sanctuary of Sidi Mogodol, standing somewhere among the neighbouring sandhills; but a town of Mogador is shown in a map published in 1608,[1] standing a short way north of the island, which is there marked 'I. Domegador.' As have most of those marked on the early maps, the ancient town had doubtless disappeared before the foundation of the present one, called by the Moors Soueira; but the old name must have survived in the country.

The low rocky island lying opposite to the town, and separated by a navigable channel, affords shelter from all winds except those from the SW.; but the depth of water is not great, and there are numerous dangerous reefs, so that in threatening weather steam is always kept up, and ships proceed to sea when SW. winds are expected. Although the island is shown on the oldest maps, and the channel is represented much as we now see it in the plates

[1] See Appendix C.

to Jackson's work, from drawings made about the beginning
of this century, we were positively assured that old people
in Mogador recollected the time when the island was connected with the mainland by an isthmus, over which cattle
could be driven at low water ; and this story seemed to
have gained credence with the European inhabitants.

Though it has no buildings of importance, the town is
in one respect the most habitable in Marocco, being remarkably clean, and in that respect superior to very many
seaports in Europe.   This is largely due to the efforts of
two intelligent French physicians, who have at various
periods visited Mogador, but especially to the exertions of
Dr. Thevenin, who has resided there for many years.

The Governor and other officials, with the European
consuls and merchants, all reside in the Kasbah—the chief
of the three quarters into which the town is divided.
Here are several narrow but regularly-built streets; the
houses are mostly of two stories, enclosing a small court-
yard, which is entered by a low and narrow doorway from
the street.   In the Moorish town, inhabited by natives of
the lower class, the houses are of one story, and poor in
appearance ; but the practice of whitewashing within and
without once every week makes them look clean, and, no
doubt, has much to do with the remarkable immunity of
this place from contagious and endemic diseases.   The
Jewish town is much overcrowded ; but we were assured
that even here the modern gospel of soap and water has
made much progress.

In the afternoon we sallied forth with our portfolios ;
but in deference to public opinion, which could not endure
that strangers of consequence should be seen trudging on
foot, we rode for about a mile out of the town.   Its sur-
roundings are not prepossessing.   The low tertiary limestone rock, on which it is built, and which doubtless extends
inland for some distance, is covered up to the city walls
by blown sand, driven along the shore before the SW.
winds, forming dunes that cover the whole surface ; and

in most directions one may ride two or three miles before
encountering any other vegetation than a few paltry
attempts at cultivating vegetables for the table within
little enclosed plots, whose owners are constantly disputing
the ground with the intrusive sand.   The chief break in
the monotony of the sand ridges is due to the small stream
of the Oued Kseb (called Oued el-Ghoreb on Beaudouin's
map), which reaches the sea little more than a mile away
on the south side of the town.   Much of the water being
diverted, the current is not strong enough to keep a
channel through the sands, but forms at its mouth a marsh,
where many of the most interesting plants of the neighbourhood are to be found.   The drip from the small
aqueduct that supplies water to the town suffices to give
nourishment to other less uncommon species.

Mogador has long been tolerably well known to botanists.   It was visited by Broussonet at the latter end of
the last century, and was for some time the residence of
Schousboë.   More recently the neighbourhood has been
explored by the late Mr. Lowe and by M. Balansa.   We
could not, therefore, reasonably expect to find here anything
new to science; but our short excursion was nevertheless
full of interest, though not altogether of an agreeable
kind.   We here saw for the first time a district recently
ravaged by locusts; and while we acquired a lively sense
of the amount of mischief effected by these destructive
creatures, we also found out how it happens that the
damage is confined within tolerable limits; how, in short,
they fail to turn the country into a desert.   When one
reads the reports of credible eye-witnesses, who describe
the arrival of swarms of locusts that devour every green
thing, one asks oneself how it can be possible for man or
animals to survive such destruction.   In the first place, it
may be remarked that, like most other sweeping statements, these are not strictly true.   The locusts do not, in
point of fact, devour every green thing.   In the spots
where they were most destructive we always remarked

that certain plants escaped untouched. The result of this immunity would naturally be to substitute the latter for the species destroyed by the locusts, were there not some very efficient agency for repairing the damage and maintaining the life of the species, if not of the individual. An important element in considering this question is the season at which the mischief is effected. The young locust grows very fast, and it is mainly during the period of growth that it consumes vegetation. When once the animal has attained its full size, it becomes comparatively inert, and its capacity for destruction is vastly diminished. If the swarm of young locusts arrives before the middle of April, when the rainy season is not quite over, the first showers revive the plants that have been devoured almost to the root with surprising rapidity. Perennial species throw out new buds, and are soon again covered with leaf and flower; and the same often happens with annuals, unless these have already shed their seed, and then a new crop soon reappears. It may be supposed that the vast amount of decaying animal matter left on the surface, even in the most barren spot, contributes not a little to the vigour of the vegetation, and thus compensates for the destruction effected at an earlier stage. It is when the swarms appear late, and attack the wheat or maize after the flowers are developed, that the consequences to the population are very serious, and famines result that periodically affect large districts.

In the present year it was clear that rain had fallen since the locust invasion, and although much damage had been done, tolerable specimens of many plants here seen for the first time were to be found. A few of these are common to the Canary Islands and this part of Africa ; others are not yet known except on this coast. The most curious of them is the *Senecio* (*Kleinia*) *pteroneura*, whose succulent almost leafless branches, as thick as a man's finger, bear a few heads of flowers that differ little, save in their larger size, from those of the common groundsel.

Well pleased with our first glance at the South Marocco flora, we returned to our comfortable quarters, and spent a pleasant evening in discussing our future movements, and in drawing upon our host's ample stores of information respecting the country and its inhabitants.

We were now for the first time brought into contact with the primitive stock of this part of Africa, one main branch of the Bereber race, which is distinguished by speaking some dialect of the Shelluh (Shleuh) language.[1] The affinity of this people with the Berebers of the Lesser Atlas—including under that name the Kabyles of Algeria, with the Riff tribes of North-west Marocco—has been denied, but does not appear to be open to reasonable doubt. The type is physically the same, excepting among some of the tribes south of the Great Atlas, where the intermixture of Negro blood has introduced new and very diverse elements. The languages now spoken among these tribes doubtless exhibit marked differences, especially to the ear of a foreigner. Jackson long ago denied the relationship between the Shelluh and the Bereber, while Washington, in the paper already quoted, came to a contrary conclusion. It may now be considered as beyond question that the differences between the Shelluh and the Kabyle are merely dialectic.[2] The value of linguistic evidence in ethnological inquiries has of late been questioned by eminent critics, and it must be conceded that such evidence, when it merely rests on lexicographical coincidences, is of less value than when it is derived from grammatical structure; yet, after all deductions, the facts remain to be accounted for, and, in the absence of proof to the contrary, it goes far towards proving community of origin. It must be remembered, that unlettered races are subject to far greater and more rapid changes of dialect

---

[1] The usage of preceding English writers is hereafter followed by writing the name, Shelluh; but to our ears the native pronunciation is more accurately given by the spelling Shleuh or Shloo.

[2] See Appendix H.

than those who preserve in sacred books or popular poetry
fixed standards of correct speech; add to this, the chances
of error when a traveller, communicating with a native
through an interpreter, and contending with sounds unusual to his ear, attempts to form a vocabulary. These
causes, acting together, tend to increase the difficulty of
recognizing linguistic affinities that really exist.

In the absence of any indication of the intrusion of a
conquering race that can be supposed to have imposed its
language on the previous population, it seems most probable that the native races of North Africa, between the
Libyan Desert and the Atlantic coast, including also the
Canary Islands, all belong to a single stock, which may
best be called Bereber. The two main branches are both
mountain peoples. To the north we have the tribes of
the Lesser Atlas, extending from the gates of Tetuan to
the hill country of Tunis, who may best bear the common
name of Kabyles—to the south-west the population of the
Great Atlas, from the neighbourhood of Fez to the coast
between Agadir and Oued Noun, broken up into numerous
tribes, but all speaking some dialect of the same language,
and thence called generically Shelluhs. Of the scattered
fragments of the Bereber stock that have spread far
through the oases of the Great Desert, till they have
come into contact with the Negro tribes from the south
of that barrier, our information is still most imperfect.
In constant conflict with each other, and with the Arab
and Negro tribes who dispute with them the scanty means
of subsistence that Nature here provides, they appear on
the whole to predominate over their competitors. The
Touarecks, scattered over a territory as large as half of
Europe, from Algeria to Soudan, form a separate branch
of the same stock; while we learn from Gerhard Rohlfs
that the predatory tribes of the desert south of Marocco
are merely Shelluhs who have changed their habits and
manner of life to suit altered conditions of existence.

The character of the Bereber has scarcely received

justice at the hands either of ancient or modern writers.
They have been inconvenient neighbours for those who
have sought to encroach on their territory, and they are
justly dreaded by the traveller through the Great Desert as
the most active and enterprising of the human enemies he
must confront or evade. Comparing them with the Moor
and Arab population of South Marocco, our report agrees
with that of Jackson, who probably knew them better
than any other European has done. They are decidedly
superior in intelligence, in industry, and general activity
to their neighbours. Two of our retinue, selected by Mr.
Carstensen among the mountaineers who resort to Mogador
to pick up a living about the port, distinguished themselves over all the rest both in physical and mental qualities; and one of these especially, who became Hooker's
personal attendant, showed an amount of general intelligence and unfailing cheerfulness that made him a favourite
with the entire party.

On the morning of the 27th we made an excursion to
the island.   It is formed of an irregular, low, knobby
mass of very friable tertiary rock, which seems to yield
rapidly to the erosive action of the heavy waves that
almost constantly break on its seaward face, where the
overhanging cliffs are hollowed into caverns. At the time
of our visit it appeared to be uninhabited. Two or three
heavy pieces of cannon, honeycombed with rust, lay near
the highest point, but seemed never to have been placed
in position. A small building was said to have been sometimes used for the custody of State prisoners, but otherwise
there was no indication here of the presence of man. In
such a spot we expected to find the coast vegetation fully
developed, but we counted without the locusts. Nowhere
else did we observe such complete destruction. A good
many plants growing on the rocks, within constant reach
of the sea-spray, had escaped; but on the rest of the island
scarcely a green leaf remained, and it required a patient
search to discover a few fruits of some leguminous plants

that appear to abound in this locality. Of the seaside
rock-plants three were supposed to be peculiar to this
single spot. *Andryala mogadorensis*, of Cosson, a very
showy species of an unattractive genus, has been well
figured in the 'Botanical Magazine' for 1873; *Frankenia velutina*, the most ornamental species of that
variable genus, appeared at first quite distinct, but we
were afterwards led to suspect it to be a local form or sub-
species of the widely spread perennial *Frankenia*, so
common in the Mediterranean region. Both of these we
afterwards found on the coast near Saffi. Of the third
plant—*Asteriscus imbricatus*, of Decandolle—but a single
stunted specimen was found by Ball, and as yet it has no
other known habitat. We here saw for the first time a
plant which turned out to be rather common in South
Marocco, and which was taken by us, as it had been by
preceding botanists, to be the *Apteranthes Gussoniana*,
of Mikan, first described by Gussone as *Stapelia europœa*,
and in truth closely resembling in habit and appearance
some of the South African species of *Stapelia*. The fruit,
which we afterwards found in abundance, did not appear
different from that of Gussone's plant; but when the
specimens carried to England by Maw flowered two years
later, the structure of the flower showed that it should be
recognised as a distinct species of the group which has
received the generic name *Boucerosia*, and it was accordingly published by Hooker, in the 'Botanical Magazine'
(No. 6137), under the name *Boucerosia maroccana*.

In the course of the day we called on Monsieur
Beaumier, the French Consul, in company with Dr.
Thevenin, an intelligent physician, who has spent several
years at Mogador, much to the advantage of the inhabitants whether Christian or native. M. Beaumier not
only received us with the proverbial courtesy of his
country, but showed a warm interest in the success of our
journey, and kindly supplied us with many items of
information, along with manuscript notes prepared by

himself during his residence in South Marocco. His premature death, from an illness contracted during a visit to France in 1875, has been a serious loss to the country which he had made his second home.

Amongst other items of information, we owe to M. Beaumier a series of meteorological observations carried on at Mogador with a single interruption for nearly nine years, and supplying all requisite particulars for eight complete years. The results are so remarkable that they have attracted the attention of many physicians, and may probably lead at some not distant date to the selection of this place as a sanitarium for consumptive patients.

Dr. Thevenin mentioned several facts of much interest in their bearing on this question. In the first place, phthisis is all but completely unknown among the inhabitants of this part of Africa; while in Algeria cases are not rare among the natives, and in Egypt they are rather frequent. In the course of ten years he had met but five cases among his very numerous native patients, and in three of these the disease had been contracted at a distance. He further mentioned several cases among Europeans who had arrived in an advanced stage of the disease, on whom the influence of the climate had exercised a remarkable curative effect.

An examination of the tables, showing the results of M. Beaumier's observations, and especially those for temperature, may help to explain these facts, as they certainly show that Mogador enjoys a more equable climate than any place within the temperate zone as to which we possess accurate information.

It should be mentioned that these observations were made with good instruments, sufficiently well situated on the shady side of the open court-yard of the French Consulate, about thirty feet above the sea level. The hours of observation were 8 A.M., 2 P.M., and 10 P.M.—not perhaps the best that could be selected, but sufficient in a climate where rapid transitions are unknown.

A few of the results here stated in Fahrenheit's scale
are derived from M. Beaumier's tables as continued to the
end of 1874 :—

Mean temperature during eight years	=	66.9°
Do. for the hottest year (1867)	=	68.65
Do. for the coldest year (1872)	=	65.75
Mean of the annual maxima	=	82.5
Mean of the annual minima	=	53·0
Highest temperature observed	=	87.8
Lowest temperature observed	=	50.7

More striking still is the comparison between the temperature of summer and winter. The following results
show the monthly mean temperature, derived from eight
years' observations :—

Summer	{ June	=	70.8
	{ July	=	71.1
	{ August	=	71.2
Winter	{ December	=	61.4
	{ January	=	61.2
	{ February	=	61.8

showing a difference of only 10° of Fahrenheit's scale
between the hottest and coldest months. It has not been
possible to ascertain accurately the daily range of the
thermometer, as there were no self-recording instruments
employed; but there is reason to believe that this would
exhibit a still more remarkable proof of the equability of
the climate. So far as the observations go they show an
ordinary daily range of about 5° Fahr., and rarely exceeding 8° Fahr. It may be added, that in the course
of six weeks from our arrival on April 26 to our departure on June 7, the lowest night temperature observed
at Mogador was 61° Fahr., and the highest by day 77°
Fahr.

If the climate of Mogador be compared with that of
such places as Algiers, Madeira (Funchal), and Cairo,
which have nearly the same mean winter temperature, it
will be found that in each of those places the mercury is
occasionally liable to fall considerably below 50°, and that

the summer heat is greatly in excess of the limits that
suit delicate constitutions, the mean of the three hottest
months being about 80° Fahr. at Algiers, about 82° at
Funchal, and 85° at Cairo. It will help to complete the
impression as to the Mogador climate to say, that rain falls
on an average on forty-five days in the year; and that,
per 1,000 observations on the state of the sky, the proportions are

Clear 785; Clouded 175; Foggy 40:

the latter entry referring to days when a fog or thick haze
prevails in the morning, but disappears before mid-day.
The desert wind is scarcely felt at Mogador. On an
average it blows on about two days in each year, and on
these rare occasions it has much less effect on the thermometer than it has in Madeira, doubtless owing to the
protective effect of the chain of the Great Atlas.

These remarkable climatal conditions have been mainly
attributed to the influence of the north-east trade wind,
which sets along the coast, and prevails, especially in
summer, throughout a great part of the year; the average
of north and north-east winds being about 271 days out
of 365. West and south-west winds blow chiefly in winter
on about fifty-seven days in each year, and variable winds
from the remaining four points prevail on an average of
thirty-seven days. The north-east breeze, increasing in
force as the sun approaches the meridian, maintains the
exceptionally cool summer temperature already indicated
as characteristic of the Mogador climate—a privilege which
is not shared by Saffi or Mazagan, where the summer heat
is sometimes excessive. It must be noted that although
the summer temperature of the interior of Marocco is
much higher than that of Mogador, it yet falls far short
of what is found in places lying in the same latitude in
North Africa or Asia. This is evidently owing to the
influence of the Great Atlas chain, with its branches that
diverge northward towards the Mediterranean, which

screen the entire region from the burning winds of the desert, and send down streams that cover the land with vegetation.

When one comes to consider how it happens that a place possessing such extraordinary natural advantages has not become frequented by the class of invalids to whom climate offers the only chance of recovering health, or prolonging life, the obvious answer is, that invalids cannot live on air alone, and that few persons in that condition have the courage to select a place where they may reasonably expect much difficulty in procuring the comforts and even the necessaries of life, competent medical advice, and some reasonable opportunities for occupation or amusement. The difficulties under the first two heads are perhaps not very serious. Lodging and food may apparently be procured on reasonable terms, and for many years past there has always been a competent French physician residing here. The resources of the place in point of society are of course limited, and must vary with the arrival and departure of the few European residents ; but any one fortunate enough to be interested in any branch of natural history would find constant occupation of an agreeable kind in a place where there are not half a dozen days in the year that may not be agreeably passed out of doors.[1]

A special subject, to be earnestly recommended to any competent inquirer, whether invalid or not, who may pass six months at Mogador, is the language and ethnology of the Shelluh branch of the Bereber race. Many of these mountain people come to seek a living at Mogador, and

---

[1] Those who are interested in the subject should consult a pamphlet entitled ' *Mogador et son Climat*,' par V. Seux, Marseille, 1870, and a paper in the Bulletin of the French Geographical Society for 1875, by Dr. Ollive, now residing at that place, styled ' *Climat de Mogador et de son influence sur la Phthisie.*' There are some errors in the tables included in the latter paper, and especially in that headed ' Tableau comparatif des Températures moyennes de diverses stations hivernales.'

from our experience it would not be difficult to find one who would become a useful servant.

In the course of the day we visited the extensive stores of Messrs. E. Bonnet & Co., who export large quantities of olive oil from the neighbouring provinces. By increased care in the preparation and subsequent purification of the oil, its quality has been much improved. The cultivation of the vine has of late rapidly increased, and wine of tolerable quality has taken a place among the products which Marocco supplies to England.

Notwithstanding all that we had heard of the excellence of the climate, we had to confess that at this season Mogador is not a paradise for the botanist. The NNE. winds come saturated with vapour, and charged with minute particles of salt from the breaking of the Atlantic waves on the reefs near the town; and, as the temperature of the land is scarcely higher than that of the sea, the air has little or no drying effect on paper and plants. The consequence was that Mr. Carstensen's kitchen was used both by day and night to save our specimens from destruction by damp.

As our interpreter, besides the cook and one or two more of our retinue, were Jews, it was decided that, in order to spare their feelings and those of the Jewish community in Mogador in respect to the Sabbath, we should despatch them along with our heavy baggage on April 28, while we should follow on the succeeding day to the spot where they were to await us. Later in the day, after completing the arrangements for our journey, we went by invitation to dine with the Governor. We found that our host had had a table prepared with chairs for Mrs. Carstensen, who with two European ladies graced the entertainment. Beside them a carpet was spread for Mr. Carstensen and our party; while the Governor himself, with three native functionaries, sat in their usual fashion, cross-legged, on another carpet several yards distant. The first preliminary was the washing of fingers. One attendant bore water, another a brass bowl or basin, and a third presented to each in turn an embroidered towel. This process is always repeated at the close of dinner, and is common to all classes in the country. The feast then began, as every well-ordered Moorish banquet must do, by green tea.   Three cups, carefully prepared in the presence of the guests, in a silver teapot half filled with sugar, were handed in succession to each, and then fresh tea, with mint leaves added, is again prepared, and of this decoction the natives usually take one or two cups more. The serious part of the repast then followed. A large dish of coarse earthenware, covered with a conical cap of fine straw, twice the size of a beehive, is laid on a low wooden frame in the centre of the circle of guests. On the present occasion duplicate dishes were prepared for us, and for the Governor and his native friends. When the cover was removed, we were introduced to the national dish which was destined to be our frequent acquaintance during our journey in the South. The basis of *keskossou* is coarse wheaten, or sometimes millet flour, cooked with butter, for which oil is occasionally substituted. To this is added mutton, lamb, or fowls, cut up into pieces, with various vegetables, either laid on the farinaceous sub-stratum or mixed up with it. Numerous dishes succeeded each other, but they appeared to be all variations on the same gastronomic theme. The cookery on this occasion was better than we often found it; but the pervading flavour of rancid butter, long kept in great earthen pots, is repulsive to European stomachs, and few strangers are ever fortunate enough to be able to enjoy Moorish feasts. To some of us this was the first occasion for practising the art of eating with our fingers, and it was lucky that our host was not at hand to observe the awkwardness of our first essays. We improved somewhat with practice, but never could approach the dexterity and neatness with which the natives accomplish the operation, using only the fingers of the right hand. Conversation was completely drowned during dinner by the native music provided in compliment to the distinguished guests. Four men, squatting on the ground, struck the stretched metal strings of an instrument somewhat resembling a very rude Tyrolese *zither*, and kept up a constant chant or recitation in loud nasal tones, very different from the slow mono- tonous almost always melancholy songs of the Arabs in the East. These men, on the contrary, declaimed the words with unflagging energy, as though determined that the hearers should understand the story; and it was a moment of intense relief when at the end of dinner the deafening clang of strings and voices ceased. The fingers were again washed, green tea again served, courtesy re- quiring that each guest should take at least three cups, and then the Governor and his friends advanced and joined our party.

Mr. Carstensen had asked permission to bring some wine for our use during dinner, and afterwards naturally took the occasion to invite the Moors present to take a share. With very slight show of reluctance, they accepted; and, though the quantity consumed was but trifling, the effect was unmistakable. The conversation became very lively, and jokes passed which excited peals of laughter, though most of them evaporated in the process of transla- tion. One of the Moorish guests—Director of the Tobacco monopoly, as we were told—from the first struck us as a man of jovial temperament; and on him the extra glass or two of wine had a potent effect, the jollity culminating in an extemporised dance, reminding one of the dancing bears, once the delight of our youth, that have disappeared since the era of Zoological Gardens. The copious doses of green tea did not prevent some of the party from sleeping; while others sat up till near morning, engaged in the almost hopeless endeavour to get large piles of botanical paper thoroughly dry, before we finally started on our journey into the interior.

CHAPTER V.

Departure from Mogador—Argan forest—Hilly country of Haha—
Fertile province of Shedma—Hospitality of the Governor—Turkish
visitor—Offering of provisions—Kasbah of the Governor—Ride to
Aïn Oumast—First view of the Great Atlas—Pseudo-Sahara—Tomb
of a Saint—Nzelas—Ascend the ' Camel's Back '—Oasis of Sheshaoua—Coolness of the night temperature—Rarity of ancient
buildings—Halt at Aïn Beida—Tents and luggage gone astray—
Night at Misra ben Kara—Cross the Oued Nfys—Plain of Marocco
—Range of the Great Atlas—Halt under Tamarisk tree.

THE morning of Saturday, April 29, was fixed for our
departure from Mogador, and about 7 A.M. all were ready
to start.

Mr. and Mrs. Carstensen, with a rather numerous
party of the European residents at Mogador, had arranged
to escort us for a distance of some seven miles ; and it was
agreed that, instead of following the direct road to the
city of Marocco, which runs about ENE. from Mogador,
we should make a detour nearly at right angles to that
direction, or about SSE., so as to gain a fuller acquaintance with the Argan forest.

Our course lay in the same direction that we had
chosen in our first short excursion from the town. Between
the belt of sandy shore that is daily washed by the tide,
and the sand dunes that rose in undulations on our left,
we rode past the mouth of the Oued Kseb, and then began
to ascend over sandy dunes, whereon the prevailing plant
is *Genista monosperma*, the R'tam of the Arabs, whose
slender silvery branches wave in the slightest breeze.
Several of the peculiar plants of this coast occurred at
intervals, such as *Cheiranthus semperflorens*, *Statice
mucronata*, a curious and somewhat ornamental species,

and two or three kinds of *Erodium*. As the track rises
and recedes a little from the coast, the tertiary calcareous
rock that underlies the sandhills crops out here and there,
and the first Argan trees begin to show themselves. As
we advanced, the trees grew larger and nearer together,
and as we approached our intended halt, at a place called
Douar Arifi, they formed a continuous forest.

The Argan tree is in many respects the most remarkable plant of South Marocco; and it attracts the more
attention as it is the only tree that commonly attains a
large size, and forms a conspicuous feature of the land-
scape in the low country near the coast. In structure and
properties it is nearly allied to the tropical genus *Sideroxylon* (Iron-wood); but there is enough of general resemblance, both in its mode of growth and its economic
uses, to the familiar olive tree of the Mediterranean
region to make it the local representative of that plant.
Its home is the sub-littoral zone of South-western Marocco,
where it is common between the rivers Tensift and Sous.
A few scattered trees only are said to be found north of
the Tensift; but it seems to be not infrequent in the hilly
district between the Sous and the river of Oued Noun,
making the total length of its area about 200 miles.
Extending from near the coast for a distance of thirty
or forty miles inland, it is absolutely unknown elsewhere
in the world. The trunk always divides at a height of
eight or ten feet from the ground, and sends out numerous
spreading, nearly horizontal branches. The growth is
apparently very slow, and the trees that attain a girth of
twelve to fifteen feet are probably of great antiquity.
The minor branches and young shoots are beset with stiff
thick spines, and the leaves are like those of the olive in
shape, but of a fuller green, somewhat paler on the under
side. Unlike the olive, the wood is of extreme hardness,
and seemingly indestructible by insects, as we saw no
example of a hollow trunk. The fruit, much like a large
olive in appearance, but varying much in size and shape,

is greedily devoured by goats, sheep, camels, and cows, but refused by horses and mules; its hard kernel furnishes the oil which replaces that of the olive in the cookery of South Marocco, and is so unpleasant to the unaccustomed palate of Europeans. The annexed cut, showing an average Argan, about twenty-five feet in height, and covering a space of sixty or seventy feet in diameter, with another, where goats are seen feeding on the fruit, exhibits a

ARGAN TREES.

scene which at first much amused us, as we had not been accustomed to consider the goat as an arboreal quadruped.[1] Owing to the spreading habit of the branches, which in the older trees approach very near to the ground, no young seedlings are seen where the trees are near together, and but little vegetation, excepting small annuals; but

[1] For fuller particulars as to the Argan tree and its economic uses, see Appendix D.

H

in open places, and on the outer skirts of the forest, there
grows in abundance a peculiar species of Thyme (*T.
Broussonnetii*), with broadly ovate leaves and bracts that
are coloured red or purple, and the characteristic strong
scent of that tribe. It is interesting to the botanist as an
endemic species, occupying almost exactly the same geographical area as the Argan. As we afterwards found,
it is replaced in the interior of the country by an allied,
but quite distinct, species. Its penetrating odour seems
to be noxious to moths, as the dried twigs and leaves are
much used in Mogador, and found effectual for the preservation of woollen stuffs.

Not many flowering plants were seen in the shade of
the Argan trees ; the only species worthy of note being a
very slender annual Asphodel (*A. tenuifolius*), and *Carum
mauritanicum*—a plant somewhat resembling our British
pignut.

Meanwhile carpets had been spread under the shade
of one of the largest Argan trees, and a copious breakfast
was displayed. Fully an hour had been consumed between eating and conversation and the parting cigar,
when, bidding farewell to our friends, we finally started
on our road for the interior, under the guardianship of the
worthy old Kaïd who commanded our escort. Separated
from our interpreter and our luggage, we felt ourselves
at first strangely isolated ; but thanks to the cheerful
readiness of our Shelluh attendants, and especially of
Omback, who had been specially assigned to Hooker, this
impression soon wore off. Our men had been engaged
in unloading cargo from English ships in the port of
Mogador, and had commenced the study of the English
tongue by picking up about a dozen words from the
sailors. They at once showed themselves anxious to add
to their store, and the result was that all, but especially
Omback, gained such a smattering of the language as
served our purpose for many of the ordinary purposes of
life. ' Catch him flower ' became the ordinary way of

desiring a man to gather some plant by the wayside, and many similar phrases soon passed current between us. The only term of disapproval in use with our men was 'bloody dog,' and this was not seldom applied to the mules whenever they gave trouble, as those creatures are wont to do.

As we rode on, the Argan forest grew thinner, the trees were gradually intermixed with other species, amongst which we noted a few specimens of *Callitris quadrivalvis* —the *Arar* of the Moors—and before long we gained, from the brow of a low hill where the forest ceased altogether, a rather wide view over a country not altogether unlike some parts of England. The hills of the province of Haha rise in successive undulations as they recede from the coast in sloping downs, relieved at intervals by clumps of trees, and elsewhere broken by masses of low shrubs. The calcareous rock, which seems never far from the surface, is thinly covered over with red earth; and patches of cultivation, chiefly barley or wheat, the former now nearly ripe, here and there indicated the presence of man somewhere within reach, but seemed to show that he plays a subordinate part in fashioning the appearance of the country. The prevailing bush or small tree is *Zizyphus Lotus*, whose double sets of thorns—one pointing forward and the other curved back—were destined to plague us throughout all the low country of South Marocco. The *Zizyphus* was often quite covered over by climbing plants, that rise ten or twelve feet from the ground. The most frequent of these, an *Ephedra* and an *Asparagus*, do not appear to require any special organs of attachment. Probably the intricate branches and complex spines of the *Zizyphus* render these superfluous.

Soon after this we first met bushes of one of the peculiar plants of South Marocco, then little known, and of which we were not able to learn much by personal inspection. The *Acacia gummifera* of Willdenow is one of a group of allied species of which the remainder inhabit Upper Egypt and Nubia, while one, at least, is widely

spread throughout Eastern Africa and Arabia. The tasteless gum known as the gum-arabic of commerce is probably
produced by several of these species. Like its allies, the
South Marocco plant flowers late in the year, after the
first autumn rains, and ripens its pods during the winter.
Hence, as seen by us in spring, without flower or fruit,
there was little to distinguish this from several of the
other forms of this group.[1]

Among herbaceous plants that attracted our notice
was *Glaucium corniculatum* (here always orange, and
never crimson as it is in Palestine), with *Campanula
dichotoma*, only just coming into flower, whilst two or
three degrees farther north, in Palestine and Syria, it
usually flowers three weeks earlier. More interesting,
as being one of the few local plants common to South
Marocco and the Canary Islands, was the *Linaria sagittata*
(*Antirrhinum sagittatum* of Poiret), very unlike any
other toadflax in the form of its leaves and its much
branched twining stems that spread far and wide over the
low bushes.

Although the air was cooled by a pleasant breeze, the
direct rays of the sun were very powerful, and we were
glad to make a short halt for luncheon near a well, where
a small ruined building of rough masonry gave a narrow
fringe of shadow. Resuming our route, we soon after re-
crossed the sluggish stream of the Oued Kseb, whose banks
were fringed with *Vitex Agnus castus*, and with *Cyperaceæ*
not yet in flower. We took this at the time for one of
the branches of a river shown on the French map as falling
into the Atlantic north of the Djebel Hadid, some twenty
miles from Mogador ; but we afterwards came to the conclusion that no such river is in existence.

At or near the ford is the boundary of the province

---

[1] It may be hoped that the plant will now become well known to
botanists, as our friend M. Cosson has obtained a good supply of seed,
which he has liberally distributed among many of the chief botanic
gardens of Europe. See Appendix D.

of Shedma, much less extensive than that of Haha, but
apparently more fertile.   The soil now sensibly improved,
and there were indications of more careful husbandry.
At the same time the larger portion of the surface remained in a state of nature, and gratified our botanical
appetites by a display of many novelties.   The varied
species of *Genista,* that are so conspicuous in North
Marocco and the Spanish peninsula, were here little
seen, but are replaced by several allied genera.   *Cytisus
albidus* and *Anagyris fœtida* are especially prominent.
Withania fruticosa, a curious Solanaceous shrub, which
we had already seen near Casa Blanca and during the
morning ride, here became extremely common ; but what
most interested us was *Linaria ventricosa* of Cosson, a
large species, with stiff erect branches three or four feet
in height, first found in the adjoining province of Haha
by M. Balansa, and which we afterwards saw to be widely
spread through South Marocco, and one of the characteristic features of the flora.

The dwarf fan-palm (*Chamœrops humilis,* or palmetto
of the Spaniards), much less common in Marocco than it
is in the hotter parts of Southern Europe, was here rather
abundant, perhaps because it is one of the few plants that
the locusts are unable or unwilling to devour.

As we rode onward, gradually ascending over a gently
undulating country, this became constantly more productive.   In two or three places the people were cutting
tolerable crops of ripe corn ; the olive, fig, and pomegranate became frequent, and for the first and last time
we saw the former tree cultivated with care, pruned, and
apparently manured.

The sun had just set when we at length reached our
camp outside the large castle of the Governor of Shedma,
and found our interpreter and other attendants anxiously
awaiting our arrival.   The tents were already pitched,
and our heavy luggage was in its place.   We should have
been glad to eat a moderate repast in peace, lay out the

plants collected during the day, and retire to rest; but
that would have been nowise suitable to the dignity of a
party travelling under the especial protection of the
Sultan, and whose importance had doubtless been exaggerated to the utmost by the inventive talents of our
interpreter. In the absence of the Governor, his son, a
stout overfed man of forty, welcomed us on our arrival,
and invited us to dine in the *kasbah*, and of course
courtesy required us to accept the invitation. After a
brief toilet, we proceeded to enter the castle, and were
led through open spaces to the inner building, which
forms the dwelling of the Governor, and then through a
court, with flower-plots in the centre, to a large and handsome hall, where we were to be entertained. As usual,
there was little furniture, save several showy Rabat carpets, but we noticed three or four ornamental French
timepieces in a recess where it would appear that the
Governor or his son were used to sleep. Besides our
host, there was present a grave man whose features differed much from the ordinary Moorish type. He turned
out to be a Turk who had already passed several months
as a guest in the Governor's castle. We never understood
accurately what had brought him so far from Istamboul;
but we were led to believe that he had come on some
informal mission, and that its traditional jealousy of
foreigners, nowise confined to Christians, had led the
Moorish Court to interpose obstacles in the way of his
advance into the country.

After a quarter of an hour's interchange of civil
speeches, conversation began to flag; but the Governor's
cook, who perhaps wished to display his professional skill
on the occasion, was yet far from completing his operations. Quite an hour passed, we were tired and sleepy,
and our fat host showed no talent for conversation, so
that the time hung heavily enough until the usual pre-
face to dinner, green tea, was introduced. Doubtless the
entertainment was everything that a Moorish connoisseur

would have thought refined and exquisite. Orange-flower water was provided for washing the fingers, and incense was burned at the beginning of the repast. Our host was attentive enough to pick out and present to us choice pieces of meat or vegetable from the dishes that followed each other in slow order, but he fortunately did not think it necessary to show the utmost mark of hospitable attention by taking an especially delicate morsel from his own mouth and thrusting it into that of a guest. It was quite ten o'clock when, after further potations of green tea, we returned to our tents. Presently Hooker was requested, through Abraham, our interpreter, to receive the *mona*, or offering of food, which, in accordance with the Sultan's order, was to be provided at each place where we stopped on our journey. The *mona* on this occasion befitted the dignity of the Governor of an important province rather than the wants of three travellers who had just been abundantly fed, and whose retinue could not, with the best intentions, consume one half of the articles supplied.

Opposite the door of our large tent a number of the Governor's servants appeared, the whole group being lit up by torchlight. First, five live sheep were dragged forward, then twenty fowls, then followed a large hollow dish filled with eggs. To these succeeded a very large earthen jar of butter, and another of honey, a package of green tea, four loaves of sugar, candles of French manufacture, which are largely imported, and finally corn for our horses and mules. As if all this were not enough, there then advanced a procession of men, carrying the usual large dishes with beehive covers, each of which in turn was laid down before Hooker. It may be here mentioned that the presentation of *mona* was henceforward a daily ceremony, repeated every evening, some time after our arrival in camp. The requisition was made by the soldiers of our escort upon the local authority, whether a governor or a mere village sheik; and this was a part of their duties which they performed

with unfailing zeal and punctuality.   On such an occasion
as the present we had no fear of pressing too hardly on the
donors of the *mona*; but in poor places, and especially in
the valleys of the Great Atlas, we had an unpleasant feeling that the exorbitant demands of our rapacious escort imposed a heavy tax on the limited means of the population.

Struggling against sleep, we diligently worked at our
plants till long past midnight, and then, at length, sought
rest after our first day's journey in South Marocco.   On
the morning of April 30, we were up betimes, and had an
opportunity of viewing the *kasbah*.   It is a large pile of
building, enclosed by a high wall, within which there is
space for great numbers of horses, camels, and domestic
animals of all kinds, with dwellings for the numerous
retainers and rooms for guests, all separate from the central block which forms the residence of the great man,
his family, and personal attendants.   Except that it is
mainly built of *tapia*, or blocks of mud, rammed into
square moulds and hardened in the sun, this and other
similar buildings in Marocco differ little from the castles
which the semibarbarous feudal chiefs inhabited throughout
a great part of Europe in the so-called ages of chivalry,
and down to the beginning of the last century.   A more
extended acquaintance with the country afterwards showed
further points of comparison.   There is not one of these
kasbahs that has not been the scene of atrocious deeds of
cruelty and treachery, such as we find in the records of
most of our mediæval strongholds.   When we shudder at
tales of Moorish atrocities we are apt to forget that they
merely disclose an anachronism, no way surprising in a
country that has stood altogether aloof from the influences
that have brought Europe to a condition of relative civilisation.

The *kasbah* of Shedma is well placed, on nearly flat
ground, at the summit of one of the highest of the undulating hills that intervene between the coast and the great
plain of Marocco, standing, by our measurements, 1,430

feet (436 mètres) above the sea level. The view over the
gently heaving surface of the lower hills to the south was
very pleasing. The slopes covered with short herbage, the
green now beginning to turn brown and yellow, are studded
with trees, chiefly Argan, olive, and fig, sometimes in
clumps, sometimes dotted over the surface. Close to us,
adjoining the gate of the *kasbah*, were several very fine
Argan trees just coming into flower.

We were rather late in this morning's start, and it was
near 9 A.M. when, after the tents and luggage were packed,
we got under way, accompanied by our host of last night,
the Governor's son, who volunteered to show us his father's
garden, of which he was evidently proud. We rode down
the hill, and soon reached a place called the 'Tuesday
Market' (Souk el Tleta), beside which we were to inspect
the first example we met of Moorish horticulture. The
enclosed space, about an acre in extent, was divided into
oblong beds, in which the only cultivated flowers were roses
and marigolds, growing amidst an abundant growth of
weeds. Along with these we noticed several beds of mint,
which is in constant requisition for mixing with green tea.

At the open space of the 'Tuesday Market,' our host
took leave of us. We had not thought it necessary to
make him a present, but he had no hesitation in asking
for such small articles as caught his fancy. Maw had
beguiled the tedious hour of waiting for dinner last night
by exhibiting the combustion of magnesium wire, and complied with a request to that effect by giving up a small
portion of his store. The Moor had spied a small lens in
the hands of Crump, Hooker's servant, and now asked for
that. He next begged for some trifling European article
belonging to Abraham, our interpreter, and finally for a
box of fusees, the last possessed by Ball.

In a country where shops are unknown, except in a few
large towns, the only chance for obtaining anything which
the peasant cannot raise on his own ground is at the nearest
market. These are held at some selected spot throughout

the inhabited parts of the country, not always near a village, and the place takes its name from the day of the week on which the market is held. We found this place to be 1,183 feet (360·3 m.) above the sea level.

Our way now lay for some distance amidst enclosed and cultivated land, through green lanes bordered by shrubs covered with climbing plants. As the enclosures came to an end, and we again found ourselves in an open country dotted with trees, we observed the Argan gradually becoming more scarce, and the *Zizyphus* more frequent, until the last of the former were seen about ten miles east of the *kasbah.* Among the smaller shrubs *Rhus pentaphylla* was prominent. The genus *Teucrium* is especially characteristic of South Marocco, as may be inferred from the fact that four new species were found by M. Balansa, besides many of those common about the Mediterranean. We here met one of the peculiar Marocco species (*T. collinum*); and the ever varying *T. Polium* constantly recurred throughout our journey, from the coast up to over 4,000 feet above the sea.

After several brief halts, requisite for collecting new and rare plants by the way, we rested for half an hour in a shady spot near a well. Up to this point our course since morning had varied between due E. and SSE.; but for the remainder of this day's journey our general direction was about ENE. The track slowly wound its way upwards amongst hills covered with *Retam*, till it reached the brow of a rounded eminence that overlooks a wide expanse of treeless plain extending eastward to the horizon, except where some low flat-topped hills were seen in the dim distance. We had now accomplished the first stage of our journey. We had traversed the zone of hilly country lying between the coast and the great plain of Marocco, on the verge of which we here stood. Leaving out of account a few prominences to be spoken of hereafter, the plain appears to the eye quite horizontal; but in fact there is a very perceptible inclination of about forty feet

per mile from south to north, as it slopes from the foot of
the Great Atlas towards the river Tensift, and a further
slighter dip of about ten feet per mile from east to west,
between the city of Marocco and Sheshaoua. The deficiency of water at once explains the great change in the
vegetation, which was speedily perceptible in detail, but
obvious to the eye from the first view of the country newly
opened before us. Corresponding to this is a considerable
change of climate, arising from the rapid heating of the
surface by day, and the no·less rapid cooling by radiation
at night. We are already far from the equable climate of
Mogador; and although the air in the shade is only pleasantly warm, we are happy to have the protection of pith
helmets covered by turbans between our heads and the
direct rays of the sun.

The verge of the great plain over which we rode this
afternoon is far less barren than the portion which yet lay
before us; and we found several species characteristic of
similar situations in Spain and Africa, along with some
others, hitherto undescribed, that appear to be characteristic of this part of Marocco. Thus *Artemisia Herba alba*
became conspicuous, in some places almost covering the
surface. Of the more noticeable herbaceous plants here
seen were *Matthiola parviflora*, *Gypsophila compressa*,
Ebenus pinnata (rather common throughout the low
country), *Onobrychis crista galli*, an *Elæoselinum*, near to
E. meoides, and numerous *Compositæ*, of which *Cladanthus
arabicus* is one of the most conspicuous. We did not notice
the fragrant odour which some travellers have found in the
flowers of this species. To the same natural Order belong
several undescribed plants, which became more abundant as
we advanced into the interior of the country, belonging to
the genera *Anacyclus*, *Matricaria*, *Anthemis*, and *Centaurea*.

About half-past four we reached our appointed camping
place, at Aïn Oumast, one of the few wells of drinkable
water found in the region we had now entered. In the coast

zone it would appéar that in ordinary years the rainfall is sufficient to enable the natives to raise grain crops wherever the soil is suitable for the purpose ; but in the interior, cultivation is limited to the tracts that are capable of irrigation from the streams descending from the Great Atlas, or else to the immediate neighbourhood of wells. The ground around Aïn Oumast had borne a scanty crop of grain, and the rough surface, now baked hard by the sun, was not very comfortable for sleeping upon, even with the intervention of a mattress of cork shavings.

For a short way before our arrival, the main chain of the Great Atlas had for the first time been in view, dimly apparent at a distance of some sixty miles ; but as the sun declined towards the horizon, the outlines became clearer, and we naturally watched with increasing interest every feature of that mysterious range seen, even from a distance, by few civilised men, whose recesses we hoped to be the first to explore. We discussed eagerly the question whether some patches of lighter colour represented snow, or merely surfaces of whitish limestone rock; and, as usual, the only effect of discussion was to confirm each in the impression first formed, which it was impossible to verify or disprove unless, by viewing the range from the same direction under similar conditions at a later season, we could discover whether the appearances in question should have altered or disappeared.

The *mona* presented by the Kaïd or sheik of the place was naturally less profuse than that offered at Shedma, but yet abundant for the needs of our camp. As almost everywhere, save in the remoter valleys of the Atlas, green tea and a quantity of white sugar formed a main feature in the entertainment, and doubtless the most expensive to the pcor people who had to provide it.

The day had been warm, though not oppressive, the thermometer probably standing at about 80° Fahr. in the shade, and the fall of temperature during the night was very sensible. Even after the sun had risen on May 1—

soon after five A.M.--the thermometer marked only 54°
Fahr., but by six A.M. it reached 67°. The observation for
altitude gave 1,132 feet (345·5 m.) above the sea; probably
too low by fifty or sixty feet, owing to the local effect of
radiation in depressing the temperature of the air in
contact with the surface.

We were on our way soon after six; and, on leaving
behind the bushes and small trees that grow on the skirts
of the irrigated ground, we entered on a wide bare plain,
stretching unbroken as far as the eye can reach, which
forms the most singular feature in the aspect of this part
of Marocco. The surface is covered with calcareous rough
gravel, mixed in places with siliceous concretions. The
scanty vegetation was already nearly all dried up, and it
was not without difficulty that we secured specimens of
most of the few species that can endure the parching heat
and drought. Conspicuous among these was *Peganum
Harmala*, forming at intervals green patches amid the
general barrenness. *Stipa tortilis* was frequent, but
mostly dried up, and here and there occurred tufts of
a meagre variety of *Avena barbata*. More interesting
than these were a diminutive annual species of *Echium*
(*E. modestum*, Ball) and two species of *Centaurea*—one
hitherto known as *Rhaponticum acaule* of Decandolle,
the other, before undescribed (*C. maroccana*, Ball). In
its general aspect, and in the character of its vegetation,
this region bears a striking likeness to the stony portions
of the Sahara, and we were not sorry to include this among
our Marocco experiences, though well pleased that the
acquaintance was not to be much prolonged.

Some six or seven miles east of Aïn Oumast we passed
a short way north of Sidi Moktar, the tomb of a saint
much venerated in this region, and the last spot where for
a long distance water is to be found at all seasons. This
is one of the halting-places, called *Nzelas*, frequented by
ordinary travellers who follow this road. The *Nzela* is one
of the peculiar institutions of this country deserving of

some notice. The Marocco Government recognises, at
least in theory, the duty of protecting travellers from violence to their persons and goods; for without some provision for the purpose the small amount of trade now existing
between the interior and the coast could scarcely continue
to exist. As well as all other executive functions, the
sovereign commits this to the Governor of each province,
who accordingly stations a few armed men at the places
where travellers are accustomed to halt. Such a post is
a *Nzela*. It does not imply the existence of any shelter,
and still less of any supplies for the sustenance of men
and cattle. In a country where the sparse population
lives in tents or temporary sheds, the traveller must provide such things for himself; but at a *Nzela* the wayfarer
may count on security from violence, and the guards are
entitled to a trifling payment for each beast of burden
that is committed to their protection. From any demands
of this nature, as well as from the tolls that are levied on
passing from one province to another, we were declared
by our escort to be free, as personages travelling under
the direct authority and protection of the Sultan. The
boundaries of the three provinces of Shedma, Mtouga, and
Ouled bou Sba met at Sidi Moktar; but such places in
Marocco are proverbially unsafe, because they are the frequent resort of robbers and outlaws. In case of a robbery
or murder being committed, the people of each tribe throw
the blame upon their neighbours, and the men of one
province are very shy of attempting to pursue malefactors
who take refuge within the boundaries of another. After
the commission of many outrages at this place, it was
found necessary to transfer a portion of territory to the
Ouled bou Sba, at the same time making the Governor
of that province and tribe responsible for the safety of
those whom business or piety lead to the sanctuary of
Sidi Moktar.

As we rode onward the Great Atlas chain remained in
view, but dimly seen through the haze that increased with

the increasing heat of the day, and ahead of us rose some flat-topped hills of singular aspect which have attracted the attention of all travellers in this region. Some of these hills extend for a considerable distance, while others form small isolated masses; but they agree in two respects —all are flat-topped, and all show a steep escarpment especially on their westward faces. We afterwards saw reason to believe that they all rise about 450 feet above the portion of the plain near at hand, and reach nearly the same height as the plain surrounding the city of Marocco. The general appearance suggested the probability of a former wide extension westward of the latter plain, and subsequent erosion by marine or fluviatile action. As we approached the most conspicuous of these isolated hills, we were struck with the singular appearance of the stunted bushes of *Zizyphus Lotus*, which form the only arborescent vegetation of this region. From a little distance they looked as if covered by some white-flowered climbing plant, or else laden with white fruit. This appearance was due to the extraordinary number of two species of snails (*Helix lactea* and *H. explanata*) that completely covered the branches. We frequently noticed the same appearance afterwards, but nowhere so markedly as here.

Towards the foot of the first and most conspicuous of the hills above mentioned, which bears the inappropriate name Hank el Gemmel (Camel's back), the plain rises gently rather more than one hundred feet in all; above this the slope of the hill becomes steep, and finally exhibits an almost vertical face at the top. At the foot of the steeper slope, about four hours' ride from Aïn Oumast, our track passed by an ancient well, now almost dry, and often completely so; and here, under the imperfect shade of a lotus tree, we made a short halt. The direct rays of the sun being very powerful, we were somewhat surprised to find the temperature of the air to be only 77° Fahr. Leaving our escort, we ascended the low but steep hill

above the well. The scarped face exhibited a section of
the yellowish-white limestone that appears to underlie
nearly the whole of the low country between the coast
and the base of the Atlas.   No fossils were found ; and in
the present state of our knowledge, or rather ignorance,
of the whole region, it seems impossible to fix its position
in the geological series.   The level summit is capped by
a thin layer of coarse chalcedony, in which we recognised
the origin of the siliceous fragments scattered over the
plain below.   This layer would offer resistance to superficial denudation, and account for the tabular forms of the
hills, but where these were attacked from below by marine
or river action the covering would necessarily be broken
up and the fragments scattered over the plain below.
With reference to the opinion expressed by Maw in his
paper in the Proceedings of the Geological Society, and
in the Appendix to this volume, as to the origin of the
tufaceous coating of the plain between Aïn Oumast and
Marocco, the only difficulty that presents itself arises from
the presence of these siliceous fragments on the surface
along with the disintegrated tufa.   If, as he and other
geologists believe, such a superficial coating is due to
evaporation from the underlying mass of water charged
with carbonate of lime, it seems hard to account for the
diffusion of the chalcedony fragments, unless we suppose a
submergence of the plain subsequent to the formation of
the tufa layer, and a renewed supply of such fragments
by further erosion of the hills that formed the sea or river
coast line.   To confirm this conjecture, we may note the
fact that the fragments of chalcedony became progressively rarer as we advanced from the lower portion of the
plain over which we this day travelled to the upper level
surrounding the city of Marocco.

The summit of the hill was found to be 1,648 feet
(502·4 m.) above the sea and 303 feet above the well at its
base.   It was barren, yet supplied a few additional plants
to our collection.   *Frankenia revoluta* was abundant, as

was also a lavender somewhat intermediate in appearance between *Lavandula multifida* of the Southern Mediterranean shores and *L. abrotanoides* of the Canary Islands. We also found a form of *Cotyledon hispanica* of Linnæus (*Pistorinia hispanica* of Decandolle), with pale yellow flowers, intermediate in some respects between the common plant of Southern Spain and *P. Salzmanniana* of Boissier and Reuter.

Resuming our journey, we bore somewhat south of east over a country similar in character to that traversed in the forenoon, but not showing such a complete dead level surface. On the way we noticed for the first time *Cucumis Colocynthis*, one of the characteristic plants of the desert region, extending from Arabia and Southern Palestine across the entire of Northern Africa, but rarely approaching the littoral zone. Here, as near Suez and elsewhere, so far as we have observed, this plant is curiously infrequent. Growing as it does in a region where it has few rivals to contend with, and the surface is remarkably uniform, one yet finds but one or two individuals scattered at comparatively wide intervals over the stony plain. The fruits are used in Marocco to preserve woollen clothing from moths, but their purgative qualities do not seem to be known to the native doctors.

Here and there in this part of our route we encountered small blocks of volcanic rock—trap or basalt—as to the origin of which we have no information. We have no grounds for supposing eruptive action to have occurred in this region within a period so recent as that subsequent to the formation of the tufa which covers the whole surface of the lower country, and it is not easy to account for the transport of these blocks from a distance after its formation.

The direct heat of the sun was great in the afternoon, and the way barren and monotonous, so that it was with thorough satisfaction that, on reaching the summit of a slight swelling rise on the plain, at near 5 P.M., we saw

I

before us a green shallow basin, at the farther end of which
our eyes rested gladly on the abundant foliage of gardens
and orchards. A stream from the Great Atlas, diverted
into numerous slender irrigation channels, is the source
of this apparent fertility, but so much of the water is
taken up in this way that only a trifling surplus remains;
and, save after heavy rains, it seems that a mere streamlet
flows northward to join the Oued Tensift, the chief river
of South-western Marocco. The green that gladdened our
eyes seemed to have given but deceptive promise, for we
at first entered on a scrub formed exclusively of Cheno-
podicaeous bushes, including *Arthrocnemum fruticosum,*
Caroxylon articulatum, Suœda fruticosa, and *Atriplex*
Halimus.

The same thing happens here that may be noticed in
the neighbourhood of the freshwater canal in the Isthmus
of Suez. Where the soil contains a quantity of soluble
salts, the first effect of admitting moisture by irrigation is
to form a salt marsh, which becomes covered with its own
characteristic vegetation; but if the surface is so disposed as to allow the percolation of fresh water, the salts
are gradually carried off, the salt marsh is converted into
fertile land, and the ugly *Chenopodiaceœ* disappear.
Accordingly, after traversing a broad belt of scrub, we
soon found ourselves amidst luxuriant vegetation, and
saw our tents, which had preceded us, pitched under the
shade of tall fig-trees, in one of the orchards belonging to
the village of Sheshaoua. This place is a true oasis, and
an abundant growth of fig, olive, pomegranate, apple,
plum, and apricot, with an undergrowth of grasses and
herbaceous plants, affords a striking contrast to the desert
tracts surrounding it.

The vegetation of the irrigated land, excepting a few
tall palms, was almost exclusively European; and not
without pleasure we gathered many common English
species, such as our common bramble, dandelion, charlock,
Sisymbrium Irio, Geranium dissectum, Hypochœris

radicata, Sonchus oleraceus, Lycopus europœus, Plantago major, Rumex pulcher, Carex divisa, and *Scirpus Holoschœnus.*

The usual *mona* was sent soon after our arrival; and the local governor, a deputy of the Governor of Marocco, paid a visit of ceremony in the evening. He was a black of nearly pure Negro type, and in all probability originally a slave. We were not then familiar with the fact that slaves frequently rise in Marocco to the highest posts in the State. The body-guard of the Sultan is exclusively recruited among the black population, either voluntary immigrants, or slaves imported young from Timbuctoo. These form the only troops in the country that can be relied on to repress internal disorder, though in case of war with a European Power there is little doubt that the whole Moorish population would respond to an appeal to their patriotism and fanaticism. Whether the same would hold good as to the Bereber tribes of the Great and Lesser Atlas may be much doubted. With these the sentiment of national, or rather tribal, independence is the predominant feeling, and so long as an invader kept aloof from their native valleys they could not be easily moved to action. It naturally happens that an absolute ruler, too conscious of his slight claim on the affections of his own people, is led to prefer men whose prominent virtue is that of the dog—attachment and fidelity to him who feeds them. When it is considered that, in addition, the Negro often possesses far more energy than the Moor, united to at least equal natural intelligence, it may be believed that the rulers of Marocco have shown no want of policy in favouring this section of the population.

The thermometer about sunset stood at 72° Fahr., while in the water flowing beside our camp it marked but 62°. At 1 A.M., when we had concluded our nightly task in laying out our plants, it had fallen to 52°, and rose only to 57° an hour after sunrise, when the barometer was recorded, and gave an estimated altitude of 1,141 feet

(347·8 m.), or almost exactly the same as that of Aïn Oumast. The coolness of night temperature throughout this region of Northern Africa doubtless contributes to make the climate not only healthy but favourable to human activity; and it was impossible for us not to speculate at times on a possible, though remote, future, when this may become the home of a prosperous and progressive community.

Early rising does not always mean an early start, and many delays occurred on the morning of May 2, before our caravan was fairly under way at about 9 o'clock. On leaving our encampment, we perceived, on rising ground close at hand, the remains of an ancient town, with stone houses, for the most part in ruins, but some of them still inhabited, and a *kasbah* or castle of somewhat imposing appearance. We failed to obtain any information as to these buildings, which may probably be of considerable antiquity. It must be remembered that throughout the portion of Marocco inhabited by an Arab population permanent houses are unknown, excepting in the coast towns and the royal cities of Marocco, Fez, and Mekines. The country people live in *douars*, which are merely groups of rude dwellings, half hovel half tent, usually formed of branches, over which a piece of camel's hair cloth is stretched, and leaving no wreck behind when choice or necessity leads their inhabitants to remove from one spot to another. Even the Governor's *kasbah*, though often a pile of large dimensions, rarely survives a single generation. The great wall and massive towers surrounding it, as well as the building itself, are constructed of unbaked bricks or of blocks of mud half dried in the sun; and save in cases where a son succeeds his father in power, the custom of the country is to level the whole structure to the ground on the death or removal of the occupant. A few seasons complete the work, and nothing remains but a few mounds of clay to mark the site. Thus it happens that in a country of which the greater part is naturally fertile, the stranger may travel long distances without perceiving a trace of human habitations, or any other buildings than the *zaouias* and *koubbas,* which are scattered over the country at unequal intervals. By these names are designated the tombs of persons, who, when alive, attained a reputation for sanctity, differing only in the rank which they hold in local estimation. The person over whose remains a *zaouia* is constructed may be regarded as the patron saint of the tribe or province, while the *koubba* marks the resting-place of a saint of less renown.

We soon left behind us the irrigated ground, and entered on a barren region, less absolutely sterile than that of the preceding day's journey, and having a more varied vegetation. Blocks of black volcanic rock were more frequent, and of larger size, indicating that we were nearer to the place of their origin, wherever that may be. In some spots *Artemisia Herba alba* was the predominant plant, but we met several new species not before seen. One of the most curious of these is a white-flowered *Picris* (*P. albida*), afterwards seen at intervals in the low country, whose ligules wither so rapidly that we failed to secure any satisfactory specimens. Without becoming hilly, the surface lay in slight heaving undulations, the upward slope being always longest towards the east; and the same remark applied throughout the day's ride. In about three hours we reached Aïn Beida, where a copious spring of excellent water fertilises a tract of about a square mile. We turned aside from our track to halt beneath a very fine pistachio tree,[1] fully forty feet high and two feet in diameter. The sun was very hot, though the temperature of the air was not more than 80° Fahr., and we were assured that our halting place for the night was only four hours' distant; and so it happened that between luncheon,

---

[1] This was apparently the *Pistacia atlantica.* The true Pistachio tree (*P. vera* of Linnæus), so extensively cultivated in the warmer parts of the Mediterranean region, was not seen by us in Marocco.

and rest, and short excursions into the blazing sunshine to
botanise in the surrounding corn-fields, we did not resume
our journey until 3.20 P.M.   The baggage train as usual
had gone on ahead; and as the evening light was fading
fast, about 7.20 P.M., when we expected to be near our
night quarters, some inquiry from our escort revealed two
disagreeable facts : first, that we were still nearly two
hours' ride from Misra ben Kara; and secondly, that the
baggage train had taken a different road.   It is not surprising that such intelligence coming suddenly on three
hungry and tired Englishmen, with the further prospect
of passing the night without food or shelter, led to a
vehement row, in which strong, if not intelligible language
was discharged at the head of the worthy Kaïd El Hadj,
the commander of our escort.   The whole affair had probably arisen from some misunderstanding ; but it was settled
by sending two of the escort to ride at full gallop after the
missing baggage train, while we jogged on sad and silent
towards our destined quarters for the night.   Being pressed
for time, we had abstained from botanising by the way
from Aïn Beida; but at one place we stopped to gather
some extraordinarily fine specimens of *Phelipæa lutea*,
which caught our eyes in the failing light.   This is the
king of the broomrape tribe; the stems stood four or five
feet high, with sceptre-like spikes of large yellow flowers,
nearly two feet long, but it was quite too dark to ascertain
on what plant this curious parasite had attached itself.

The stars shone down with marvellous brilliancy on the
desolate tract over which we rode in single file, always
ascending slightly, and the chain of the Great Atlas stood
out more definitely than we had yet seen it, when, at past
9 o'clock we reached Misra ben Kara, and found to our
relief that the baggage train had just preceded us.   About
11 P.M. some food was prepared, and, being fairly tired, we
soon lay down for the night after a frugal meal.   But
not to sleep, for the furious barking of the dogs from the
adjoining village, or *douar*, and the clatter kept up by

our own people, did not let us close our eyes till the night
was far spent.

On this, as on many another occasion, we were forced
to admire the extraordinary endurance of the common
people of this country. It was not mainly the amount of
work they are able to accomplish, but their high spirits
and cheerful demeanour under hardships and difficulties.
Four of our men travelled on foot, walking or running at
a jog trot under a burning sun, and on arrival in camp
the same men were always ready for work in setting up
tents, moving heavy luggage, and attending to the various
wants of their employers.   Having often to wait till midnight for their food, they would pass the time in lively
talk, and after the stimulus of a draught of green tea,
their renewed spirits generally broke out in the form of
songs or chaunts that seemed interminable.   Then, after
three or four hours' sleep, they were ready to begin again
next morning with the same unflagging energy and spirit.
During the day the men on foot resorted to a curious
expedient for diminishing the effect of heat, by thrusting
a stick down the back between the skin and their scanty
woollen garment, and thus securing ventilation.

We were up soon after daybreak on May 3. Our camp
was close to the wretched village of Misra ben Kara, a
large collection of mere hovels put together with mud
and dried branches, and enclosed, as the *douar* generally
is, within a sort of rampart formed of the dried stems and
branches of the *Zizyphus Lotus*, piled up to a height of
eight or ten feet, through which a single opening gives
admission to the inhabitants and their domestic animals.
It stands at a short distance from the Oued Nyfs,[1] one of

---

[1] The spelling here adopted is that used by M. Beaumier in his
sketch-map of the route from Mogador to Marocco, but it is extremely
difficult to fix the sounds expressing many of the native names. Sometimes this sounded to our ears as Oued enfisk, sometimes as Oued
enfist ; the latter, it will be remarked, is merely a slight anagram of
the name Oued Tensift, belonging to the main river flowing westward
on the north side of the city of Marocco.

the chief streams flowing northward from the Great Atlas.
We started about 7 A.M., and soon reached the banks,
fringed with magnificent oleanders in full flower, below
which the shallow stream runs in a deep bed. Like all
the rivers of this country, this is liable to great oscillations;
and though it seemed nowhere two feet deep when we
crossed it, travellers are said to be sometimes detained for
days, owing to the impossibility of fording the stream in
rainy weather.

We found here a few plants not hitherto seen, but
were especially pleased with an undescribed *Statice* (*S.
ornata*, Ball), not found elsewhere on our journey, whose
numerous bright amethyst blue flowers were scattered on
long, slender, much-branched panicles.

On the east side of the river we fairly entered on
the portion of the great plain immediately surrounding
the city of Marocco, extending some thirty miles from
west to east, and southward to the base of the Great Atlas.
This is inclined upwards from west to east, and still more
decidedly from north to south; but to the eye it appears a
dead level, and the hills represented on Beaudouin's map
as approaching near to the city on the south and east
have no existence in fact. The north-western border of
the plain is, on the other hand, marked by prominent
rough hills of a ruddy hue, as seen from a distance, which
rose on our left as we advanced towards the city.

Some portion of these hills, seeming to form an interrupted range, extending along the north side of the Oued
Tensift and parallel to its course, was traversed by Washington on his route from Azemor to Marocco in December,
1829. He estimates their height above the plain at from
500 to 1,200 feet, and describes the rock as schistose, with
veins of quartz, the line of strike from north by east to
south by west, and the dip 75°. To us it appeared
that the higher summits, which perhaps do not lie near
Washington's track, must rise fully 2,000 feet above the
plain. On the southern side of the Oued Tensift, and

nearer to the city, are some lower hills, very similar in appearance to the others, and probably of similar geological structure. One of these, visited by Maw, is described as formed of very hard, dark, grey rock, with knotted white concretions elongated in the line of stratification, the strike from north-west to south-east, and the dip south-west, varying from 50° to 80°.

Our attention, commonly fixed on the vegetation of the country, was on this day chiefly engaged by the great range of mountains, no longer very distant, that bounded the horizon to the south. We had expected to find no difficulty in singling out the peak of Miltsin, described by Washington in the first volume of the Journal of the Royal Geographical Society, as the highest peak of the Atlas visible from the city of Marocco, and the altitude of which, as determined by a rough trigonometrical measurement, he fixes at 11,400 English feet. Approaching the city by a very different route from that of Washington, we soon convinced ourselves that there is no summit visible in the main range much surpassing its rivals in height, and we subsequently came to the conclusion, that Miltsin, which appears somewhat higher than its neighbours in the view from the city, is situated somewhat on the north side of the watershed, and therefore nearer to the observer than any other lofty summit of the range. It may fairly be inferred from Washington's account that he had no opportunity for measuring a base-line—such as could allow him to determine accurately the height of distant summits. The conclusion to which we now came, and which was confirmed by our subsequent observations, was that the part of the main range within sight of Marocco and its neighbourhood is remarkably uniform in height. There are many prominent points that probably approach the limit of 13,500 English feet, and no depressions that fall more than about 2,000 feet below that height. This, as will be seen hereafter, does not apply to the westerly part of the chain lying west of the sources

of the Oued Nyfs, but this is only imperfectly seen from
the neighbourhood of Marocco.

The day was hotter than any we had yet experienced,
the temperature in the shade being about 85° Fahr., and
the breeze which usually rises during the hottest hours
was scarcely felt.   But the vicinity of lofty mountains
usually determines strong currents in the heated air, and
these must have been at work, though unfelt by us.   As
we looked towards the mountain chain, we noticed lofty
columns of sand or dust, remarkably uniform in shape,
that travelled steadily westward across the plain in the
opposite direction to the breeze, so far as this could be
detected.   At one time as many as three of these were
seen at the same time, each moving independently.
These miniature cyclones, arising from the interference of
opposite currents of air, are not uncommon in the plains
on the south side of the Alps, but are rarely to be seen on
so great a scale as here.

About two hours after starting, the great tower of the
chief mosque came into view, and one of our soldiers
rode on ahead to announce our approach.   Not long afterwards we met a courier bound for Mogador with letters
for Mr. Carstensen, and we took the opportunity of
reporting progress and sending him a few details as to our
journey.   In default of regular postal communication,
which is not to be thought of in such a country, the facility
for forwarding letters in Marocco is far greater than could
be expected.   For a few shillings a native is easily induced
to make a journey of many days, and take care of letters,
which always reach their destination.   The reverence with
which Mohammedan people generally regard all written
communications—which may perchance contain the name
of Allah—serves as a protection so effectual, that the loss
of letters and despatches is scarcely ever heard of.   These
couriers travel forty or even fifty miles a day, and after a day's
rest are ready to return to the place whence they came.
The chief object of Mr. Carstensen's letter to Marocco had

been to recommend us to the good offices of some wealthy and influential Moors, correspondents of English mercantile houses, and we were not long before experiencing the benefit of this piece of kindly attention.

The heat of the sun was much felt as we rode over the open plain, and it was suggested that we should do well to halt awhile, and await the return of the soldier who was to report to us the state of affairs in the city. The only spot on the way affording the slightest shelter is under the reclining trunk of a fine tree of *Tamarix articulata*, which had apparently been blown down, though still adhering to the ground by its roots, and throwing out vigorous shoots and branches. The remaining portion of the trunk was 24 feet long, and at 8 feet from the roots the girth was 7 feet 7 inches (2·32 m.) We saw no other specimen of this tree, characteristic of the semitropical region of Northern Africa; but our opportunities for exploring the country surrounding the city were very limited, and it seems probable that it is here indigenous, though the extreme scarcity of fuel may have led to its partial extermination. The slender twigs into which the branches are divided gave no protection from the sun; but, by throwing a carpet overhead, we extemporised a serviceable roof, whose shade was most welcome. Though bare to the eye, this part of the plain produced many small herbaceous plants, such as *Notoceras canariensis*, our native *Coronopus Ruellii*, *Mesembryanthemum nodiflorum*, and *Schismus calycinus*. The *Mesembryanthemum* is as common here as it is in the East; but the last-named grass, so characteristic of the skirts of the desert in Egypt and Arabia, seems to be rare in South Marocco.

Throughout our morning's ride, as well as on the journey between Sheshaoua and Misra ben Kara, we noticed the apparently unaccountable way in which certain social species prevail over a considerable tract, and then suddenly give place to others, without any apparent reference to the composition of the soil. Where Chenopodiaceæ,

such as *Suæda* and *Caroxylon* prevail, it is reasonable to
conjecture the presence of nitre, gypsum, or other salts in
the superficial layer; but such plants as *Artemisia Herba
alba*, *Genista monosperma*, and a local form of *Helianthemum virgatum* will sometimes take almost exclusive
possession of the surface, though this in some places is
mainly composed of siliceous sand, in others of disintegrated calcareous tufa, and in others of decomposed volcanic rock, nowhere seen by us *in situ*, but derived from
scattered blocks of various sizes. In the plain near the
city siliceous sand predominates, and, as a consequence, the
vegetation is more meagre than elsewhere.

We hereabouts first saw the only works of public
utility which we encountered during our journey. What
first struck the eye were long lines of irregular earthen
mounds traversing the plain in a north and south direction,
and we soon ascertained that these were watercourses
rudely arched over. The streams from the mountains
south of the city are distributed through irrigation canals
over a large part of the plain, and thus render it fit for
cultivation. Early experience must have taught the people
that by protecting these canals from evaporation, they
could be made available to a much greater extent; and it
is probable that the construction of these covered water-
ways, some of which were in a ruinous condition, goes back
to a remote period. In point of fact, the whole drainage
of three considerable valleys, whose torrents we afterwards
crossed, appears to be intercepted by this irrigation process,
and absorbed by the vegetation of the plain. It is probable that by the skilful extension of the same system
wide tracts, now barren, might be made productive.

CHAPTER VI.

Approach to the City of Marocco—Pleasant encounter—Halt in an olive garden—Interior of the city—Difficulty as to lodging—Governor unfriendly—Camp in the great square—Negotiations with the Viceroy — Successful result— Palace of Ben Dreis — Diplomatic difficulties—Gardens of Marocco—Interview with El Graoui.

AFTER vainly waiting nearly two hours for the return of our soldier, we determined to push on towards the city of Marocco, though somewhat uncertain what the character of our reception might be.   This was not merely a matter affecting our personal comfort during our short stay, but was certain to have an important effect on the success of our subsequent journey, and it was a most fortunate circumstance that Hooker's long experience in contending with the jealousies and suspicions of the native authorities in Nepal and the border States of North-eastern India, and his thorough knowledge of the character of people, who, though far removed, very much resemble the Moors in their ideas and maxims of policy, had prepared him to deal with them successfully.   We had gained the first essential condition for exploring the unknown valleys of the Great Atlas, by obtaining the consent of the Sultan ; but it was impossible to guess the precise tenor of the orders forwarded by him to the local authorities, and, whatever these might be, the ultimate result would largely depend on the good or ill faith of the latter in carrying them into effect.   The person whose favourable dispositions it was chiefly important to secure was El Graoui, who exercises under the Sultan a wide authority as Governor of nearly all the tribes of the Great Atlas that recognise the imperial authority, extending over a considerable

portion of the country at the foot of the mountains inhabited by a Shelluh population. El Graoui, as we knew,
lived in the city of Marocco, but outside his own jurisdiction. The Governor of the city and its immediate
neighbourhood was Ben Daoud, a man notorious for his
dislike to strangers and especially to Christians, and it
was to him that we had to look for our reception, on our
arrival.

From whatever side it be approached the city of
Marocco presents an imposing appearance. The western
side presented an outline about a mile and a half in
length. Massive walls, some thirty feet in height, with
square towers at intervals of about 120 yards, completely
enclose it, and on two sides at least it is girdled by a wide
belt of gardens in which the date-palm, olive, and fig are
the most conspicuous trees. Here, as elsewhere, the date-
palm flourishes in a sandy soil where the roots plunge
into a more compact subsoil kept moist by infiltration.
On the north-west side the palm groves, which we passed
on our left, are so close and continuous as to give the
effect of a forest, while along our route they alternated
with other cultivated trees. The effect of the scene as
we approached was peculiar and new. The luxuriance of
the vegetation that at intervals screened from view the
great range of the Atlas, the majestic old olive trees, the
rough trunks of the tall palms on which stood many a
motionless chameleon ; the walls and towers of the great
city seen at intervals as we wound among the gardens
and groves, combined to form a striking and highly picturesque scene. It was near 4 o'clock when, as we were
drawing near the walls, we were startled by the sudden
appearance of a party of well-mounted Moors in flowing
haïks, whose horses leaped in succession through a gap in
an adjoining fence and advanced to meet us. Our surprise was increased when the foremost of the party greeted
us in English with a friendly welcome. This gentleman,
whose outward appearance was quite undistinguishable

from that of his companions, was Mr. Hunot, brother of the British Vice-Consul at Saffi, and representative at Marocco, of the house of Messrs. Perry & Co., of Liverpool. His companions, grave and courteous-looking Moors of considerable local influence, were Sidi Mohammed Hassanowe [1] and Sidi Boubikir, with two or three more of less note. We were invited into an adjoining garden, where carpets were spread under large olive trees, and a refection in the Moorish style, consisting of green tea, cakes of wheaten flour and kabobs, was speedily prepared.

Mr. Hunot had spent several months in the city, and in the absence of M. Lambert, a French merchant, who has lived there for many years, was the only European resident. Although his thorough familiarity with the language, customs, and ideas of the natives made his position less difficult, he found it practically so irksome that he was then on the point of returning to Saffi, his ordinary residence; and we owe it to his kind desire to assist us during this critical period of our journey that he postponed his departure for a day. The time passed very pleasantly, and we listened with satisfaction to the opinion expressed by our hosts that no obstacles would be interposed to our intended journey in the Atlas; when the first sign of rocks ahead was disclosed on the return of the messenger despatched to ascertain what sort of lodging had been prepared for us. When the messenger acknowledged that a very small house with but two rooms had been provided, it became clear that, so far as the city authorities were concerned, there was no disposition to show us much courtesy. Hooker at once sent back a message to the Governor, that we should require a much larger house, or else an enclosed garden in which to pitch our tents. After about half-an-hour's delay, the messenger again appeared, saying that a much larger house, with four

[1] The designation *Sidi*, equivalent to the Italian Signore, given to persons of consideration, forms no part of the name. In conversation it is abbreviated to *Si*.

rooms, and more adjoining if required, was ready to receive us. Hereupon we resolved to enter the city, and in
company with our new friends, whose ample *haïks* pic-
turesquely draped must have contrasted favourably with
the ugly *jellabias* that we wore over our European dress,
we defiled in a long cavalcade, followed by the mules,
camels, and donkeys of our train, through the gardens that
on this side approach close to the city walls.

Before the gate we found an officer, evidently of inferior position, with some ten or twelve ragged fellows on
foot, armed with rusty matchlocks, posted there to receive
us, and to conduct us to our quarters ; and with this sorry
escort, we made our entry into Marocco. It is impossible by
any language to convey the sense of utter disappointment
and disgust which overpowered us on our first arrival ; and
though these feelings soon became subordinate to others
connected with our personal position, they are those which
predominate in our subsequent recollections.

After passing the gateway we had before us a wide
road, with a lofty mud wall on one side and a lower mud
wall on the other. The high wall on the right forms
part of the enclosure of the Sultan's palace ; over that on
the left branches of shrubs or trees appeared, showing that
gardens or orchards lay behind. On either side of the
road rose accumulations of refuse and filth that looked as
if they might have been the growth of centuries, and the
farther we went the greater became the piles of abomina-
tion, until it seemed as if these would block up the passage. We had passed a fine Moorish arch of wide span,
that forms the chief entrance to the palace enclosure, and
following this as it makes a sharp turn to the right, there
were still no signs of dwellings. The mud walls on either
side, on which many storks built their nests, were often in
a ruinous state ; and here and there it seemed as if people
had burrowed beneath so as to make something between
a den and a hovel. At length we turned into a sort of
lane, and soon emerged into what appears to be one of the

main streets. Hitherto we had met very few passers-by;
but we now found ourselves in a rather crowded thorough-
fare, encountering a good many men on horseback, and a
large number of foot passengers, many of them veiled
women. The street displayed nothing but mud walls,
about twenty feet in height, without a single window, but
with openings at frequent intervals leading into short and
narrow passages or lanes. The behaviour of the people
as we passed was singular. Some of them cast scowling
looks at us and muttered words, certainly not of welcome,
which may very likely have included some unflattering
references to our grandmothers; but the great majority
went by without seeming to heed us in the least, as though
European costume, which had probably not been seen
within the walls since M. Beaumier's visit in February,
1868, were a familiar sight calling for no remark.

At length the head of our escort turned suddenly into
one of the narrow lanes, barely wide enough to let a laden
camel pass; we followed and, after passing the entrances
of five or six other houses, reached a low door at the end
of the lane. Stooping through the mean entrance and a
short passage, we found a small open court, about fifteen
feet square, on each side of which was a narrow room,
receiving no light except from the court. A very brief
inspection showed that the whole place was swarming with
insects of every kind; and as Hooker turned round to
express his opinion and his intentions, it was found that
the officer with his rabble escort had decamped the
moment he saw us safe into the house, thinking no doubt
that we had thus no option but to remain there. When
Hooker announced in very decisive terms his resolution
not to sleep in such a house, Mr. Hunot and our new
Moorish friends, foreseeing a row between us and the
Governor, urged that we should put up with the house for
that night, and on the following morning negotiate for a
more suitable dwelling. As we were holding council
together as to what should be done, a number of men

K

appeared on the scene, each bearing one of the customary large beehive covered dishes, as a *mona* for our evening meal. They were instantly ordered to carry their dishes back to Ben Daoud, and inform him that we refused to stop in that house. They said they could not take away the food, as their orders were to leave it for us; but on the order being repeated in more imperative tones, they departed, most likely settling the difficulty by appropriating the *mona* between themselves and their friends. At this point Hooker's knowledge of the Oriental character was conspicuously shown. If it be often true in the West that people are taken at their own valuation of themselves, this becomes an invariable rule among Eastern people. It was absolutely necessary for our eventual success that it should be understood that we were persons travelling by the express authority of our own Government, and entitled to all respect from the officers of the Moorish Sultan, however high their position might be. Were we to allow ourselves to be treated as mere private persons recommended to their good offices, there was an end to all hope of breaking down the barriers by which national prejudice and ancient tradition had closed the interior of the country against the intrusion of strangers. Had we even given way for a single day, the ingenuity of the natives would have found abundant pretexts for evading our demands; it was much easier to refuse the proffered lodging at once than to give a good reason why we could not spend a second night in a house where we had passed the first.

A messenger was despatched to Ben Daoud. 'Tell the Governor,' said Hooker, 'that my Sultana gives me a large house with a garden to live in; hospitality would require that the Governor of Marocco should provide me —the guest of his Sultan—with a better house; but, in any case, I shall not live in a worse one.' In a short time the messenger returned with the answer: 'The Governor has no better house to give the Christians;

but Marocco is large, and they are welcome to provide
for themselves!' It was then immediately decided to
camp out for the night; and the better to mark our
sense of the reception given us, it was at first proposed
to pitch our tents outside the walls. From this, however, we were dissuaded. In such a position, apparently
deprived of the protection of the local authorities, we
should certainly, it was said, be attacked by robbers, from
whom our Mogador escort might not prove a secure defence. It was finally decided that our best course would
be to encamp in the great open space beside the chief
mosque and tower of the Koutoubia. Our men had been
ordered not to unload the baggage, so we were immediately under way. In the twilight the filth and abjectness
of all that met the eye were not so glaringly prominent as
before; but as we rode through more streets and lanes and
open spaces, we saw no single building of the slightest pre-
tension, until we entered the great square, at the farther
end of which is the tower of the Koutoubia, the solitary
specimen of architecture of which the ancient capital can
boast.

The daylight was fading fast, but enough remained to
show that the spot of our encampment was anything but inviting. Go where we would the ground was covered with
refuse of every kind, full of scorpion holes, and swarming with
insects, of which the most abundant and unsightly, though
the least mischievous, were very large black Coleoptera,
distant relatives of our European cockroaches, and the
whole space was bordered by a ditch or open drain that
reeked with foul exhalations. Meanwhile, we had sent the
captain of our escort with a message to Muley Hassan,
the Viceroy, informing him of our resolution to encamp in
our own tents, until a suitable house had been provided
for us. A civil answer was returned, expressing a wish
that we should not camp out, and saying that he had sent
orders to Ben Daoud immediately to provide us a suitable
residence. Soon after came a polite message from El

Graoui, expressing his regret that we had not been lodged to our satisfaction, and forwarding a letter that he had addressed to Ben Daoud on the subject. Nearly an hour passed, when at last a final answer came from Ben Daoud, saying that it was then too late to comply with our wishes, but that on the following morning we should have a good house with a garden.

During all this time we had remained grimly sitting on horseback, no way anxious, until it should be quite necessary, to commit ourselves for the night to the unpleasant accompaniments that surrounded us; but there was no longer any choice, and the order was given to pitch the large tent and unload the baggage.

During the interval we had been much struck by the demeanour of the people, who had from time to time passed by as we stood grouped together in this most public place of the city. Whether in obedience to orders, or from a spontaneous desire to display their utter indifference as to the doings of the infidels, no one paid the slightest attention, or even turned a head to notice us or our proceedings. Even the very boys engaged in some rough play when we first arrived on the square showed the same ostentatious disregard—in striking contrast to the eager curiosity of the Arab children in Algeria and the East, who will sit for hours together watching every movement of European travellers. By this time, however, the night had closed in, and the great square was silent and dark before our large tent was pitched and the baggage brought under cover.

Soon after there appeared a train of men bearing dishes—the evening *mona*—along with twenty-four soldiers sent by El Graoui to guard our camp during the night, and about half-past 9 we were able to get the rather long delayed evening meal and discuss our further prospects and proceedings. While thus engaged the sound of angry voices outside the tent interrupted conversation; it was evident that a violent altercation was going on, in

which many voices took part. When all was again quiet,
we ascertained the cause of the row. Ten soldiers sent by
Ben Daoud as a guard for the night had come to take
their places round the camp, when they found the ground
already occupied. El Graoui's men warned them off,
telling them they had no business there, and when the
others insisted on remaining to carry out their master's
orders the first comers threatened to thrash them if they
did not immediately depart. Peace was re-established
when Ben Daoud's men retired to the farther end of the
square behind the great mosque.

When we came to talk over the varied experiences of
the day, we first of all agreed that, old travellers as we all
were, and familiar with the squalor of Oriental cities, we
none of us had ever known, or even imagined, the existence of a large town so expressive of human degradation,
so utterly foul and repulsive, as this wherein we found ourselves. Of all the places commonly visited by travellers
Jerusalem is perhaps that which at the first moment approaches nearest to the same impression ; but, not to speak
of the numerous important buildings and the associations
connected with them, nor yet of the modern structures
that have arisen during the last half-century, the poorest
quarters of Jerusalem are far from rivalling the universal
squalor and hideousness of all that meets the eye in Marocco. A ruinous house, with windows closed by weather-
beaten rickety lattice-work, is not a beautiful object, but
it may be sometimes picturesque, and, at the worst, is far
better than a dead wall of crumbling mud, such as here
meets the eye on every side. It would seem as if the
most miserable suburbs of all the other towns of North
Africa and Western Asia had been collected together and
enclosed within a lofty wall, so that seen from without the
whole might be palmed off on mankind as the effigy of a
great city.

On deliberating over the events of the evening in relation to our own future prospects, we found reason to think

that what had happened did not necessarily bode ill for
the objects of our expedition. A fierce rivalry, as we knew,
already existed between El Graoui and Ben Daoud, men
whose power and influence in the State were supposed to
be of equal weight. Whether to gratify his own feelings,
or because he so understood the intentions of the Sultan,
Ben Daoud had showed himself unfriendly, while El
Graoui had clearly declared himself on our side. But as
Ben Daoud had no authority whatever in the Atlas valleys,
his enmity could do us no real harm; while El Graoui,
whose opposition alone was to be feared, might easily be
carried farther than he would have otherwise gone on our
behalf for the mere pleasure of thereby spiting his rival.
In this way our visit came to play a certain part in the
interior politics of Marocco, and the serio-comic development of the story acquires a share of interest from the
light it throws on native character.

Some time after midnight, after finishing our customary task in laying out our collections of the day, which
had been much smaller than usual, we sallied out to view
the surrounding scene. The moon stood high in the
cloudless sky, wherein there was so little vapour that the
stars seemed scarcely dimmed by her brilliancy. The
great tower, stark against the black vault, appeared gigantic in its proportions as it looked down on the strange
scene below. The noises of the city—even the howling
of the dogs—were for the moment completely stilled;
our camels, horses, mules, and asses lay resting after their
day's work, and amongst them the sleeping figures of our
men, wrapped up in white *haïks* or *jellabias*, looked weird
and ghostly in the moonlight. The distinctness with
which we heard the occasional whispers of the guards
around our camp served only to make the deep silence of
the night the more impressive.

On this night the advantages of a tent of what is
known as the Alpine Club pattern, where the floor is made
of canvas continuous with the sides, and the entrance is

closed by a flap rising about a foot from the ground, were shown in a striking way. In the great tent, where the ground underfoot was pierced with scorpion holes and swarming with insects, Hooker and Maw did not venture to undress, and had to pass the night perched upon the baggage, while Ball was able to spread his mattress regardless of the creatures that might be moving about under the canvas floor. When his tent was struck next morning the ground underneath was absolutely covered with a continuous mass of creeping things, yet not a single insect entered the tent.

When we all rose betimes on the morning of the 4th, we felt that this must be a decisive day in our contest with the Moorish authorities. At an early hour Hooker despatched two messengers, one to the Viceroy, requesting an interview, the other to Mr. Hunot, begging him to use his local knowledge and influence to make sure that the request should reach the Viceroy. Soon after arrived a message from Ben Daoud, saying that we were at liberty to pitch our tents in an adjoining garden. If that offer had been made on our first arrival, it is most likely that it would have been accepted; but, as it was now clear that Ben Daoud was intent on yielding as little as possible, Hooker wisely resolved to insist on the demand which he had made on the previous evening, and returned an answer in nearly the same terms as before.

At 8 A.M. a morning meal of wheaten cakes and milk came from El Graoui, and throughout that and the following days he continued to supply our wants and those of our followers on the most liberal scale. Bésides a light breakfast, three copious meals with meat and vegetables cooked in the most approved style, accompanied by dates and oranges, were regularly furnished; and the addition of a mule-load of oranges that came later in the day furnished in abundance the most acceptable luxury that nature affords in this region.

It was clear that the question debated among the

Marocco authorities as to the best way of dealing with
the troublesome Christian visitors was considered a rather
knotty one, for fully two hours elapsed before our Mogador
kaïd returned with the Viceroy's answer. We were welcome to Marocco, he said, and he had ordered the palace
of Ben Dreis, with the adjoining garden, to be prepared
for our reception. That building belonged, it was added,
to his father, the Sultan, and not to the Governor of
Marocco, so that we should consider the use of it as a
mark of the personal favour and friendship of the Sultan.
The request for an interview was evaded, probably to avoid
any further demands that may have been apprehended ;
but we had obtained a complete victory, and had nothing
more to ask so far as our stay in Marocco was concerned.

Although the sequel of the story was not unfolded
till a day or two later, it may as well be here given.
Si Boubikir, one of our Moorish friends who had interested himself on our behalf, was sent for by the Viceroy,
and at the same time Ben Daoud was also summoned.
The latter was addressed by the Viceroy in the coarsest
terms : ' You dog ! you slave ! you son of a slave ! how
have you dared to neglect my father's orders ? Were you
not ordered to provide a suitable residence for these English gentlemen ? ' With further additions of threats and
abuse. On the following day (after we had paid our visit
to El Graoui) a person sent by Ben Daoud came to Abraham, our interpreter, to express a hope that we should
also pay a visit to the Governor of the city in token of
reconciliation. He was to assure us that Ben Daoud was
no way to blame for anything that had happened, as he
had acted throughout by the express orders of the Viceroy,
who had desired him to begin by offering the smaller
house, then one somewhat larger, and to leave it to the
Viceroy to meet our demands, if we persisted in asking
for a house with a garden. It was quite impossible to
guess how much or how little truth there might be in this
tale, and how far the scene got up before Si Boubikir was

a mere comedy; but it is characteristic of the country, that it should not be considered improbable. Hooker decided that it was not expedient to overlook the affronts of which Ben Daoud was either the author or the instrument, and his message was met by a curt refusal.

The house or palace of Ben Dreis, which we were to inhabit, originally belonged to a powerful minister, whose property, after the custom of the country, had been confiscated by the sovereign. In 1864 it was occupied by Sir Moses Montefiore; and a correct sketch of it is given in Dr. Hodgkin's narrative of that gentleman's mission of benevolence to Marocco.

We were told that a short time would be required to prepare the house for our reception, and it turned out that the first requisite step was to knock down the wall that stood where the entrance had formerly been. A house in which Jews or Christians had lived was regarded as unclean and unfit for the dwelling of a true believer, and accordingly after the departure of Sir M. Montefiore, the entrance had been walled up, and the house had so remained ever since. When the way had been cleared, an escort of soldiers, despatched by the Viceroy, accompanied us to our new dwelling, which stands inside its walled garden very near to the Bab Roub—the gate by which we yesterday entered the city. We were agreeably surprised when we approached by far the finest house which we any-where saw in this country, a massive square building, entered by a Moorish arch. As usual, the ground-floor rooms, with the central court, roofed in, contrary to the usual practice, were fit only for servants, or for stabling animals and storing goods, and the best apartments were on the upper floor. These were, of course, destined for us. But the first glance showed that in a country where animal, and especially insect, life is so active, the rooms in their present state would be no pleasant habitations. This, however, was foreseen and provided for, and, before many minutes were over, a crowd of

men, including our own attendants, were hard at work
carrying up large vessels of water from the irrigation
stream in the adjoining garden, and armed with rough
brooms, with which to complete the work of cleaning the
premises. Water was turned on in such abundance as to
stand ankle deep in most of the rooms, and pour in a
copious stream down the staircases and other openings.
When all was done the blazing sun soon dried all up, and
during our stay we suffered no inconvenience, and scarcely
saw any insects, except a few harmless beetles. When the
rooms had been thus cleansed and dried, we proceeded to
instal ourselves in our new quarters. There was a large
central room, open to the sky in the middle, with roofed
bays or recesses around, and several adjoining, which served
as bedrooms. The terrace roof, overlooking the trees of
our garden and the city wall, commanded a magnificent
view of the Great Atlas range, and in the early morning,
and towards sunset, afforded an unfailing attraction.

During this and the following days much time was
consumed in long discussions respecting our future plans
and arrangements. During the remainder of his stay in
Marocco Mr. Hunot was kind enough to devote most of
his time to us, and in his visits he was generally accompanied by Si Mohammed Hassanowe and Si Boubikir, who
sat gravely by, rarely taking any part in the conversation.
One of the subjects requiring mature consideration related
to the manner in which the objects of our journey might
best be made intelligible and satisfactory to the Moorish
authorities. The matter had already been under discussion at Mogador and during our journey, but its importance was now much more obvious when it was clear that
our farther progress would depend on the view that El
Graoui might take of our character and intentions. We
were well aware that anything so simple as the statement
that the object was to gratify our curiosity as to the vegetation of the Great Atlas, would at once be set aside as a
false pretext, intended to cover some sinister design. That

one man should be crazy enough to make a long journey for such a purpose might have been thought within the range of possibility; but to suppose that three should all at once be smitten with the same form of insanity was plainly too ridiculous. To endeavour to explain that Hooker, as Director of a great national establishment such as Kew Gardens, should be anxious to enrich it by the introduction of new, rare, or useful plants, was not likely to be more successful. The existence of any public institution having a claim to attention apart from the personal will or caprice of the sovereign could not be made intelligible to the native mind. It was clear that if we did not present ourselves as persons in some way carrying out the direct orders of the Queen of England, we should have no claim to respect, and should be regarded as adventurers prompted by some motive we did not care to avow.

Of course, we felt a natural reluctance to use the Queen's name in an unauthorised way; but, without entering into subtle discussion as to the extent to which the acts of ministers are to be regarded as those of the sovereign, the fact that the Foreign Secretary had, through the Queen's representative, applied for the Sultan's permission for Hooker's journey, undoubtedly justified him in assuming a position different from that of an ordinary traveller. It is true that the knowledge and personal interest which Her Majesty has always shown in matters relating to Art have never been publicly displayed in reference to natural history; but it would certainly not be straining the truth to let it be understood that such a unique institution as the Royal Gardens at Kew is regarded by her with sympathy and favour. The most natural way of conveying this to the Moorish mind seemed to be to say that the Sultana of England had great gardens, in which were plants from all the countries in the world, excepting the Great Atlas, and that she had sent Hooker and his assistants to collect and send home whatever new plants they could find there. To this suggestion, a serious objection was made. It would appear unworthy of a great ruler, we were told, to trouble herself about anything so frivolous as a garden : 'Her thoughts must be engaged in the government of her vast dominions, and above all in the management of her armies and fleets, and not on mere matters of amusement.' 'But,' as it was urged, 'there is one use of plants that every one can understand. Cannot you say that you are seeking for herbs useful to cure diseases, and are charged to bring these home to the Queen of England?' Of course it was true that if by any chance such new plants as we might find possessed medicinal qualities, they would thereby acquire additional interest, and, therefore, in our numerous subsequent communications with the authorities, Hooker stated that his commission was to collect and bring home the plants of the country, and especially those useful in medicine. It is pretty certain, however, that the imagination of our interpreter enlarged upon this text, to what extent we could not of course say. How much that we afterwards heard was serious, and how much more play of fancy, it is hard to guess ; but there is no doubt that the current belief among our own followers was that the Sultana of England had heard that there was somewhere in Marocco a plant that would make her live for ever, and that she had sent her own *hakim* to find it for her. When, in the course of our journey, it was seen that our botanical pursuits entailed rather severe labour, the commentary was : ' The Sultana of England is a severe woman, and she has threatened to give them stick (the bastinado) if they do not find the herb she wants !'

It was impossible to decide on our future route until after an interview with El Graoui ; but whatever that might be, it was certain that we should require a number of animals to convey ourselves and our baggage ; and we yielded to the general opinion of the country in preferring mules for this purpose. Camels are unfit for the rough mountain paths; and the mule is decidedly superior to the horse in endurance of prolonged fatigue, inferior food, and vicissitudes of climate. In a journey of some length it is decidedly economical to purchase horses and mules rather than hire them; and we resolved to supply a part of our wants in that way, Mr. Hunot being good enough to undertake to choose eight mules, for which on the following day we paid 8*l*. each.

The current coin in South Marocco we found to be French five-franc pieces (called by Europeans, dollars) for all except small transactions. These are carried on by Moorish silver pieces, worth respectively something less than four pence and two pence, and little coins of an alloy of copper and zinc, called *flous*, of which about fifteen go to an English penny. It was necessary to provide our interpreter, Abraham, with bags of these coins to defray the trifling expenses of our journey. It being understood that the provision of food for our followers and the animals of our train would be undertaken by the local authorities, wherever we should go, the only serious expenses we had to provide for were the purchase or hire of mules, and such gratuities as we might think proper to distribute amongst our escort and our servants on our return to Mogador. For presents to governors, sheiks, and others whom it might be desirable to conciliate or reward, we had brought with us a supply which turned out to be more than sufficient for the purpose.

To pass a quiet evening in our own house, free from any immediate cause for trouble, and with the prospect of a good night's rest, such as we had not known since we left Mogador, was an enjoyment keenly felt; and though our quarters were absolutely devoid of furniture of any kind, the mere sense of quiet and freedom from intrusion made them seem to us perfectly luxurious. The position of our dwelling was indeed admirably chosen. Completely separated from the inhabited quarters of the city, with their noises and their stenches, by large gardens and high walls,

the only building within our view was the great tower of
the Koutoubia.   Some idea of the effect as seen through
one of our windows is given in the accompanying woodcut.
It is very similar in design and dimensions to the Giralda
at Seville and the great tower at Rabat, and like these is
said to have been built by Christian captives.   Including
the lanthorn at the top, the height is about 270 feet.   It
is a singular proof of the deficiency of the Moors in con-
structive faculty, that the only stone structures in this, the
ancient capital of the country, once the abode of wealth
and barbaric luxury, should be this tower, and the great
arch forming the entrance to the Sultan's palace, of which
the carved stones were transported piece by piece from
Spain.

The morning of May 5 presented the unusual appearance of heavy clouds covering the sky and concealing from
view the range of the Great Atlas.   This did not last
long.   The sun soon reasserted his dominion over the
plain, though the clouds still hung round the higher peaks.
The direct heat of the sun was already great at this season, but the air was always relatively cool.   In the shade
of our rooms the thermometer marked about 80° Fahr.
during the warmer hours of the day, and fell to about 70°
at night.

It was a matter of some interest to us to study the
spontaneous vegetation of the gardens of Marocco.   We
could without difficulty have obtained permission to visit
the very extensive gardens that occupy the larger part of
the enclosed space surrounding the palace of the Sultan ;
but we decided that we should be able to work more
effectually, and without risk of exciting the suspicions of
the natives, by confining ourselves to the rather large space
surrounding our own dwelling.   To the English reader it
may be well to remark that, in Marocco, as in most Eastern
countries, a garden means something very different from
what we understand by it at home.   So far as any idea of
enjoyment is connected with it the paramount object is

J. D. H. & J. B. delt.

TOWER OF THE KOUTOUBIA AT MAROCCO

shade and coolness. Trees, and running water, without which in this climate few trees will grow, are therefore the essential requisites. Beyond this the Moor, if he be rich and luxurious, may plant a few sweet-scented flowers, of which the rose, violet, jessamine, and *Acacia Farnesiana* are most prized ; but beyond this, no mere pleasure of the eye is ever dreamt of, and here, as elsewhere, there seems to be among the natives a complete want of the sense of beauty.

To the Moor the chief object of a garden is not pleasure but profit. In this admirable climate nearly all the vegetable products of the temperate and subtropical zones may be had in profusion wherever water is attainable, and of this the Great Atlas provides an unfailing supply to the city and its neighbourhood. Even at the low prices of the country, fruits are the most profitable of all crops ; and it is asserted that the Aguidel Garden—the largest of those within the palace enclosure—containing about forty English acres, produces on an average 20,000*l.* a year.

The fruit-bearing trees planted in our garden were the date-palm, orange, olive, fig, pomegranate, apricot, almond, pear, apple, and mulberry, along with a few vines. Besides these were cypresses, willows (*Salix babylonica*), aspens, *Robinia*, *Melia*, and *Celtis*.

There were several tall bushes of *Acacia Farnesiana*, just coming into fruit, and of white jessamine. The only cultivated flowers were the rose, Mirabilis, and hollyhock, and a large-leaved variety of sweet violet, which has also been found in Madeira. Of wild arborescent plants we noted only *Zizyphus Lotus* and *Lonicera biflora* ; the latter species (peculiar to North Africa) we observed here and there throughout our journey. Although the list of wild herbaceous plants includes few that are not common throughout the Mediterranean region, it may interest some readers to give it in full. Specimens of nearly all the species enumerated were preserved by us.

List of Plants growing wild in the Garden of Ben Dreis, in the City of Marocco. Those marked (*) are British species.

* Papaver Rhœas, L.
* Fumaria parviflora, Lam.
  ———— agraria, Lag.
* Sisymbrium Irio, L.
  ———— erysimoides, Dsf.
  Brassica geniculata, (Dsf.)
* Capsella Bursa-pastoris, L.
  Lepidium sativum, L.
* Viola odorata, L., var.
  Frankenia pulverulenta, L.
  Spergularia diandra, (Guss.)
  Lavatera cretica, L.
  Malva parviflora, All.
  ——— nicæensis, All.
* Erodium moschatum, (L.)
  ———— malacoides, (L.)
* Oxalis corniculata, L.
  Zizyphus Lotus, L.
  Medicago pentacycla, D.C.
* Trifolium resupinatum, L.
  Lotus arenarius, Brot.
  Lythrum flexuosum, Lag.
  Bryonia acuta, Dsf.
* Conium maculatum, L.
* Apium nodiflorum, (L.)
  Ammi majus, L.
  Carum Petroselinum, (L.)
* Caucalis nodosa, (L.)
* ———— infesta, (L.)
  Lonicera biflora, Dsf.
* Galium Aparine, L.
* Gnaphalium luteo-album, L.
  Anacyclus Valentinus, L.
  Chrysanthemum coronarium, L.
  Senecio gallicus, Chaix, var.
  Calendula stellata, Cav. ?
  Carduus myriacanthus, Salzm. ?
  Onopordon illyricum, L.
  Picris pilosa, Del.
  Leontodon Rothii (Thrincia hispida, Rth.)

Leontodon hispidulus, Boiss.
Sonchus maritimus, L.
* ———— oleraceus, L.
* Convolvulus arvensis, L.
  Solanum villosum, Lam.
  Withania somnifera, (Link.)
  Hyoscyamus albus, L.
  Cynoglossum pictum, Ait.
  Verbascum sinuatum, L.
  Linaria græca, Chav.
  Scrophularia auriculata, L., var.
* Veronica anagallis, L.
* Marrubium vulgare, L.
* Verbena officinalis, L.
* Beta vulgaris, Moq.
  Chenopodium ambrosioides, L.
* ———— murale, L.
* ———— album, L.
* Rumex crispus, L.
* ——— pulcher, L.
* Polygonum aviculare, L.
  Euphorbia pubescens, Vapl.
* ———— peplus, L.
* Mercurialis annua, L.
* Urtica dioica, L., var.
* Parietaria officinalis, L., var.
* Typha angustifolia, L. ?
* Cyperus longus, L.
  Phalaris minor, Retz.
  Piptatherum multiflorum, P.B.
  Agrostis verticillata, Vill.
* Cynodon dactylon, L.
* Poa annua, L.
  Kœleria phleoides, Pers.
  Cynosurus aureus, (L.)
* Festuca rigida, (L.)
  Brachypodium distachyon, (L.)
* Bromus madritensis, L.
* ——— mollis, L.
  ——— macrostachys, Dsf., var.
* Hordeum murinum, L.

This list affords a fair illustration of the general uniformity of what may be termed the ruderal vegetation

throughout the Mediterranean region. Of 81 species enumerated there are but four (*Brassica geniculata, Lonicera biflora, Picris pilosa,* and *Leontodon hispidulus*)
that do not extend to Southern Europe; fully one-half
are found in Northern France and Germany; and 37 are
included in the British flora.

About 4 P.M. we started by previous appointment to
pay our visit to El Graoui. The usual course in this
country is to make the first visit to a man in authority
one of pure ceremony, wherein presents are made that are
intended to prepare the way for any serious business, which
is reserved for a second interview; but the necessity for
deciding without delay on our future plans, which depended altogether on the consent and assistance of El
Graoui, forbade this dilatory mode of proceeding, and it
was decided that we should go at once thoroughly into
the subject of our intended journey. This was a proud
day for our interpreter, Abraham. While, in spite of some
concessions made to Sir Moses Montefiore, his coreligionists in the city are forced to put off their shoes when
they leave their own enclosure, Abraham, rejoicing in a
pair of gamboge leather boots, could enter with head erect
the presence of the most powerful subject in Marocco.
Preceded by several mounted soldiers, we passed, by the
same filthy roads as before, through the great gate leading
to the palace. Perhaps the sight of a stately procession
irritated the feelings of the people; certainly there was
on this occasion no doubt as to the disposition of the
bystanders, made sufficiently clear by muttered curses and
spitting towards us, and even by a few stones thrown in
the same direction, though scarcely intended to reach us.

The walled space which surrounds the dwelling of the
sovereign appeared to be a nearly regular square, of which
the sides measure about half a mile. Besides the residence of the Sultan, rarely inhabited of late years, and
the extensive ranges of mean irregular buildings used by
his attendants and body-guard, another block of buildings

L

served as the dwelling of his son, the then Viceroy ; and a
third group, to which we proceeded, was the home of El
Graoui, who thus avoided the inconvenience of inhabitiug
a place subject to the authority of his rival, Ben Daoud.
We saw no building of the slightest architectural preten-
sions, or at all comparable to the house in which we
ourselves were lodged. Through narrow tortuous passages,
amidst low buildings, scarcely more than ten feet high,
and of the meanest appearance, we reached a whitewashed
building of two floors, and through a narrow door and
passage were ushered up a short flight of steps into a small
room, wherein sat a stout man of completely black complexion, whose broad countenance gave the impression of
considerable energy with an habitual expression of good-
humoured ferocity.   The room was decorated with wood-
work, cut into elaborate geometrical patterns and painted
in bright colours, the only form of decorative art known
to the Moors, and lighted by a lanthorn overhead through
small bits of coloured glass.   We seated ourselves on the
carpet-covered cushions ranged on either side, and a few
of the ordinary phrases of courtesy, familiar to all readers
who have made a tour in the East, were exchanged.   Presently, on a signal from the great man, the inevitable
green tea was served in English china tea-cups, followed
by a slight refection, the air of the room being meanwhile
perfumed with the heavy scent of incense burned over
charcoal.

    After this, we, without further preface, commenced
conversation as to the object of our journey, taking care
to adhere as nearly as possible to the line of discourse
previously fixed upon.   Having in general terms explained that we wanted to collect the plants of the high
mountains, Hooker was careful to add that we did not
care about stones or minerals.   We had been warned that
the belief in the existence of precious metals in the Atlas
is traditional in the country; and though no ruler of
Marocco is known to have made any effort to search

for or work mines, extreme jealousy is felt lest strangers should be attracted by the prospect, and attempt, in consequence, to establish themselves in the country. In point of fact, we were obliged to refrain from any overt attempt to collect even the commonest rocks, and the fragments which we carried away were picked up and pocketed—as if by stealth—when removed from the watchful eyes of our followers.

El Graoui, with apparent frankness and good-humour, said that he should have much pleasure in carrying into effect the intentions of the Sultan in our regard, and that we should have full liberty to go where we pleased in the portions of the Great Atlas where the authority of the Sultan is recognised, orders being given that the local authorities should provide food for us and our followers. This brought us at once to a question of pressing importance. Up to this moment we had no notion as to the limits that might be set to our journey by the Marocco authorities, though too well aware that Hooker's engagements in England would not under any circumstances allow us to carry our explorations very far. In answer to our inquiries, we were told that we were free to travel on the northern slope of the mountains that send their waters to the plain of Marocco. The names of various districts were mentioned, several of which were strange to our ears, and not to be found on any map; but we retained those of Glaoui, Ourika, and Reraya, the latter two being afterwards familiar to us. Further than this, we were informed that we might travel eastward through the provinces of Demenet and Ntifa, both apparently high mountain regions, whose waters run northward to the Oum-er-bia. We were emphatically told that we must not attempt to cross the chain southward in the direction of the Sous valley, and we gave a distinct promise to abstain from doing so. The actual decision as to our future route was far too important a matter to be disposed of on the spot by men so imperfectly prepared as we necessarily were, and it was arranged that Hooker should, in the course of the next day or two, acquaint El Graoui with the route which we might adopt.

The main point being thus settled, some further conversation ensued as to the arrangements for our journey. El Graoui informed us that he would send an escort of five soldiers under the command of a Kaïd, and further suggested that the escort that had already accompanied us from Mogador should remain with us throughout the journey. As it seemed desirable to humour the great man's fancies, the latter arrangement was at once agreed to, and in the sequel we found it decidedly advantageous. The interview was brought to a close by Hooker presenting to El Graoui a pair of excellent, though not showy, rifled pistols, which he had brought from England under the impression that they might be useful for personal defence. We had an unnecessarily large store of articles intended for presents—silver watches, musical boxes, opera glasses, cutlery, and the like ; but, on reviewing our stores, Mr. Hunot decided that such objects would be thought too paltry for a man of El Graoui's importance, and that serviceable fire-arms would be far better suited to his taste.

CHAPTER VII.

On returning from our interview with El Graoui, we felt
that our most urgent want was reliable information about
the districts mentioned by him as within the possible
range of our expedition.   Hooker had already ordered
Abraham to make inquiry for some one who had actually
travelled eastward from the city into the mountain country; but such a person was by no means so easily found
as a stranger might suppose.   The few Moors who ever go
into the interior are cut off from communication with the
natives by ignorance of the language; and, besides this, the
Moor is usually too incurious and intellectually sluggish
to carry away information about anything not directly
concerning his own business or pleasure.   Had our stay
in Marocco been prolonged, we might perhaps have been
able to collect some details as to the interior provinces
from natives of the mountain valleys who must from time
to time resort to the city; but it is very doubtful whether
a Christian stranger could obtain anything reliable in
this way.   People constantly forget how wide the gap
is that separates the mind of a modern European from
that of the inhabitant of a barbarous country, where the
conditions of society are such that apprehension of danger

to life and property becomes the predominant feeling. The notion that a man can care to acquire knowledge of any kind for its own sake is not for a moment admitted, and suspicion is necessarily the first feeling aroused by any inquiry, however apparently harmless. Bearing this in mind, we often felt astonishment at the share of success that has been attained by some geographers, and especially by Captain Beaudouin, the author of the French War Office map of Marocco. It is true that some large portions of that map are quite unreliable, and that it contains many grave errors as to the direction of the mountain ranges and valleys; but, considering that the greater part of it was compiled by the comparison of itineraries and descriptions furnished by a large number of separate native informants, the wonderful thing is that in many districts it should approach so near to accuracy as it does; and it undoubtedly shows a very remarkable degree of care, patience, and intelligence on the part of its author.

In the course of the afternoon, Abraham brought to us an elderly Jew, named Salomon ben Daoud, described as a man employed by the merchants trading with the interior, and familiar with all the roads leading to those parts of South Marocco with which the people of the city have any intercourse. The contrast between the appearance of this man and that of the Moors was complete. He had something of the downcast, long-suffering expression common among his coreligionists in this country, but an unmistakable air of intelligence that at once made him interesting. It was easy to understand that, although despised and often ill-used by the governing race, these people by their superior brain-power have contrived to make themselves indispensable to their masters, and that all people in authority, from the Sultan to the deputy-governor, are forced to rely upon them. Although Salomon was able to answer readily most of our questions respecting the several routes leading from Marocco into the neighbouring portions of the Great Atlas, it was inevitable that the information given by an uneducated man should fall very far short of what we should have desired; and the abundant catalogue of names of places—very few of them possible to be identified on the map—which he glibly enumerated, rather tended to confuse than to clear up our understanding of the country. With a view to mature deliberation on a point of such importance Salomon was requested to write down the chief particulars which he had given us verbally; and his memorandum, written in Hebrew, was afterwards translated by Abraham. This translation, checked by the memoranda taken down at the time by Ball, is printed in Appendix B, and affords a slight contribution towards the topography of a portion of Marocco hitherto completely unknown to Europeans.

The morning of Saturday, May 6, found us still in doubt as to our future course; but, on a careful review of the whole matter, we adopted a decision of which we saw no subsequent reason to repent. It appeared that if we decided on pushing forward into the interior of the country we might be able to reach the head of the valley of the Tessout—the main western branch of the Oum-er-bia—lying probably about 120 miles due east of Marocco. The portion of the Atlas chain whence that stream flows is in all probability as high as the range seen from the city, and perhaps somewhat higher, and the district through which we should travel was and still remains a complete *terra incognita*, as to which Beaudoin's map is almost certainly incorrect.[1] Against these strong inducements we had, however, to set many weighty reasons in favour of the alternative course, which consisted in at once directing our steps towards the main chain south-east of the city, and thence travelling gradually towards the Atlantic coast, penetrating in succession to the head of as many of the chief valleys as circumstances should allow.

The first course was open to the objection that, under

[1] See Appendix C.

any circumstances, it would involve a considerable amount
of travelling through a comparatively uninteresting country—at least four days, and probably six, for the route to
and fro between Marocco and Demenet, and four days at
least for returning to Mogador ; and further, that if difficulties should arise to prevent us from reaching the head
of the Tessout valley, we might possibly miss altogether
the main object of our journey by failing to reach the
higher region of the Great Atlas. On the other hand,
by coasting along the northern skirts of the chain, and
penetrating as many valleys as might be found practicable,
we should avoid altogether the need for retracing any
part of our course, and might reasonably expect to reach
a part of the chain whence a couple of days' ride would
carry us back to Mogador. The strongest argument was,
however, the consideration that by choosing the latter
route we should have numerous chances of accomplishing
our desire to reach the upper part of the mountain range,
and that if we should find impassable obstacles in one
or another valley, we should yet have an unimpaired
chance of succeeding elsewhere. Hooker's strong impression that our future course would not be unopposed
gave especial weight to the latter view, and the sequel
will show that we were well advised. It was therefore
decided to apply to El Graoui for letters to all the Kaïds
of the valleys subject to his authority in the range extending from Tasseremout to the borders of Haha, while,
with a view to a possible change in our course, he was also
requested to write to the Governor of Demenet.

In one way or other the days passed in Marocco were
so fully occupied as to leave no leisure, and Maw alone
was able to afford time for an excursion to one of the
low hills on the south side of the Oued Tensift, seen
on the left of the track by which we had approached the
city. The nearest of these—a rough mass of metamorphic
rock, rising nearly 800 feet above the level of the plain—
is only about three miles distant from the walls.

Our regret at not having been able to accompany
Maw on this excursion was much increased when towards
evening he returned with a number of interesting plants,
several of which proved to be additions to the Marocco
flora. The most notable of these was an undescribed
species of the tropical and subtropical genus *Boerhavia*,
and a curious *Reseda*, seemingly distinct from any described species. Besides these, he had gathered a variety
of *Forskåhlea tenacissima* described by the late Mr.
Webb as *F. Cossoniana*, *Andropogon laniger*, a fine
grass whose leaves have the scent of Verbena, *Echinops
strigosus*, and one or two more species characteristic of
the flora of Southern Algeria.

Maw also visited some of the bazaars, and described
them as miserable and repulsive, and we preferred to
let the carpet merchant who had been recommended to
us bring his goods to our quarters. The carpets made
here are not considered equal to those of Rabat, but they
are comparatively cheap and durable. On inquiring how
our purchases could be forwarded to Mogador, we found
that the hire of a camel with his driver for this weary
four or five days' journey amounted only to about seven
shillings of our money, and that the risk of robbery was
considered too trifling to be worth mentioning.

The comparison of five observations, taken on as many
successive days, gives for the altitude of our station in
Marocco a height of 511·9 mètres, or 1,679 English feet.
Allowing for the difference of level, the height of the
great square may be taken to be very closely 500 mètres
or 1,640 English feet. The observations were calculated
on the assumption that the barometer at sea level at
Mogador stood at 760 millimètres, and hence it is not
surprising that the results of each day's observation vary
form the mean, in some cases as much as fifty feet; but, as
settled weather prevailed at this period, the mean adopted
is probably very near the truth. Most of the results of
our observations (see Appendix) agree well with the few

observations previously made in the interior by M. Beaumier and M. Balansa; but in this instance there is a
difference of seventy mètres (or 230 English feet), M.
Beaumier's result being 430 mètres above the sea. No
particulars are given by him as to the instruments used,
or the methods of observations and reduction, and we feel
no hesitation in provisionally adopting the height resulting from our own observations.

As may well be supposed, the object that most fully
and constantly engaged our attention during our stay in
Marocco was the view of the Great Atlas range, for which
the terraced roof of our house afforded every needful facility. The interest attaching to an almost unknown chain
of great mountains was enhanced by the hope of penetrating its recesses. We were often tantalised by finding
clouds hanging about the flanks, or clinging to the higher
peaks, as happened during the latter days of our stay, but
there was always enough to be seen to reward our attention. We were able to identify the mountain, Miltsin,
which Washington took to be the highest peak in the
Atlas chain visible from Marocco; but we had already,
during the last day's journey before entering the city,
satisfied ourselves that there is no summit visible from the
plain of Marocco that can claim any marked predominance over its neighbours. Travelling, as we were, nearly
parallel to the main chain, we were far more favourably
placed than Washington, who approached the city from
the NNW., and always viewed the chain from nearly the
same direction. The crest is undoubtedly more sinuous
than it appears as laid down on the map, or when seen
from a distance; some of the projecting summits are
therefore nearer to the eye than others; but it appeared to
us then, and our subsequent experience only strengthened
that belief, that most of the peaks or prominences in
the higher portion of the chain seen from the plain of
Marocco, in a distance of fifty or sixty miles, attain to
very nearly the same height.

Washington speaks of a base-line of seventeen miles which served him for his trigonometrical observations, but it is obvious from his narrative that this cannot have been measured so as to admit of much accuracy in his results. As a matter of fact, it appears from his map and accompanying section, and from the narrative of his excursion to Tagherain, that Washington considerably underrated the distance from the city to the crest of the Great Atlas. His Miltsin is doubtless a summit near the head of the Ourika valley, which apparently stands some short way north of the axis of the chain.   According to the scale of his map Miltsin is twenty-seven and a half geographical miles distant from the house which he occupied, whereas it is impossible to estimate the true distance at less than thirty-three geographical miles.   If we allow for the height of Marocco above the sea level, and increase the estimated height of Miltsin in the ratio of its true distance to that assumed by Washington, we get for the height of the peak 13,352 feet (4,069·6 m.) above the sea, which is perhaps somewhat above the true measurement.

Owing to the prevalence of clouds during the latter period of our stay in Marocco, we failed to secure a satisfactory outline of the Atlas chain ; but, through the kindness of Sir J. D. Hay, we are enabled to insert a copy of a drawing made by the late Mr. William Prinsep, who accompanied the mission to Marocco in December, 1829. The view of the same range given in Jackson's generally accurate work must have been done from description or from imperfect recollection, as it bears no resemblance to nature.

We had been many times struck by the demeanour of the wild birds during our journey from Mogador.   They seem in this country to be quite free from what we look on as the instinctive fear of man, which in truth is an inherited tendency only in those countries where the human population habitually pursues them.   As we rode along, the turtle doves, which abound wherever there are trees or bushes, scarcely seemed to notice our passage, and would remain perched upon a bough close beside the track. Here in the city of Marocco a small bird about the size of a sparrow, but much more elegant in shape and attractive from its green-grey plumage—the *Fringillaria Saharæ*—displayed still greater boldness. During our meals, which were always taken in the central saloon open to the sky, they would boldly alight beside us, and pick up the crumbs that were sometimes purposely scattered for their benefit.

We saw nothing of the harmless serpents, one of which at least is said to inhabit every house in Marocco, and which the natives consider it unlucky to destroy or drive away. Probably they find the food that suits them only in inhabited houses, and ours had been so long untenanted that they had deserted the empty rooms.

Sunday, May 7, was fully employed in completing the arrangements for our journey, and packing up the botanical collections already secured. Our men had doubtless enjoyed the rest, and were gratified by a distribution of new shoes, or rather slippers, which replaced those pretty thoroughly worn out on the journey from Mogador. In a country where the surface is generally stony, and the soil abounds in plants beset by thick sharp spines, the thin slippers universally used by the people are very soon consumed.

Abraham appeared to-day gorgeously arrayed in a new suit, with dark yellow boots such as are worn by Moors of the better class. He had grown much in importance during the last few days, since, in his capacity as our interpreter, he, a Jew, has sat with his slippers on in the presence of El Graoui, the most powerful subject in Marocco.

With regard to the position of the Jews, there can be no doubt that the benevolent efforts of Sir Moses Montefiore, backed up by the representatives of England and other civilised States, have produced some permanent

effect.   In the coast towns, under the eyes of European
consular agents, they seem to enjoy security from violence,
and even from insult.   In the city of Marocco, where
they inhabit a separate quarter, walled in and accessible
only by two gates, they are safe so long as they keep
within those limits; but they are still forced to walk
barefoot when they pass into the city, and are exposed to
derision and insult against which they dare not protest.
In the remoter parts of the territory, where their scattered
communities are found here and there, their condition is
apparently still worse, and they are frequently subjected
to brutal ill-usage; but even there their superior intelligence and skill in industrial crafts, for which the Moor
is incompetent, secures them a certain degree of consideration.[1]

We this day made acquaintance with Kaïd el Hasbi,
the captain of the escort of five men, who, along with our
Mogador guard, was to travel with us through the Atlas.
Nature had given him a disagreeable countenance with a
forbidding expression, and our subsequent experience fully
confirmed the first unfavourable impression.

It had been arranged that our first day's journey from
Marocco was to be a short one, and accordingly our final
start on Monday, May 8, was delayed until 8 A.M.   Our
large tent, too heavy for mules, had been sent back to
Mogador; but, nevertheless, our baggage formed a very
sufficient load for nine mules.   Not counting our interpreter and Hooker's European attendant, we had nine
followers engaged in various capacities, besides twelve men
in charge of the hired animals, making up altogether, with
the escort, who numbered nine privates and two officers,
a party of thirty-seven men and thirty-three horses and
mules.   We wound slowly through the filthy lanes of the

---

[1] The reader who may desire a more detailed account of the city
of Marocco than we can afford is referred to the Description and Plan
of the City, by M. Paul Lambert, in the *Bulletin* of the French Geographical Society for 1868.

Jews' quarter, and went out by the south-east gate of the city, having on our right the high wall that encloses the vast gardens attached to the Sultan's residence. Having entered the city through groves of the date-palm, the foliage of which is too tough for the teeth of the locust, we had scarcely noticed these pernicious creatures on that occasion; but in the well-irrigated tracts south and west of the city which are devoted to tillage they had this year been more than usually destructive. It is in their young condition, while still active on the wing, that their voracity is greatest; but in that stage it is practically impossible to contend with them. When they have attained their full growth they become unwieldy, and at length nearly torpid; and it is then that the natives endeavour to exterminate them, with a view to prevent the females from depositing their countless eggs and leaving to the district a legacy of future devastation.

It seemed that El Graoui, for his own reasons, wished to give us a parting testimony of good-will and favour, without at the same time committing himself too glaringly for native ideas. It was not, doubtless, by mere accident that about a mile outside the walls we found him close to our track, with a train of mounted attendants, superintending the process of locust slaughter, and were informed that he wished to bid us farewell. Mounted on a splendid black charger, the old man, in spite of his unwieldy figure, had a commanding appearance. His manner was quite friendly; and, as the brief conversation proceeded, he rode along with us for a couple of hundred yards, and then shook hands with many good wishes for our safety and success.

The process of locust destruction which El Graoui was supposed to superintend was of the rudest description. The bodies of the bloated sluggish insects are swept into heaps with rough brooms, and a fire of twigs is then lighted over each heap. On the way from Sektana to Mogador, Maw afterwards saw another more expeditious process adopted in a part of the country intersected by open irrigation channels. Rough screens made of reeds are set up along one side of the watercourse, as shown in the annexed cut, and the inactive insects, being driven against them, fall into the water and are drowned. Some effect may doubtless be produced by these contrivances; but it seems very doubtful whether, if every locust that reaches the inhabited districts were destroyed, the plague would be materially abated. In a region including wide tracts almost without population there are unlimited opportunities for depositing the eggs; from these arise countless swarms, which, in their active condition, are capable of traversing wide spaces in search of nourishment. The suppression of the locust plague probably

awaits the spread of some creature to which their eggs would afford suitable food.

In great measure influenced by Washington's account of his visit to Tasseremout, and his conviction that from that place the highest ridge of the Atlas might be reached in a single day's excursion, we had decided on making that our first halting-place; and, as the distance can be little over twenty miles, we reckoned on reaching it by evening. Our way lay about due south-east through the district of Mesfioua, which is under the rule of a Kaïd, or sub-governor, subject to the orders of El Graoui.

To the eye the country seems a dead level; but the rapid flow of water in the covered channels and smaller

open rivulets showed that the slope of the ground from
south to north is very decided.   Along the smaller watercourses we noticed in abundance what appeared to be a
new *Pulicaria,* but was afterwards found to be the same
as an eastern species (*P. longifolia*) described by Boissier.

We rode along in high spirits, delighted to leave the
city, and still more with the near prospect of setting foot
on the mountain chain whose unknown recesses had so long
been a fascination for us; and the only drawback on our
enjoyment was the shifting veil of clouds that hung about
the higher summits, only now and then allowing some rugged peak to stand out for a few moments. As we gradually
drew nearer, our attention was more and more fixed on
the remarkable line of flat-topped bluffs, conspicuous in
the view from the city, that extends for a distance of fully
twenty miles along the base of the Atlas chain, and on the
east side seems to jut out in a northern direction.   From
a distance the face appears quite precipitous and almost
vertical, and there is but one conspicuous break in its
continuity.   This, as we afterwards found, is caused by the
stream running under Tasseremout, which has cut a deep
channel through the barrier.

After riding about three hours we approached an inhabited place, which we were told was the residence of
the Kaïd.   We had left behind us the tract of country
ravaged by locusts, and the general aspect of things was
here much brighter than we had beheld since leaving the
coast region.   The more brilliant green and more vigorous
growth of herbaceous plants led us to infer that, irrespec-
tive of the influence of irrigation, the zone extending
round the base of the mountain region must receive at
least some share of the more frequent rains that occur
there at seasons when the low country in general is con-
demned to utter drought.   On reaching the *kasbah* of the
Kaïd, which showed as a low but substantial building, with
walls sloping outward, we were accosted by an official
deputed to apologise for the absence of his chief, who

was to return towards evening, and to invite us to halt there for the remainder of the day. As it was now about noon, this proposition was met at once by a decided negative, when the chief of our Marocco escort intervened and, with an air of dogged insistance, urged the necessity for a halt. There ensued the first of many an altercation with the same disagreeable person, in which it is needless to say that Hooker's decision and firmness prevailed, and the order went forth to continue our journey.

Amongst the bushes near at hand we for the first time gathered a curious, but no way ornamental, Cruciferous plant, first found by M. Balansa, which exhibits the only distinct generic type yet found in the interior of South Marocco, and has been described by M. Cosson under the name *Ceratocnemum rapistroides*. It here grew four or five feet high, with long slender branches; but in open places we afterwards found it in comparatively stunted condition—a foot, or less, in height.

The country, after quitting the *kasbah*, gradually changed its character. Scattered blocks of moderate size showed themselves with increasing frequency, and seemed to be of very varied composition. Some were formed of a coarse-grained sandstone or fine conglomerate, others appeared to be granitic, though deficient in mica, while others looked like porphyry. The restrictions by which we were bound prevented us from undertaking any close examination, and still more from attempting to carry away specimens. The predominance of silex in the soil was made apparent by the vegetation. We had already often admired the pretty little rose-coloured *Spergularia diandra*, common on sandy soil throughout Southern Marocco; but this here became a conspicuous ornament; its numerous delicate flowers forming large cushions of bright colour on the surface of the soil. Among other characteristic species not before seen were *Aïzoon canariense* and a new species of *Anthemis*; but the predominant element in the vegetation was furnished by the *Leguminosæ*, and espe-

M

cially by the genera *Trifolium, Medicago, Ononis,* and *Lotus.* Of the first two genera we found in the lower region none but the common Mediterranean species, while the others displayed many local forms. One *Ononis* here found was altogether new ; and a *Lotus,* not elsewhere seen in our journey, seemed identical with an Oriental species not hitherto found west of Greece.

As we advanced, the upward slope of the ground towards the foot of the great range became much more perceptible, though still very gradual. At near 4 P.M., we arrived at another house belonging to the Kaïd of Mesfioua, and were informed that that functionary was waiting to receive us, and expected us to halt there for the night. The instinctive feeling of an Englishman who has made up his mind to accomplish a certain distance in his day's journey is to close his ears to any suggestion of delay, and all the more decidedly when there is reason to think that other people are scheming to oppose him ; so at first it seemed as if we should have further altercation with our escort. But as prudence pointed out that, whatever the feelings of the local authorities might be in our regard, it would be injudicious to do anything to give offence, and as at the same time the appearance of the country near at hand promised good botanising, we speedily decided on making a virtue of necessity, and with sufficiently good grace agreed to pitch our tents. By this time the Kaïd had come out to receive us, but retired after a brief salutation, it being understood that conversation was reserved till evening.

Without loss of time, we sallied forth with our portfolios, attended by one of the soldiers who was supposed to watch over our safety, and directing our steps to a dry river-bed that winds through the plain close by, were rewarded for our self-denial by finding a number of interesting plants not before seen. The most conspicuous of these was a *Tamarix,* which in some places grew thickly near the banks. It is remarkable for the bright pink colour of the seed vessel, and differed much in general aspect, though not widely in structure, from the common *T. gallica.* The river-bed is probably the natural channel of the stream that flows below Tasseremout, part of which is diverted into irrigation channels, but during rainy weather resumes its original course.

As the sun declined the clouds cleared away from the higher ridges of the Atlas, of which we enjoyed the finest view we had yet attained. Nearly due east and thence bearing towards ESE., was a group of high summits which, to judge from several large patches of snow, must be quite as lofty as that nearer to us. Between this, which belongs as we believe to the district called Glaoui, and the nearer range it was clear that a considerable valley runs deep into the chain. The drainage of this valley must flow to the Oued Tensift; but whether that be the main eastern branch of the river, or an affluent not indicated on the maps, is as yet uncertain.

After dinner, we adjourned from our tent to pay our promised visit to the Kaïd, who, according to custom, had green tea served in a small low room, which was reached through intricate passages. As a matter of course, the object of our journey was the chief topic of conversation, Among other plants we spoke of the *furbioun*,[1] or *Gum Euphorbium*, which we knew to be produced by a cactoïd *Euphorbia* that grows about the base of the Great Atlas, east of the city of Marocco. Concurrent native testimony fixes the province of Demenet as its chief home; and it must be very rare, or altogether absent, in the districts traversed by us, as it is scarcely possible that it should have been overlooked. Apparently conciliated by some trifling presents, the Kaïd informed us that he had in his garden some plants brought from Demenet, and offered one of these, which was safely forwarded to Kew, as a gift to Hooker.

[1] See Appendix D.

M 2

When nearly ready to start on the morning of May 9,
we were informed that the Kaïd meant to accompany us
on the way to Tasseremout.[1] This caused a slight delay,
which was not unpleasantly occupied in looking around us.
The morning air was delightfully cool (58° Fahr. at 6 A.M.),
although the day before had been hot, and at 11 P.M. the
thermometer had fallen only to 73°. The position of the
Kaïd's dwelling was in itself very beautiful, in the midst
of a fertile country encircled by hills, and these backed by
a majestic range of lofty mountains. The line of escarpment skirting the base of the Atlas, already distinctly seen
in the view from Marocco, was conspicuous on the eastern
side, but towards the south was partly concealed by fine
olive groves. The mean of two nearly concordant observations gives for the height of this place 2,399 feet
(731·1 m.) above the sea.

At about 7 A.M. we moved, and, crossing the dry river-
bed, very soon began to ascend among low hills, apparently
formed by erosion from an upper plateau that surrounds
the base of the mountains. We often rode along hollow
ways between high banks or lofty hedges formed of tangled
shrubs and climbing plants, in which were mingled some
familiar forms with several altogether new to us. A dog-
rose, scarcely distinguishable from the common British
Rosa canina, was common in some places, along with
profuse masses of *Ephedra altissima* and other southern forms. Climbing high over all these was a beautiful
Coronilla, with very large white and lilac-blue flowers.
We knew that a fine species of this genus, first brought
from Marocco by Broussonnet, had been formerly cultivated in England, though long since lost from our gardens;
but the *Coronilla viminalis* figured by Salisbury shows
yellow flowers, and is placed amongst the shrubby species
of the genus. It was clear that in the plant before us the

---

[1] As constantly happened, it was very difficult to fix the sound of this
name. The accent sometimes fell on the last syllable, sometimes on
the penultimate.

stems die down nearly to the root every winter; and our
belief that this was an entirely new species only yielded to
subsequent careful examination, which proved that it is no
other than Broussonnet's plant.

The date-palm had disappeared soon after we entered
the hills; here, and elsewhere on our route, it seems to be
confined to the lower region, rarely attaining the level of
3,000 feet above the sea. Its place was here supplied by
the palmetto (*Chamærops humilis*), which seldom forms
a trunk, perhaps because it is not allowed to attain a
sufficient age. As we advanced, the vegetation constantly
offered a more varied and attractive aspect; and one of
our first prizes was a new species of thyme (*Thymus
maroccanus*, Ball), somewhat like the species of the Argan
zone, but with oblong leaves and uncoloured bracts. Of
comparatively familiar forms there were *Cistus monspe-
liensis*, and *C. polymorphus*, the first species of that
genus that we had seen in South Marocco, the pretty
little *Cleonia lusitanica*, with many other Labiatæ. Of
plants new to our eyes by far the most interesting was the
curious *Polygala Balansæ*. To those who know only the
milkworts of Europe and North America, it must seem
strange to hear of a large shrubby Polygala, with branches
that end in a sharp point, few small leaves, so quickly de-
ciduous that it generally appears quite leafless, and large
flowers of a showy purple-red colour. In truth, although
there is great variety of form in this large genus, the
species which is common throughout the lower valleys of
the Great Atlas is very distinct from all its congeners.
In Arabia and South Africa there are some species forming
dwarf bushes with spinescent branches, but in other respects very different. When full grown this is six or eight
feet in height; and the round, green, almost leafless stems
give it, when the flowers are absent, much the appearance of *Spartium junceum*, the large broom of Southern
Europe.

After riding some way up a rather steep stony track,

we reached a grove of very fine olive trees, and our escort
came to a halt. We had reached Tasseremout. For some
time we had seen a large pile of solid masonry which
crowned the hill immediately above the olive grove. This
seemed to deserve a visit; but, on the other hand, the
attractions of the surrounding vegetation were irresistible
to botanists. The matter was settled by Hooker proceeding to visit the castle with the Kaïd, while Ball botanised,

FORT AT TASSEREMOUT.

and Maw secured living specimens of some of the more
interesting plants.

The castle of Tasseremout is one out of a large number of similar buildings standing on the northern outworks
of the Great Atlas chain that will afford interesting matter
for inquiry to future travellers when the country becomes
more accessible, and the lessened jealousy of the natives
will make a thorough examination of them less impossible
than it would be at present. The natives vaguely attribute

their construction to Christians or Romans, the same word
conveying either meaning; but the Jews often explain
this to mean Portuguese.   The general character of these
buildings, as far as our information goes, is tolerably
uniform.   The walls are of great thickness and built of
rough hewn stone: the arches are always rounded and
the lower chambers vaulted; and they are evidently
places of defence.   There is little reason to believe that
the Portuguese, who held at one time or other most of the
Atlantic coast of Marocco, ever established a firm footing
inland, and still less that they had such a hold on South
Marocco as would be implied by the erection of a chain
of forts along the foot of the Atlas.   On the other hand,
the history of Mauritania during the long period of the
decline of Rome, and preceding the Saracen conquest, is an
almost complete blank, save for a few apocryphal stories.
It is certain that the lower country was once completely
subject to Roman power and Roman institutions, and it remains to be ascertained how far an organised government
survived the weakening of the central authority.   That
the independent tribes of the Atlas may have been inconvenient neighbours to the half-Romanised inhabitants of
the plain is more than probable, and that the forts should
have been erected to hold the former in check seems the
most likely conjecture as to their origin.   Excavation,
whenever that may be practicable, will scarcely fail to tell
something of the original occupants of these buildings,
and to diminish our ignorance of a dark period of past
history.

   As to the question which interested us most nearly,
the Kaïd had at first been reserved; and when it became
necessary to decide, his language was decidedly unfavourable.   It was impossible, he said, to reach the high mountains with snow on them from Tasseremout.   Any one
attempting to do so would pass beyond his district, where
he could not protect us, and he could not allow us to incur
such a risk.   We remembered Washington's account of

his winter excursion from this place; and, what was more
curious, we found that a tradition of the visit of Christians
who have gone up the mountains here many years before
survived among the people. When, in December, 1829, the
late Sir J. Drummond Hay was received at Marocco with
great distinction by the then Sultan, it was arranged that,
after taking leave of the sovereign, the party should enjoy
two or three days' hunting towards the foot of the Atlas, and
they accordingly encamped somewhere below Tasseremout.
Washington and some other officers attached to the mission resolved to take the opportunity for ascending the
mountains as far as possible. At starting they evidently
thought it practicable to attain the higher peaks from
this place by a continuous ascent, and appear to have been
surprised to find, after several hours' climbing, when they
had reached and somewhat passed the limit of the winter
snow, that the 'highest peaks were still far beyond their
reach.' To one familiar with high mountain countries, the
natural course for attaining to the backbone of a considerable chain is by penetrating to the head of one of
the deeper valleys; and the course taken by Washington's
party would appear no more promising than the attempt
to scale Monte Rosa from the plain of Piedmont by ascending the mountains behind Ivrea. The mountain
stream that flows below Tasseremout seems to come from
the SE., where the range presents no conspicuous summits; whereas the higher points visible from our camp
at Mesfioua lay nearly due south. We were therefore not
inclined to insist on carrying out our original design of
making Tasseremout our base of operations; and when we
were told that the valley of Ourika, lying some distance
to the west, led to the snowy mountains, we at once decided
on moving thither in the course of the afternoon. To console us for our disappointment, the Kaïd invited us to a
repast which, like the food supplied at Mesfioua, was much
better cooked than usual. We especially appreciated
some cakes, or bannocks, of wheaten flour that made an

agreeable change from the biscuit to which we were often reduced.

Our impressions on this our first acquaintance with the outer region of the Great Atlas were very agreeable. The country appeared populous and fruitful. There was, indeed, little space for tillage, and that was of the rudest kind; but besides the olive, which attains a great size, the carob (*Ceratonia Siliqua*) and walnut, both growing to perfection, combine beauty with economic value. The common *Opuntia*, or Indian fig, also grows luxuriantly, and supplies an item in the diet of the natives.

Before we started, about 3 P.M. a body of miserable-looking Jews presented themselves, and offered a *mona* of olives, chilis, cakes of repulsive appearance, and some terrible spirituous liquor served in a battered tin teapot. When we excused ourselves on the ground that we had but just finished eating, they insisted that we should, at least, partake of the liquor. Abraham explained that we could not possibly drink out of a vessel so indescribably foul as the earthenware cup presented to us; whereupon one of the women lifted the skirt of her filthy petticoat, and proceeded to polish the cup to her own satisfaction. There was something pathetic in the abject air of these poor people, of whom there are many communities in this part of the Atlas. Born to suffering and oppression, they yet contrive to hold together, and even increase their numbers, thanks to superior intelligence and skill which make them almost indispensable to their neighbours. They are forced by law or custom to wear none but black outer garments, and the older men have high brimless cylindrical hats, tapering somewhat towards the top. They had taken it into their heads that Christian strangers travelling with a large escort must be persons of influence and authority, and had come to implore our favour and protection. The men concluded by kissing the skirts of our *jellabias*; and, as we were riding off, the women, who stood in a group behind, advanced and kissed our knees,

in true Oriental fashion. We were assured by our interpreter, who naturally sympathised with the people of his
own race, that they often suffer from ill-usage, for which
there is absolutely no redress; but it does not appear that
their condition is practically as bad as that of the same
people in Roumania and some other so-called Christian
States. In some respects, indeed, they are better off than
their Mohammedan neighbours. Not suspected of wealth,
their head-men are not liable to be 'squeezed,' and, living
apart, they are not engaged in the intestine feuds of
adjoining tribes, and not often victims of the cruelties
that accompany them.

During our afternoon ride from Tasseremout to the
Ourika river, our course lay to the SW., along the base
of the escarpment which had so much attracted our notice
from a distance; and much discussion arose as to the
origin of the vast masses of boulders that were spread along
the comparatively level shelf along which we rode, and
descended, in some places at least, to the margin of the
plain.

During the ascent from our camp of last night to
Tasseremout, we first made acquaintance with these
deposits, at a height of about 3,000 feet above the sea,
that of the olive grove at Tasseremout being 3,534 feet.
On the slope to the right of our track a mass of irregular
weather-worn blocks of sandstone lay in disorder, the
most prominent characteristic being that they were all of
large size (measuring from ten to twenty cubic yards,
according to Maw), with little or no intermixture of finer
materials. Maw was disposed to consider these as glacial
deposits ;[1] but, among other difficulties, it was urged
that the moraines of glaciers descending from a great
mountain chain always include a large proportion of finer
materials along with large blocks, and that these include
fragments of the various rocks through which the glacier

[1] See Mr. Maw's paper on the 'Geology of Marocco.' Appendix F.

flows, while it was *primâ facie* improbable that such a mountain chain as that before us should be altogether formed of the sandstone of which, so far as we could see, the blocks before us were exclusively composed. Soon after leaving Tasseremout, we came to the opening of a narrow valley or ravine cutting through the escarpment, and exposing to view great piles of boulders similar to those seen below, but on a larger scale. After this, the escarpment showed an unbroken face for a distance of about ten miles. Seen near at hand the slope, which from a distance seemed nearly vertical, appeared to have an inclination of from 35° to 45°, and rose to an average height of about 1,000 feet above its base. The upper beds, whose exposed edges were everywhere seen, seemed to consist of hard limestone with siliceous concretions; while the lower beds were of less consistent shaly limestone.

The ground over which we rode in a SW. direction, parallel to the base of the escarpment, was very irregular in form, rising in places into mounds sloping inwards towards the cliffs as well as outwards towards the plain ; and, although in great part covered with vegetation, it appeared pretty certain that the whole was composed of irregular masses of sandstone intermixed to some extent with fragments of the rocks forming the barrier beside us. To those who did not admit the probability of the boulders before seen being deposited by glacial action, the phenomena here presented offered strong confirmation. A glacier descending from a main valley necessarily flows down the slope towards the plain, and could not turn aside at right angles to its previous course, and to the line of maximum inclination, unless there had been a barrier of solid rock stopping the way, of which there was here not the slightest indication. Whether or not materials that are borne down a steep incline by sub-aërial denudation form a talus with a diminishing slope resting against the face of the escarpment, or form mounds at a greater or less distance from the base, is a question de-

pending upon the momentum with which they descend;
and this again depends on three elements—the weight of
the blocks, the steepness, and the length of the slope.   If
the greater portion of the materials consist of large blocks
launched down a steep and long incline, these will travel
to a considerable distance from the base of the cliff, and
gradually form a barrier that will stop the course of other
similar masses, until these accumulate into considerable
mounds, as may be seen in many instances of berg-falls in
the Alps.   Whatever be the origin of these accumulations
in this part of the Great Atlas, it would appear that the
conditions that gave rise to them have now ceased, as we
saw no instance of any large block that appeared to have
been recently borne to its present position.

As it was important to reach our night quarters by
daylight, we collected few plants during the afternoon
ride; a fine *Asperula*, with numerous flowers varying
from white to pink, seemingly not different from the
Spanish *A. hirsuta*, was a great ornament here, and in
several other places on the skirts of the Atlas.

As we approached the opening of a considerable valley,
it was apparent that the escarpment ridge here comes to
an end, and is not again traceable as a distinct feature in
the scenery on the west side of the Ourika river.   The
name Ourika, with which we now became familiar, appears
to be that of a district, governed by a Kaïd under the
orders of El Graoui, which includes a fruitful valley running deep into the heart of the Great Atlas.   Having
descended from the hummocky ground over which our
course lay, we struck the valley just where the stream
issues from between the hills below a village named
Achliz.   Nearly all the water was at this season diverted
from its natural bed into irrigation channels that are
carried through the plain of Marocco.   We rode some
way along one of these channels, bordered by tall reeds,
and a grand *Senecio*, fully eight feet high, but not yet in
flower, probably *S. giganteus* of Desfontaines.   The wide

bed of the stream, nearly quite dry, afforded the most convenient situation for our camp, which, by the mean of two observations, stood at 2,889 feet (880·6 m.) above the sea.

Though the more we afterwards knew of him the less we liked him, we observed, on this and some other occasions, that our disagreeable Marocco leader, Kaïd el Hasbi, shared in a quality that is common enough among uncivilised people, and especially noticeable among the Moors, of which due account should be taken by travellers. It is not the desire to please, still less real benevolence ; but a certain impressionableness, an involuntary sympathy, that makes these people thoroughly uncomfortable when they see a stranger annoyed or disappointed. In common phrase, ' they can't bear to see you put out.' An Englishman, a German, or a Swiss may travel with you the whole day, when you are suffering from annoyance, perhaps at something in his own behaviour, but will either not notice, or, if he do notice, will not heed, your humour. In this country a man who would see you killed or tortured with perfect composure, can be made more uneasy than you are by seeing you vexed or out of spirits. The disappointment we felt at our first failure to penetrate the inner recesses of the Atlas from Tasseremout, had been very perceptible during the afternoon ; and though our Kaïd was quite resolved to let us go no farther than he could possibly help, he wished to do what he could to keep us in good humour. Accordingly, we were scarcely housed in our tents when El Hasbi appeared with a supply of fruit, oranges, dates, and walnuts, that he wished us to regard as a present from himself, but were doubtless part of the ample *mona* that was obtained from the village authorities. Later on, the Kaïd of the valley made his appearance, civil, but no way cordial, and the result of the interview was not very favourable to the prospect of penetrating to the head of the valley.

We were led at the time to suppose that the more

or less overt resistance which we encountered here and
elsewhere in South Marocco, was altogether due to a
fanatical dislike to Christian strangers; but we afterwards doubted whether that feeling, undoubtedly prevalent
among the Moors, is equally general among the Shelluh
population; and as we came to know more of the practical results of our visits to these remote valleys, the
less surprised we were to know that they were unwelcome
to the inhabitants. The Sultan's order, as we learned
from El Graoui, had gone forth that we were not to be
put to any expense for the living of ourselves and our
attendants during our journey. So far as our personal
consumption went that was but trifling, as we largely relied on the provisions we had carried with us. Our
attendants no doubt consumed an ample share of food at
the one serious meal of the day, usually after nightfall,
and were ready to set to again in the middle of the night
when a good opportunity was offered; but it was the rapacity of our soldier escort that made our visit a calamity
in a poor district. Not satisfied with gorging themselves
with meat, cakes, and fruit, they demanded luxuries such
as green tea and white sugar, and in such quantities that,
as we afterwards learned, Kaïd el Hasbi used to send from
each valley in which we halted a mule laden with provisions to his family in Marocco. An altercation which we
heard this night, and which was repeated more than once
on subsequent nights, arose from our usually pacific Mogador Kaïd, who revolted at seeing the lion's share of the
spoil taken possession of by his colleague from Marocco.
On this occasion the quarrel threatened to become serious,
and the long guns were actually drawn out of the red
cloth cases; but it seemed that on one or both sides dis-
cretion overcame valour, as peace was ultimately restored.
Our interpreter, Abraham, as a prudent man, wished not
to embroil himself in these disputes, and it was only
gradually that we got to learn the real mischiefs and
hardships of which we were the involuntary occasion.

CHAPTER VIII.

Vegetation of Ourika valley—Destruction of the native trees—Our progress checked—Enforced return—Shelluh village—Ride from Ourika to Reraya—Trouble with our escort—A friendly Shelluh sheik—Native desire for medical advice—Characteristics of the Shelluhs—Zaouia of Moulaï Ibrahim—Camp in Aït Mesan valley—Excursion to the head of the valley—Reach the snow—Night travelling in the Atlas.

AT sunrise, on April 10, the thermometer stood at 60°, and in this delightful climate we were in the best spirits for undertaking the work that seemed to be ready cut out for us, by exploring the fine valley that led directly from our station to the heart of the great mountain chain. Our expectations were, indeed, somewhat damped by the discussions that had already taken place with the Vice-Governor and with Kaïd el Hasbi. They did not deny that we might travel some way up the valley, but asserted that it would not be practicable for us to 'reach the snow.'

It was impossible to leave the spot where we encamped without giving a little time for collecting some of the very interesting plants that grew close at hand. Foremost amongst these was a leguminous shrub that seems destined to become an ornament to the gardens of Europe. This—*Adenocarpus anagyrifolius,* of Cosson—was first found in 1867, by M. Balansa, and seems to be common, especially near the banks of streams, between 3,000 and 5,000 feet above the sea level throughout this part of the Great Atlas. The long racemes of bright yellow flowers were conspicuous from a distance, and we afterwards found the pods, densely covered with black glands, but nowhere containing ripe seed. Another new bush belonging to the same natural order was also seen for the first time—*Hedysarum membranaceum*, of Cosson. Unlike the other, this is rare, and seems to be limited to the lower zone at the foot of the mountains. We failed to find either flower or fruit, though M. Balansa gathered both, perhaps at a lower level, in May 1867. We also collected fine specimens of two new and very distinct species, first seen on the preceding day—*Lotononis maroccana*, Ball, and *Lotus maroccanus*, Ball.

Soon after 8 we got under way, and, after a short ride along the right bank, reached the stream above the part where the water is carried off for irrigation purposes. It was now seen to be a rapid torrent, from twenty to forty yards in width, and nearly two feet deep. For some distance the narrow floor of the valley was nearly flat, and the moist soil was covered with poplars and willows, and a dense undergrowth of grasses and herbaceous flowering plants. Among these were two large *Ranunculi*, and a gigantic orchid, growing four or five feet high, only a variety of our common *Orchis latifolia*.

The vegetation became still more interesting when we left the flat bottom of the valley, and began to ascend on drier ground, between tangled masses of bushes that formed a sort of thick hedge on either side of the track. For the first time in South Marocco, we saw two species of *Clematis*—*C. cirrhosa* and *C. Flammula*,—along with several other Mediterranean species; but our minds were especially exercised by a little bush with slender twigs and pinnate leaves, which, in default of flower or fruit, we were at first unable to refer to its place in the natural system. It turned out to be a curious species of ash, first found in the plateau region of Southern Algeria, appropriately named by M. Cosson *Fraxinus dimorpha*. As long as it remains a bush, with numerous twiggy stems, the leaflets are blunt and rounded; but when it becomes a shrub, with a stout trunk, it throws out leaves that approach in form those of the flowering ash. We nowhere

saw it in this condition in Marocco, and only by the help
of Algerian specimens could we have suspected the identity
of the two forms. Among many thorny bushes we saw
here one, first gathered the day before, near Tasseremout,
which Hooker at once pronounced to be a *Celastrus.*
This was first found in the South of Spain, and described as *C. europœus* by M. Boissier, but is in truth
one of the many forms of *C. senegalensis*, a widely-spread
tropical species, that extends from India to the west coast
of Africa.

Among other novelties, we here saw, for the first time,
a little annual stonecrop (*Sedum modestum*, Ball), that
nestles in hollow places under large stones, or about the
roots of trees; but the most curious trophy of our day's
work was a miniature bramble, lying flat on earthy banks,
with small, mostly undivided leaves, and very few minute
prickles. It is possible that the imperfect fruits that we
saw had been dried up by the sun; but it seems more
likely that this belongs to the group of dry-fruited brambles—the genus *Dalibarda* of some botanists—hitherto
known only in America and Eastern Asia.

We passed near to several villages; but, as a rule, the
valley tracks in the Great Atlas are carried on one side,
and do not approach near to the houses. The valley grew
narrower as we advanced, and the moderately steep slopes
on either hand were covered with small trees of *Callitris*,
and *Juniperus phœnicea*, none of them more than about
thirty feet in height. If this country were administered by people capable of taking thought of its future
prosperity, the former tree might undoubtedly become an
important source of wealth. The beauty of the wood, if it
were only allowed to attain a sufficient size, would always
secure a ready market, even though it never reached the
extravagant price which, under the name of citrus wood,
it obtained in the days of Imperial Rome. The only use
which it serves in Marocco is the production of gum
Sandrac, of which a small quantity is exported to Europe.

N

The destructive practice of setting fire to the brush-wood is the sole cause that prevents the northern slopes of the Great Atlas from being clothed with valuable timber. The motive is not only the desire to obtain pasture for sheep and goats, but also to deprive an enemy of cover for ambush during the frequent skirmishes that occur between neighbouring tribes. The olive, carob, and walnut, which are planted in the main valleys, and produce annual crops, are carefully protected; but the notion of looking forward to future profit after an interval of thirty or forty years would be absolutely unintelligible to a native of this country. By a very rude process, the natives extract from the trunks and branches of *Juniperus phœnicea* a sort of tar which is said to be a useful application in wounds and sores of men and animals.

As we continued to advance, the valley narrowed almost to a defile, and the track, carried along rather steep slopes, became difficult for laden animals, though not worse than one commonly meets in high mountain countries, nor nearly so bad as some that we afterwards traversed in the Atlas. Presently, Abraham announced that Kaïd el Hasbi declared the track too bad for farther progress of the baggage mules. As it was apparent that the valley widened out a short way ahead, and that the *mauvais pas* would soon come to an end, we turned deaf ears to the remark, and rode doggedly on for a short distance farther, encouraged by the view in front, which disclosed a long reach of valley, running deep into the heart of the great chain. Shortly after, those who were in front became aware that a vehement discussion was going on at the centre of our scattered line. The energy of our interpreter was taxed to the utmost in striving to render the emphatic sentences that were exchanged between Hooker and Kaïd el Hasbi, supplemented by the panto-mimic gestures of the latter.

The gist of the argument was to the effect that even if we did go some way farther, at the risk of our baggage animals rolling down into the torrent—one of them had already slipped, and had a narrow escape—our progress to the head of the valley was out of the question, as the people there were in full revolt, and would not recognise the authority of the Sultan. We were inclined at the moment, and afterwards, to believe that this was a lie trumped up for the occasion; but the story might possibly be true, and, whether it were so or not, it was clearly impossible for us to proceed against the positive and determined opposition of our escort. With feelings of bitter disappointment we dismounted, and ordered that the baggage should return to a village near which we had passed an hour before, while we climbed to the top of a projecting spur of the mountain, commanding a view of the upper valley. For the moment, our interest in the vegetation yielded to the attractions of the scenery, and our curiosity as to the nature of the great mountain chain that rose steeply before us, seamed with snow that nowhere formed wide fields of *névé*, but lay in hollows and ravines forming long vertical streaks throughout the upper zone of several thousand feet in height. About two miles ahead of us the valley forks, the main branch from SSE. receiving a tributary from the S. or SSW. On a lofty spur between the two streams stood a village, conspicuous from a distance.

To be turned back at the very moment when the main object of our journey lay before our eyes, and that on pretexts that we utterly disbelieved, was sufficiently aggravating; and it was not in the most cheerful humour that, about mid-day, we retraced our steps down the valley, and, yielding to the suggestions of our escort, halted at the olive grove beside the village which we had noticed during our ascent. This was the first Shelluh village that we had been able to inspect nearly, and it was of the same type which we afterwards found throughout the mountain region. Unlike the Arabs, the Berebers always use stone for building when it is available. The walls are thick and solid

N 2

below, but rudely constructed, and the upper portions are sometimes put together with mud and small pieces of stone. There is usually an upper story extending over the whole or part of the ground floor, and the roofs are flat and made tolerably water-tight with mortar or cement. In structure and appearance they reminded Hooker of the village houses of the mountaineers of Bhotan.

We collected a good many specimens during the afternoon, but were able to add little to the list of species already noted in the valley. The most important business in hand was to bring to book our enemy Kaïd el Hasbi—for so we began to consider him—and ascertain clearly whether he did or did not mean to carry out the orders of the Sultan, and convey us to some point within reach of the upper regions of the mountains. When pressed on this point, he distinctly declared that from Reraya—the next adjoining district to the west—we should be able to 'reach the snow;' and with that assurance we had to satisfy ourselves, and give such orders that the next evening should find us in the desired district.

The name of the village was differently noted by the members of the party; Assghin, as it is entered in Hooker's notes, by an observation taken this afternoon, with the thermometer in the shade at 72° Fahr., stands at 3,427 feet (1,044·4 m.) above the sea. Up to this our observations had been reduced on the assumption that the pressure at sea level was exactly 760 millimetres; but henceforward we had the advantage of direct comparison with observations recorded twice a day at Mogador.

The evening was fine, but flashes of lightning were seen to the SE. after dark, and during the night fresh snow fell on the higher ridges, which looked brilliant on the early morning of April 11, but rapidly melted under the mid-day sun. We started rather late, about 8 A.M., and by 10.30 had returned to the site of our former camp in the broad bed of the stream below Achliz. Here our

ill-humour was increased by a long and quite unnecessary delay. As a rule, a light luncheon was all that was
consumed at our mid-day halt, the men being content with
some fragments of the *mona* of the previous night. But
our greedy soldiers had requisitioned a further *mona*,
nominally on our behalf, from the adjoining village, and
were determined not to move forward until it was supplied.
When Hooker happened to surprise our Mogador Kaïd in
the act of secreting a quantity of tea and sugar, the old
fellow in self-defence began to narrate the misdeeds of his
colleague, and so gave us a clearer notion than we before
had of the sort of abuses that pervade the whole fabric of
Moorish administration. It is true that one is told that
the value of goods requisitioned in this way by Government officers is allowed to the villages as payment on account of taxes; but the poor country people tell a different
tale, and it is probable that any allowance made on this
head is quite inadequate.

It was 2.30 P.M. when we were at last able to start,
and, as we knew that there was still some distance to
travel, we had but very little time for botanising during
the afternoon ride. Our way lay for more than two hours
along the base of the hills, whose forms were much of the
pattern usually seen where a high mountain rises from a
plain country. The ridges dividing the main valleys gradually diminish in height as they recede from the axis,
and ultimately are weather-worn into eminences of a more
or less conical form, which project to an unequal distance
towards the plain.

Towards five o'clock we began to ascend to a low pass
connecting a long projecting spur to the right with the
main mass of the hills on our left. Up to this we had
seen a good many scattered blocks of sandstone, but nowhere forming mounds. We now came on limestone—
showing traces of fossils. The hills hereabouts were bare
of trees, with a thick growth of palmetto, bushy *Labiatæ*,
Helianthemum, and perennial grasses, except where, under

tillage, they produce good crops of red-bearded wheat.
From the first pass we descended rather steeply to cross a
narrow torrent bed, and reascend to a second somewhat
higher pass, reached at 6 P.M., which we found to be 3,590
feet (1,094·3 m.) above the sea, or just 700 feet above the
river at the mouth of the Ourika valley. The country
here appears to be fully peopled. We saw several villages,
and one or two quadrangular buildings of larger size, probably the dwellings of local sheiks. We had during most
of the way wide views over the plain of Marocco, and were
able to distinguish the city itself, with the great tower of
the Koutoubia and the extensive palm groves on its
western side. Slight undulations of the ground are not
perceptible when seen from above; but it was quite clear
to us that nothing deserving to be called a hill breaks the
uniformity of the gentle slope with which the plain subsides from the base of the Atlas to the region traversed by
us on our road from Mogador.

The monotony of the march was diversified by another
furious quarrel between the officers of our escort. Whatever may have been the pretext, the cause was doubtless
the mutual ill blood arising from the disputes over their
respective shares of the spoil obtained from the villagers.
Not content with volleys of guttural abuse, and seemingly
ferocious threats, they appeared intent on serious mischief,
and proceeded to unpack their long guns. Luckily these
were securely wrapped up in complicated covers of red
cloth, and, before the combatants were ready for action,
prudence once more restrained valour, and the storm passed
away in sullen growlings and mutterings of future vengeance. It sometimes struck us that if there were such
things as professional brigands in Marocco, we might have
been robbed or murdered with perfect safety before one of
the guns of our escort had been extracted from its case,
and made ready for use.

Indignant, as we were, at the rapacity of our escort,
we assured ourselves, when we came to know more of the

country, that there is a great deal of human nature among
the Moors, as there is amongst most of the people that
travellers make acquaintance with, and that the conduct
of the soldiers and their officers was pretty much what
might be expected from any other men put in the same
position. The pay of a captain in the regular army is
equivalent to 4s. 2d. per month, and that of the men not
nearly enough to support life, even allowing for the frugal
habits of the people. It is only natural that when the
opportunity is offered, along with the certainty of impunity, they should make the most of it, as they certainly
do. In some places, as we afterwards learned, they
were not satisfied with the large supplies that they requisitioned, but demanded and obtained money from the
village authorities.

From the summit of the second pass, the track descends about 400 feet into a broad valley, well wooded with
olive, carob, and other cultivated trees.

Soon after seven o'clock we reached a convenient spot
on flat ground, beside a rapid stream, near a village called
Tassilunt. The scenery here was very picturesque, although we had no view of the higher part of the chain.
The nearer mountains were of a deep red colour—probably sandstone—contrasting finely with the rich green of
trees and shrubs that covered most of the slopes. The
floor of the valley here, as in most parts of the range that
we visited, is chiefly devoted to olive cultivation, poor
crops of grain being raised beneath the trees.

The sheik of the village soon presented himself, and
before long an abundant *mona* was brought to the tent
door, and laid, according to custom, at Hooker's feet.
Five large dishes of cooked meat and keskossou, and piles
of wheaten cakes, were designed to stay or to whet the
appetites of the party; while a sheep, twelve fowls, fifty
eggs, and five pounds of butter formed a provision for
their future wants. When it is remembered that nearly
as much more cooked food was supplied in the morning, it

may be imagined that the tax on the resources of a poor mountain village was not trifling.

We had now entered the district of Reraya, which is under the rule of El Graoui, represented by one or more deputies. The whole population, excepting some miserable-looking Jews, is Shelluh; but here, as everywhere among the Berebers, these are divided into tribes or clans, who are often at feud, and always jealous and suspicious of each other. There is generally a superior chief or sheik, having a wide, but ill-defined authority over the whole clan; but among those that recognise the Sultan's paramount temporal as well as spiritual supremacy, this is subordinate to that of the Governor. In this part of the Great Atlas, the clans, as well as the districts named after them, preserve the Shelluh patronymic of *Aït*; but the Bereber tribes of the high ranges E. and SE. of Fez have generally adopted the Arabic *Beni*, as with the powerful tribes, Beni Mtir and Mghill. One of the many difficulties of the geographer in this country arises from the practice of naming each district from the tribe that inhabits it, and the fact that, either from compulsion, or a taste for migration, it is not uncommon for a tribe to remove from one valley to another. The next valley to that where we now were was called at this time Aït Mesan; but if the Aït Mesan should take possession of some neighbouring valley, or be driven out by a stronger tribe, the traveller who visits the country some years hence may find the Aït Mesan valley in quite another place from that which we have described.

Our chief anxiety now, was to ascertain that the promise held out, of penetrating to the inner recesses of the Great Atlas in the district of Reraya, was to be realised. We were told that in the next adjoining valley we should reach a place only two hours' journey from the snow, and that the sheik of that valley had been summoned, and would arrive on the next morning. At the same time, whether from ignorance or a design to mislead us, El

Hasbi's language was decidedly vague and confused, and, after the experience of the last few days, there was no reason to feel the slightest confidence in his assurances. Hooker therefore decided on bringing matters to a point by informing El Hasbi that if any further difficulty was made, he should despatch a courier with letters for El Graoui and the Viceroy at Marocco, saying that the commander of our escort has failed to carry out the Sultan's orders, and requesting that another should be sent in his place. As this was the last thing to suit El Hasbi's book, he became profuse in assurances of devotion to our wishes, and for some days, at least, we had no reason to suspect him of further machinations to defeat our plans.

In point of fact, a courier was sent on the following day with a letter for El Graoui. It was desirable to obtain the direct sanction of the authorities for our intention to remain several days in the district we had now reached, and to make it understood that this would be essential for the object of our journey.

A further topic requiring some previous arrangement arose from Maw's desire to return to England as soon as possible, after effecting the desired ascent to the higher region.

Our camp, which stood at 3,160 feet (963·1 m.) above the sea, was in a pleasant and sheltered position, and the temperature was thoroughly enjoyable. The relative coolness of the nights was not, indeed, so remarkable here as in the plain, for the actual temperature was pretty much the same, while that of the shade of day, which at this season, there usually rises somewhat over 80° Fahr., rarely exceeded 70° in the lower zone of the Great Atlas, however powerful might be the direct rays of the sun for several hours in each day. Towards dawn the air was usually keen, often almost cold; and at the hour which, when possible, was selected for observation—about one hour after sunrise—Fahrenheit's thermometer, as well in the plain as in the main valleys of the Atlas, usually ranged from 56° to 60°. It will be seen that this country, when made accessible to civilised Europe, will supply the nearest approach yet discovered to the perfection of climate, whether for health or enjoyment.

The morning of May 12 was in every sense a busy one. During our afternoon ride of the day before, one of the baggage mules, while following a narrow track along the bank of a watercourse, had slipped into the stream, and a large parcel of Hooker's plants had been thoroughly soaked. Several hours of the night, and the early morning, were consumed in repairing the damage, by laying the plants in dry paper, and drying in the sun that which was wet. Then came the important affair on which our hearts were mainly set. The sheik of the Aït Mesan valley had arrived, and it was necessary by a judicious combination of compulsion and conciliation to secure his co-operation in our undertaking. It was true that the orders of his superior, El Graoui, if duly conveyed by our escort, should alone have sufficed for our purpose; but we had already learned that, by a mutual understanding between the treacherous El Hasbi and the local authorities, our progress could at any moment be effectually barred. How were we to detect and expose the falsehood of the stories that were daily trumped up, and were seemingly accepted for truth by our own attendants?

The sheik appeared sufficiently cordial, especially when he was made to understand that, in case we were satisfied, he should receive a handsome present; and it was arranged that our next camp should be fixed at a spot within reach of the snow.

When the time for our departure drew near, a fresh, but not quite unforeseen, cause for delay presented itself, by the appearance in our camp of a crowd of native applicants for medical advice. Before leaving England we had been advised not to neglect the surest means for conciliating the good will of an African population, and had fortunately provided ourselves with a sufficient stock of

common medicines. Even an ordinary traveller, with no
more knowledge of medicine than the elementary notions
possessed by most educated persons, may safely apply simple
remedies in many of the cases of sickness that commonly
occur among uncivilised people; but in our case there was
no occasion for rash experiments, as Hooker's medical
knowledge and skill were more than sufficient for the
needs of the patients who flocked in considerable numbers
to ask for advice.   From this time forward, except in one
or two places where the people were withheld by the bigotry
of the authorities, this became one of the daily demands
upon his time and patience.

To judge from our own observation of the Shelluh
people, and the experience of French travellers among the
Kabyles, it seems probable that a traveller having some
knowledge of the Bereber language, and a little medical
skill, who could once make his way among the independent tribes of East Marocco, might safely explore the unknown portions of the Great Atlas.   The first condition
would be, that he should be able to overcome or evade
the obstacles that would be put in his way by the Moorish
authorities; and the second, that he should avoid treating
any case that was likely to have a fatal termination.   The
position of an infidel stranger who might be supposed to be
accessory to the death of a native of one of these wild tribes
would doubtless be very perilous.   The only branch of
natural history that could be followed by a traveller under
such conditions would be botany.   In collecting plants he
would be supposed to be following his proper avocation;
whereas the slightest attention given to stones or minerals
would be set down to a search for treasure.

The Shelluh population of the Great Atlas is strikingly
different from the Arab stock, but scarcely to be distinguished in appearance from the cognate Bereber races,
the Riff mountaineers of North Marocco, and the Kabyles
of Algeria.   Long faces, of a deep sallow complexion, high
cheek-bones, eyes closely set and not so dark as those of

the Moors and Arabs, are the prevailing types. The hair
is cut short, and the use of the turban seems to be confined to the women. The men, when they use an upper
garment, wear a black cloak or large cape of goat's hair or
camel's hair, into which is inserted on the back an angular
patch of red woollen stuff. Their character seems even
more different from the Arab type than their aspect.
The Arab hates work, takes to it occasionally from necessity, but passes his time so far as he can between talk,
story-telling, and song, and dreamy contemplation, in
which he is helped by the habitual use of *kief*, prepared
from Indian hemp, the local substitute for tobacco. The
Shelluh, on the contrary, is active and hardworking. He
has some natural fitness and acquired skill in agriculture. His intelligence is readier for all practical purposes; and, in spite of difficulties of language, which
generally involved a double process of interpreting between us and the natives, we found it much easier to
obtain information on any matter of interest than from
the Arabs. Intense curiosity was always shown by them in
our proceedings, and a circle of people from the nearest
village, standing hand-in-hand, generally encompassed our
camp.

During the morning Maw amused and interested the
people by showing a little practice with a small English
rifle. The long flint-lock guns and bad gunpowder used
in this country form such ineffective weapons that the
people cannot conceive the possibility of every shot telling.
This accounts for the fact that in the frequent skirmishes
that arise between neighbouring tribes so little damage
is usually done. Several hundred men may spend the
day in firing at each other; a vast quantity of ammunition
may be consumed; but the list of casualties on both sides
seldom exceeds half a dozen killed and wounded.

At 11 A.M. we left our camping ground, and began to
ascend the valley, soon approaching the banks of the
stream, which was everywhere easily fordable. In places

where it has cut a channel through sandstone rock there was space to ride along the bed, and we here found several rock-plants of some interest. The most conspicuous was the European *Catananche cœrulea*, not before seen by us, but extremely common in the interior valleys of the Great Atlas, growing two or three feet high in the warm zone, and dwindling to a few inches in colder and exposed stations. Of greater interest was *Selaginella rupestris*, a species of club-moss that makes the round of the world in the tropics, but is very rare outside those limits.

Before long we began to ascend the slopes on the western side of the valley. The hill was covered with a dense growth of shrubs and low bushes, in great part evergreen, and had more the characteristic aspect of the region surrounding the shores of the Mediterranean than anything we had seen since leaving Tangier. But, although there were several identical species, the differences were very marked, and a single glance sufficed to show that we were far removed from the flora of North Marocco. The arbutus was the sole representative of its natural order, and no heath extends to the Great Atlas. The oak-scrub, in this and the neighbouring valleys, is all formed of some form of the evergreen oak, *Quercus lusitanica*, *Q. coccifera*, and the allied forms being all absent. The Alaternus is common to both regions, but a narrow-leaved form of *Rhamnus oleoides* is here more common. Seven species of Cistus that adorn the hills in North Marocco are on the slopes of the Great Atlas reduced to two, and those the least conspicuous. On the other hand, the number of bushy Labiatæ was here largely increased, and included many peculiar species not known to grow elsewhere ; and there were many *Umbelliferæ*, of which several were not yet sufficiently advanced for recognition. Of *Leguminosœ*, which everywhere play a conspicuous part in the flora of this region, the most striking novelty was a new *Coronilla* (*C. ramosissima*, Ball), that forms a low bush, with very numerous slender intricate branches, covered at this

season with rather small yellow flowers. In the midst of
so much that was strange to the eye, it was pleasant to
see two familiar European orchids, *Orchis pyramidalis*
and *Ophrys apifera.*

There was something comical in the effect of our long
cortège, with the escort swollen to-day by the addition of
three sheiks of the valley, winding solemnly up the slope
of the mountain, but thrown every now and then into
general excitement by the appearance of some unpre-
tending plant. The order ' catch him flower' would then
issue to the native attendants, or one or other of the
travellers would set foot to ground the better to inspect it.
But any sense of incongruity between the pomp and circumstance of our mode of travelling and the simple nature
of our favourite occupation was lost on the natives. To
them one pursuit of civilised man is as unintelligible as
another, and they can conceive no other serious occupation for men not forced to labour than war or hunting.
It is a curious instance of the survival of barbarous in-
stincts, that a good many people in our own islands, who
imagine themselves to belong to the upper classes of society, have scarcely advanced a step beyond the mental
condition of the Shelluh mountaineer.

We passed a village where we noticed some rude oil
mills; and, after an ascent of about a thousand feet, reached
the summit of the ridge dividing the valley we had left
from the long and important one, the upper part of which
is known, from the tribe that inhabits it, as Aït Mesan. It
is very difficult to trace the course of the streams that
flow northward from this part of the Great Atlas, because
they are so extensively diverted into irrigation channels
that the natural bed is often dry, except after heavy rain.
According to Beaudouin's map the streams from this and
several adjoining valleys all flow to the Oued Tensift by
the east side of the city of Marocco. This we were led to
believe an error in that map. It is probably true of the
Ourika river and its affluents; but our own observation,

confirmed by the statements of the natives, led us to think
that all the streams from the Reraya district flow north-
westward after entering the plain, and unite with those
from the districts of Gurgouri and Amsmiz to form the
river Oued Nyfs, which we had passed at Misra ben Kara ;
the same name, variously pronounced Oued Enfist or Oued
Enfisk, being applied to several of the separate torrents
above referred to. It will be remarked that the name
Oued Enfist is merely an anagrammatic form of Oued
Tensift, the main river that drains all this portion of the
Great Atlas ; and it is a question whether the natives do
not apply the same name, with the usual laxity as to the
order of the consonants, to all the affluents of the principal stream.

After descending some way on the western side of the
ridge, we came in sight of a large village perched on the
summit of a hill, on the opposite side of the stream that
ran at a great depth below us. This we soon learned to
be Moulaï Ibrahim, a *zaouia*, or sanctuary, much venerated
in all this part of Marocco, governed by a sherreef, belonging to the family of the saint whose tomb is the chief
building of the village. This semi-independent sherreef
gave permission to M. Balansa to remain in the village
for some days in 1867 ; but just as that active traveller
was prepared to attempt to penetrate into the interior of
the chain, an order from El Graoui made it necessary for
him to depart, and follow the direct way to Marocco. As
we came in sight of the *zaouia*, each of our troop, Shelluh
as well as Moor, commenced to recite prayers, and then,
after prostrating himself on the ground, with his face
towards the sanctuary, proceeded to add a stone to certain
heaps that stood beside the track. The Berebers, in
general, are said to be very lax in conforming to the
precepts of the Koran, but they are as assiduous in their
show of reverence for saints and sanctuaries as the Moors
themselves, and it would appear that this is the only
practical form in which their religion exhibits itself.

On the summit of the ridge, which may be about 4,500 feet above the sea, the rock is a grey schist, often shaly in texture, with the strike about east and west, and dipping at a high angle approaching the vertical. These beds may perhaps be identical with the schists, sometimes containing mica, and sometimes more calcareous in composition, which we afterwards found at the head of the Amsmiz valley, and with the rock, described as micaceous schist, seen by Washington in his ascent from Tasseremout. Our course now lay about due south, parallel to that of the torrent which ran at a considerable depth below us. At Moulaï Ibrahim this, according to M. Balansa, is called Oued Ghaghaia, but we never heard any similar designation for it. The difficulty of seizing the shades of more or less guttural sounds from the mouths of the natives makes it not improbable that the word Ghaghaia of M. Balansa is the same that we agreed in writing Reraya, and that the name may mean that this is the stream draining the district of Reraya.

On this ridge we found that curious grass, *Lygeum Spartum*, characteristic of Sicily and Southern Spain, where it is much used for making fine basket-work, but not seen elsewhere in Marocco. Soon after we lit upon a single specimen of a very fine plant of the artichoke family, evidently distinct from all those described, but unfortunately not yet in flower. It has been provisionally named *Cynara Hystryx* (Ball). The next find was not less interesting—an Oriental *Echinospermum* (*E. barbatum* of Lehmann) that extends from the Punjab to Asia Minor and the Caucasus, but had not before been seen in Africa.

About two o'clock we left behind us the rough irregular ground over which we had been riding, and found ourselves in a broad open valley, with a level floor, half a mile or more in width, at the head of which rose some fine snow-seamed peaks. As we advanced towards the main chain, our suspicion that the dividing ridge and the higher peaks were at once more distant and more lofty

J. B. delt.

GREAT ATLAS FROM LOWER VALLEY OF AIT MESAN

than had hitherto been supposed, was more and more confirmed; and we were soon able to certify that M. Balansa's expectation that any of the higher points might be reached in a single day from Moulaï Ibrahim was based on miscalculation of the scale of these mountains.

Our short mid-day halt was in a pleasant spot, under the shade of some very fine carob and olive trees, in view of a village with large quadrangular windowless buildings, that seemed to show that the mountaineers here are far better lodged than the people of the plain. The nearer hills, and one of the higher but nearer peaks, displayed long unbroken lines of escarpment, formed by the exposed edges of thick beds of rock (doubtless sandstone), of a deep red colour, indicated in the annexed plate. We here noticed the first indication of one prominent characteristic of the Great Atlas flora—the reappearance of many of the common field plants of Europe, which are not seen in the lower region. Among others, we gathered three species of *Ranunculus* (*R. arvensis*, *R. parviflorus*, and *R. muricatus*) beside our halting place.

We were soon again in the saddle, and every step as we advanced disclosed some new object of interest, either in the scenery that gradually opened before us, or in the vegetation close at hand. We passed close to a village where the demeanour of the people was more distinctly friendly than we had yet experienced since we landed at Tangier. The whole population—men, women unveiled, and children—turned out to see the cavalcade pass, and something approaching to a smile was seen on many a countenance. It appeared that the fame of Hooker's skill as a *hakim* had travelled before us, and during the following days his patience was often tried by the numbers who flocked to consult him. In this and the other neighbouring valleys there are a good many Jews, who appear to find life among the Shelluhs less hard than among the Arabs of the plain. True to the instinct of race, they contrive to make a living as brokers, by conducting the sale of the

o

surplus produce of the mountain country to Moorish traders, and the purchase of the grain, which must be brought from the low country for the subsistence of the people.

Some more fine plants were collected by the way. Among these were three species of *Astragalus*, one of them new, but nearly allied to *A. narbonensis*; and *Atractylis macrophylla*, of Desfontaines, a noble plant of the thistle tribe, much the most ornamental of the genus, reaching a height of three feet; but, as it flowers late, we saw only the withered heads of the previous year.

About 6 P.M. we reached the spot which was destined to be our head-quarters for several days. The site chosen was an olive grove, on a shelf of level ground about one hundred feet above the stream. The soil in the openings between the trees must have been lying fallow for some time, and was not so uncomfortably rough as the ploughed land on which we often had to pitch our tents. The two nearest villages are named Hasni and Tassghirt; but the former was taken by us as the name of the place that became to us a sort of temporary home. By the mean of four nearly concordant observations, compared with those at Mogador, the height of our camp was 4,205 feet (1,281·8 m.) above the sea.

By the time we were installed in our tents it was nearly dark, but a much longer delay occurred before the *mona* arrived from the neighbouring village. The interval was well employed in a negotiation with Si Hassan, the sheik of the valley, ending by an engagement on his part to conduct us on the following day ' to the snow.'

With eager anticipation, we rose early on the morning of May 13, and soon made our arrangements for the day's excursion. Abraham, with most of the escort, remained in charge of the camp, while three or four of the soldiers went with us, and Si Hassan with two or three wild-looking followers took charge of the expedition. During the past fortnight, our Mogador attendants, and especially

Ambak, whose superior intelligence was conspicuous, had
picked up enough of English to make the constant
presence of a regular interpreter less necessary than it
had been at first, though occasions were pretty frequent
when the attempts at. mutual understanding between us
and the Shelluhs were evidently unsuccessful.

Immediately above our camp the valley narrows
rapidly, and for some miles the torrent flows through a
mere cleft with steeply sloping precipitous sides.   To
avoid this, the upward track ascends steeply for several
hundred feet, and is then carried along the slope at a considerable height above the torrent.   After suffering from
the usual delays, we commenced the ascent about eight
o'clock.   The morning was bright, and the temperature
delightful.   The thermometer had fallen to 50° about
sunrise, but during the day it stood some ten degrees
higher in the shade in the middle part of the valley.   To
a party of naturalists it was tantalising work to ride along
the rocky track, passing at every step objects of the
greatest interest, yet unable to do more than snatch a
fragment, or hastily drag up an imperfect specimen.   The
pace over the broken ground was necessarily slow, and it
was easy for a man on foot to keep up with the horses ;
but then the temptation to linger by the way became
irresistible.   What botanist could be expected to pass by
new and hitherto unseen forms of vegetable life without
at least securing two or three specimens ? As one or other
of us yielded to the impulse, he was called to order by the
cry of his companions, ' We must lose no time—we must
keep together'—and so reluctantly remounting, he was
forced to keep time with our sheik guide, who led the way.
At a point about four miles above our camp the valley
opened a little, and near a village (Ouanzerout ?) our track
lay through a grove of large olive trees and then descended a little to cross the stream. We now found this an
impetuous torrent, with a much greater body of water than
it had showed where we crossed it the day before many

miles lower down, and a rocky bed full of deep holes through which it was not quite easy to take our horses and mules.

Throughout the valley we were struck by the proofs of native industry and skill given by the numerous irrigation channels, such as one sees in Piedmont, and in the tributaries of the Rhone valley in Switzerland, sometimes cut along steep faces of rock, sometimes maintained by high terraced banks. Where the ground is favourable, walnut trees are often planted along these watercourses, and must largely contribute to the dietary of the inhabitants. It thus appears that the drainage of the Great Atlas is, in great part, absorbed by irrigation, even before the streams enter the low country, while a further portion is there taken up for the same purpose, and but a small percentage reaches the sea in ordinary weather. This helps to account for much that at first sight appears so strange in the hydrography of Marocco. A vast mountain region, fully exposed to the currents of saturated warm air from the Atlantic, sends but four rivers to the ocean from its northern and western flanks, in a coast line of over 400 miles from El Araisch to Cape Guer; and these, at ordinary times, are all easily fordable. But when rain falls on the mountains, the irrigation channels are speedily filled to the brim, and the entire surplus reaches the rivers, which are then said to rise ten or twelve feet in the course of as many hours. As bridges are unknown, the Moors speak of travellers being detained for many days before a flooded river channel, as a common occurrence.

Above the ford, the valley was again contracted to a mere gorge, and the narrow path mounts on its eastern flank, and winds along the extremely steep rugged slopes much after the fashion of some unfrequented valley of the Southern Alps. Although the sun was already high, the mountain rose so sheer upon our left that the shadow often gave welcome protection; and the track was so narrow in places that we were not free from anxiety for the baggage animals.

The rock was now porphyry of a prevailing red colour, which, with occasional intrusive masses of diorite and dark green basalt, makes up the whole mass of the central ridge of the Great Atlas in this part of the chain. As compared with the rich and varied flora, insect life appeared, at least at this season, to be remarkably scarce, and the only butterfly noted was *Papilio podalyrius*.

The porphyry rocks appeared to be very hard, and far less yielding to erosion than those of somewhat similar character in South Tyrol. Hence the gullies and ravines cut by the water channels, round which the track wound, were not nearly so deep as those that add so much to the picturesqueness of the scenery, and at the same time to the length of the way for a traveller traversing the valleys near Botzen. After winding along the slopes for several miles, our track descended a little to approach once more the channel of the torrent. The valley was still narrow ; but the inclination of its bed was much less, and the ground on either bank left space for a track, and in places even for a strip of cultivation. The natives seem to be quick at availing themselves of every spot possible for agriculture. Rye and barley were here seen in ear, and the olive extends very nearly to 5,000 feet above the sea, or considerably higher than it does on the flanks of the Lebanon.

As our track ran along the bank of a slender watercourse, it was completely overarched by a row of elder trees in full flower, that forced us to lay our heads upon our horses' necks, one of many instances of the meeting of the common plants of Northern Europe with very different endemic forms that characterise the upper region of the Great Atlas. Some conspicuous plants of the lower country, and notably *Adenocarpus anagyrifolius* and *Linaria ventricosa*, extended thus far up the valley ; and these, together with a wild *Isatis*, scarcely different from the dyer's woad, gave a prevailing golden hue to the neighbouring slopes. A reach of the valley now opened before us, backed by a stern range of dark red bare rocky

peaks.   On our own (the eastern) flank, the enclosing
wall receded somewhat, and above a high and rather steep
convex acclivity stood a village whose people had brought
the whole slope into cultivation.   The torrent ran through
a cleft on the right of this knoll, and our course lay
directly up it, amidst fields and meadows, gay with spring
flowers, all enclosed within stiff hedges of thorny bushes,
among which our common gooseberry was abundant.   As
if because the natives would spare no space that could
be turned to profit, we soon found that on the steeper
portion of the ascent the only way was up the bed of
a brawling stream that had for irrigation's sake been
diverted from the upper course of the torrent.   The track
lay over big blocks of porphyry, with deep holes between,
over which the water leaped and tumbled, between straggling branches of spiny bushes, that left many a mark on
the faces and clothing of the passing horsemen.   Up to
this we had little experience of what the horses and mules
of Marocco can do in the way of getting over rough ground,
and it was not without surprise that we saw how successfully they managed to scramble up the slippery channel
over blocks worn smooth by the constant passage of men
and animals.   In the midst of the scramble we all dismounted, for we here saw for the first time the blue daisy
of the Atlas, growing in the shade under the bushes, or
nestling in the hollows between the rocks.

Having reached the top of the knoll at about noon, we
found a sort of shelf of nearly level ground, covered in
great part by a large village of rude but solid stone
houses.   Here a halt was called, and we were informed
that a *mona* was provided to supply the mid-day meal for
the party.   Burning with impatience, as we were, this was
anything but a welcome announcement.   The dark ridges
rising thousands of feet above the head of the valley were
still distant, and no snow was to be seen, save in rifts and
hollows of the rocks, high up and difficult of access ; but to
refuse the proffered hospitality of the mountain chieftain

would have been deemed an affront; and to insist on taking our escort on without food would have caused discontent, if not mutiny.   We made a virtue of necessity, and, while awaiting the repast, carried on a semblance of broken conversation, in which the ready wit of Ambak, our ever active attendant, supplied, it is likely, the chief materials. The name of the village, even more difficult to seize than usual, was noted by Hooker as Adjersiman.   It stands, by our observations, at 5,535 feet (1,687 m.) above the sea level.

An hour—a whole precious hour—was consumed before the meal was over, and we were again on our way. Above the village the bed of the valley rises very steeply, the central part being filled with a vast mound of huge boulders, which on further examination proved to be the undoubted remains of the terminal moraine of the glacier which once filled the head of the valley.   The principal mass of course marked the limit of the glacier during a prolonged period; but there were traces of two parallel moraines of smaller size, of which the outer marked the limit of its maximum extension.   The blocks of porphyry and other metamorphic rocks were mostly of great dimensions.

The track was carried in zigzags up the face of the rocky slope, keeping towards the top close to the edge of the moraine; and on reaching the summit of the barrier disclosed to us for the first time a full view of the head of our valley.   A few yards below us was a small miserable-looking village called Arround, the highest in this district. This stands at the meeting of two short and rather broad glens, each enclosed by the rugged masses of the central range of the Great Atlas.   The shorter of the two, which opened on our left in a SE. or ESE. direction, does not apparently reach the main watershed, and a pass from its head would in all probability lead to one of the tributary branches of the Ourika valley.   The other glen that opened right in front of us, somewhat W. of due S., was enclosed by a still loftier and more stern barrier, the rocks, since the sky had become overclouded, having passed from a dull red to a dark brown complexion. The ground for some distance behind the village was flat and swampy, showing that a small moraine lake had been filled up with gravel and silt. On the level space most of the soil was under tillage, and wheat as well as rye and barley are grown, and even maize, as we learned, is raised in this inclement position. On the low dykes that enclose the little fields we noticed *Iris germanica*, evidently planted, but whether for the production of orris-root, or for the sake of ornament, we failed to ascertain. The only large tree was the walnut, which had been planted along the skirts of the cultivated ground.

Now that we were able to pry into the inner recesses of the chain, we perceived that snow lay in abundance at a much lower level than we had hitherto supposed, but nowhere in masses of any great extent. All the higher ridges around us were extremely steep, though not cut into actual precipices; but on these snow could nowhere accumulate, save in clefts. Towards their base, however, at the foot of each narrow ravine that furrowed their faces, at many spots not much more than a thousand feet above our level, were large patches that seemed likely to maintain their position for some time. Though not without experience of mountain lands, we could none of us call to mind any spot much resembling the scene before us. Nowhere in the Alps is there anything of at all a similar character. Excluding the village, and the small fields, and the walnut trees, which, after all, filled but a small space in the view, there was something to remind one of the wilder valleys of the Northern Carpathians, but on a much greater scale. In the Tatra, as here, the rocks rise in broken masses, very steep but not quite precipitous, and the snow is seen only in clefts of the higher ridges, not because the climate forbids it to accumulate, but because the surface affords so few spots on which it can

rest. But the rocky masses of the Tatra tend to form
isolated peaks, usually of rugged and very steep conical
shape ; while in this part of the Great Atlas the depressions that separate the summits are of little depth as
compared to the great height of the range. Seen from
below, as from the spot where we now stood, many points
assume the aspect of sharp peaks ; but it is easy to ascertain, by varying the point of view, that these are mere
projecting bastions from the wall of the main chain,
rising little, or not at all, above the level of the adjoining
ridges.

The day was already far gone—nearly two o'clock in
the afternoon—when, leaving our horses at the village,
we started on foot, with our sheik as guide, descending
slightly to the level of the stream, here easily crossed, and
then mounting the slope on the west side of the main
branch of the valley in the direction of one of the nearest
of the patches of snow already seen by us. No guide was
needed, for the lower slopes on either side were easily
accessible in all directions ; but the sheik evidently wished
to fulfil in person the promise of 'leading us to the snow.'
Difficulty there was none, except that of moving onward
over ground where every step brought to view some fresh
object of interest. It was clear that we had at last
reached the threshold of the *terra incognita* that we had
so long dreamed of—the subalpine region of the Great
Atlas. There could be no doubt that in the short space
between the lower village and Arround—that is, between
the lower end of the ancient moraine and the ground formerly covered by glacier—the flora had undergone a complete change. Nearly all the peculiar species which we
had hitherto looked on as characteristic of the Great Atlas
had disappeared, and their place was occupied in part by
others peculiar to this region, and not known elsewhere ;
but more largely by species either identical or nearly
allied to well-known mountain plants of the Mediterranean region, along with some of the common plants

of middle Europe, including several familiar British field plants.

It will be more convenient to reserve details for the remarks on the vegetation of this and the Amsmiz valley, which will be found in the Appendix; but it must be owned that the general impression now made, and increased on further acquaintance, was not free from disappointment. As compared with any of the higher mountain masses surrounding the Mediterranean, already known to us, this is singularly unproductive of ornamental species attractive to the eye. The Sierra Nevada of Granada, the Lebanon chain, and the mountain ranges of Asia Minor, all exhibit at this season a multitude of bright-hued plants to delight the traveller, even though they may not rival the splendour of the higher zone of the Alps and Pyrenees to one who sees this a month or two later. Another remarkable feature was the absence of trees, and especially of true conifers. The dwarf evergreen oak that clothes the middle zone of the Atlas was no longer seen, and there was no pine, or spruce, or cedar to take its place. The solitary juniper that we afterwards saw was scarcely noticed at this, our first, visit. It is sometimes said that naturalists take no delight in the beauty of the objects of their study; but this is surely untrue of the great majority. Probably the notion has arisen from the fact that, in addition to the sources of pleasure which he shares with the rest of the world, the naturalist finds food for the sense of beauty as well as scientific interest, unsuspected by his critics, in exploring the internal structure of organised beings. Be this as it may, it is certain that the generally sombre aspect of the vegetation, harmonising as it did with that of the scenery, had a somewhat depressing effect on all the members of our party, while at the same time it only increased our desire to reach the upper part of the rugged barrier of rock that rose some 5,000 feet around the head of the valley where we stood.

Meanwhile, the afternoon was wearing away; we did

not clearly know how we were to return to our camp after dark, but remembered distinctly one awkward place—the fording of the torrent—where some daylight seemed indispensable. We hurried back, our portfolios and tin boxes fully charged with spoil, and found the horses and mules awaiting us on the flat ground below Arround. A man can usually travel over rough mountain tracks as fast as a mule; but if the man be a botanist, and the track lies among new and rare plants, it is quite certain that he will not do so; and when haste is a matter of real importance, it is necessary to submit to the restraint of riding. Hurried as we were, it was necessary to dismount and make a short halt on our return to Arround. The laws of Bereber hospitality required that the villagers should present a *mona*, and that we should at least make a show of partaking of it. There was a large dish of barley porridge, with a lake of oil in a crater-like hollow in the centre, and another of buttermilk, in which were some of last year's walnuts, as well as other unexpected delicacies. This entertainment was briefly despatched by our followers, and we proceeded down the steep track beside the moraine, and again reached Adjersiman. Here, to our vexation, another halt was called, and another slight refection was presented. Our impatience was so far successful, that the delay was limited to a very few minutes. We should, perhaps, have displayed more interest in these specimens of native cookery, if we had been acquainted with the curious passages in which Leo Africanus [1] minutely describes the dietary of the Atlas mountaineers, and the mode of preparing the identical dishes that were here presented to us.

Once more we were in the saddle, and the whole party felt that no more delay was permitted. To ride down the steep, slippery bed of the watercourse below the village seemed even a more trying affair than the ascent; but our companions seemed to take it as a matter of course, and our

[1] See Ramusio, *Delle Navigationi et Viaggi*. Venetia: 1563. Vol. i. 12.

sturdy beasts accomplished the task bravely, though not
without hard struggling, that would have strained the muscles of animals less strong and hardy.   As always happens
when the ground is looked at in the reverse direction, we
espied, on retracing our track, several plants not before
noticed, one or two of them certainly new. No botanist can
resist such a temptation, even though he were flying for his
life; and two or three times we dismounted to snatch a
specimen or two, but were soon recalled to the necessity
for pushing on.   For the first time since we landed in
Marocco, the evening sky was overcast with heavy dark
clouds, and the last of twilight was fading fast when we
reached the ford over the torrent.   The banks are here
overarched by poplars and other tall trees, and in the dim
light the rapid stream seemed fiercer, and its roar more
menacing, than when we crossed it in the morning.   The
passage was achieved; but not without a good deal of excitement among our followers, when one of the soldier's
horses slipped into a hole, and only after violent plunging
and loud shouting of the natives, scrambled to the farther
bank.

Without more trouble we ascended the slope on the
western side of the valley, and reached the olive grove,
to which we had given little attention when we passed
through it in the morning. This now unexpectedly presented the most difficult, and even dangerous, stage of our
excursion.   Such faint glimmering of light as remained
up to this disappeared under the trees, and gave place to
absolute pitch darkness.   The rough spreading boughs, all
beset with the ragged, leafless, half-dead branchlets characteristic of old olive trees, stretched out on every side,
at a height of four or five feet from the ground.   There
was no regular beaten track through the grove, but by day
it was easy for man and horse to thread a way among the
trees. The case was now very different. Our keen-sighted
Shelluh followers were as much at a loss as we were.   One
or two men on foot went first, and we then followed, the

train being brought up by the soldiers of our escort. For
a while, by moving slowly and cautiously, nothing serious
happened. The beasts seemed to understand the difficulty
of the case, and as one or another of us rode against a
branch, with head bent down to lessen the risk of mischief,
they stopped at once, and even backed a step or two.
Before long the cavalcade was separated by long gaps. A
loud cry of pain, followed by the vociferations of the
natives, brought the foremost to a full stop, and after a
while we were once more near together. It was not altogether reassuring to be told by Ambak, when we asked the
cause of the row, that one of the soldier's eyes had been
torn out. On this Ball determined to proceed on foot ;
but Hooker and Maw, after a few steps over the very
broken ground, thought it better to remount, and rely on
the sagacity of their four-footed companions.

In our awkward position the time seemed long ; but at
last we got through the olive grove, and found when we
emerged from it that the night was even darker than before.
It is well known to those who have made night excursions
in mountain countries, that anything approaching absolute
darkness, in places not overshadowed by trees or rocks, is
very unusual. It may be impossible to distinguish one
object from another ; but the outline of opaque bodies
against the sky is almost always traceable, and it rarely
happens that a path is not in some degree distinguishable
by its lighter hue from the surrounding rocks or vegetation. On this night, however, nothing whatever could be
seen ; and as we knew that the narrow track was carried
for the next three or four miles along a very steep slope,
precipitous in places, we felt that our difficulties were not
yet over. The horses and mules, however, showed themselves deserving of the confidence placed in them. Ball,
who led the way on foot, feeling his way with an alpen-
stock, had a narrow escape, as, misplacing his foot, he fell
over the edge, but was luckily stopped by a dense mass
of thorny bushes, from which he was rescued with a little

trouble. We were heartily glad when, on reaching the
spot where the track turns downward towards the river,
we at last saw the lights of our camp glimmering through
the trees.   The roof of dark clouds overhead had by this
time grown rather less dense; some faint light helped us
down the steep slope, and a little before eleven o'clock we
reached the welcome shelter of our tents.   The case of the
wounded soldier was first attended to.   It was much less
serious than we supposed ; the eye was not much injured,
but there was an ugly flesh wound on the face below it,
where a jagged stem had torn through the upper part of
the cheek. Wounds heal with remarkable readiness among
the natives of this country, and after a few days nothing
remained but a scar on the man's face.   A sheep, several
fowls, eggs, and three large dishes of cooked food were
soon forthcoming as the evening *mona*, and a rather late
supper closed the proceedings of the day.

CHAPTER IX.

WE had at last succeeded in breaking the charm that
seemed to have hitherto kept us from the inner recesses of
the Great Atlas; but we had done little more, and what we
had as yet seen and handled of the vegetation of the
higher region merely served to whet the appetite, and
increase our natural voracity.   Our talk on that night of
our return to Hasni, and our first thoughts on the following morning, turned on the possibility of making the
wretched village of Arround our base of operations for two
or three days, as it was clear that only by starting from
that point would it be possible to make a fruitful ascent of
the higher ridges.   During the day's excursion, Hooker
had ascertained a point of great practical importance.
While mounting the slopes on the west flank of the valley,
he noticed a path leading upwards towards a narrow ravine
at its head, and learned in answer to his inquiries that
this led to Sous—the great valley on the farther side of
the main range.   It was clear then, that, with Arround as a
starting point, we should have the advantage of a beaten
track as far as the crest of the ridge; and, even if this
should not be very high, we might, from that point,
ascend one of the adjoining summits.

The greater part of the following day, May 14, was devoted to putting our large collections into order; but meanwhile negotiations for carrying out our plan of sleeping at Arround were the most pressing business, and at length, after endless palavers, and discussing countless difficulties, were brought to the desired conclusion. Almost alone among the men in authority, whom we met in Marocco, the sheik of this valley seemed to have no special aversion to us as strangers and Christians. For very sufficient reasons he was longing for the moment that should see us and our escort depart from his district; but meanwhile he seemed anxious to keep on friendly terms, and do what he could to meet our wishes. We had already made him several presents; but here, as elsewhere in the country, we found that most of the articles we had provided for that purpose were little appreciated. Opera glasses, musical boxes, and even watches are of small account, unless with the comparatively civilised men who have lived in the coast towns or the great cities: cutlery is much more sought after, and some large sheath knives, of which we had a fair supply, were always highly acceptable; but fire-arms, not necessarily of modern make, are far more welcome than any other gifts. On this occasion we resorted to a strange engine of seduction. Before leaving London, Ball had happened to pick up, in an old curiosity shop, an antiquated weapon, of the size of a large horse pistol, with four barrels intended to be loaded and discharged all together. This, which we had called the 'young mitrailleuse,' had been the subject of many jokes during the journey, but was now with due solemnity presented to the sheik. The effect of our munificence was immediate and satisfactory, and the sheik was gained over to our cause. Fortunately, the efficiency of the 'young mitrailleuse' was not tested while we were in the country. It may probably have been since employed with deadly effect; but it is doubtful whether the victim would be the person against whom the four barrels may have been directed.

The sheik undertook that one of the houses in Arround should be cleared out for our reception; but, to provide for all contingencies, we arranged to take with us the two smaller and lighter tents, along with the usual supply of botanical paper and tin boxes.

Many natives of both sexes came to our camp during the day, in quest of medical treatment, as they had done during our absence the day before; and we were much amused to find that Abraham had coolly undertaken medical practice on his own account. He had provided himself with a large bottle of black stuff, containing heaven knows what nauseous ingredients, and this was doled out impartially to all applicants. It appeared to be a strong purgative, and may have answered sufficiently well in the rather frequent cases of indigestion arising from overeating. Of more serious complaints, ophthalmia was one of the most prevalent here, as elsewhere throughout South Marocco. Scrofulous sores and strictures were also common. Women desirous of offspring were brought to the camp by their husbands, and some cases of natural deformity also presented themselves. All seemed to have that great condition for remarkable cures that depends on a firm faith in the efficacy of the remedy. We had deliberately refused to follow the example of many African travellers, by including aphrodisiacs among the drugs carried with us, and all applications for such were met by a stern refusal.

We found time in the afternoon to examine the vegetation of the valley in the neighbourhood of our camp, of which we had hitherto obtained merely a passing view. The general aspect was very much what may be found at a level lower by 1,200 or 1,400 feet in Southern Spain or Calabria. In the valley bottom the prevailing trees were poplars—the common black poplar, and a small-leaved

P

variety of the white poplar—and *Salix purpurea*, with walnut, olive, and carob, the latter three being extensively cultivated. The small-leaved ash, *Fraxinus oxyphylla*, was also rather common, but does not grow to so large a size as in North Marocco. On drier ground, on the stony slopes, *Callitris* occurs here and there; and two junipers —*J. oxycedrus* and *J. phœnicea*—are rather frequent, and when allowed to reach maturity attain to the stature of small trees. The evergreen oak is the predominant tree on the flanks of the mountains, and exhibits several varieties, but rarely attains its natural size. Shrubs and low bushes, as usual in the Mediterranean region, are very numerous and varied, most of them, such as the alaternus, phillyrea, lentisk, oleander, and colutea, being widely diffused species of Southern Europe; and a beautiful honeysuckle (*Lonicera etrusca*), with large, sweet-scented flowers, was a conspicuous ornament. Along with these, several common forms of Central and North-western Europe, such as the common bramble, the ivy, the dog-rose and elder, here find their southern limit. It was not possible, however, for a botanist to look about him at any spot in the valley without being struck by abundant evidence that he had entered a region very distinct from any part of Southern Europe or Algeria. This impression was strengthened throughout our ride of the previous day, as we ascended from our camp to the foot of the ancient moraine, and everywhere saw conspicuous plants peculiar to the middle and lower zones of the Great Atlas. Our first impression was that the proportion of such endemic species was larger in this part of the valley than in the higher zone; but this was due to the fact, that so many more of those inhabiting the lower zone strike the eye by their greater size and by the brilliancy of their flowers. When we came to examine our collections with the requisite care, we found that about one-seventh of all the species found in the middle region of this valley is made up of peculiar endemic species, while the proportion of the same element in the higher region rises to one-fifth.[1]

We were especially struck by the complete absence of new generic types. There were, indeed, but two species seen in this valley belonging to North African genera—*Callitris* and *Lotononis*—that do not extend to Europe. All the rest are referable to European types, of which the large majority extend to the centrad and north-western parts of our continent. No representatives of tropical and sub-tropical types, such as are seen in Arabia, Persia, and Northern India, are here to be found.

During our absence on the 13th, a courier had arrived from Mogador, with letters from M. Carstensen. The man had first gone to Marocco, and thence, for the most part following our track, had found us in our Atlas head-quarters, and was well pleased with the trifling pay of a few shillings for the journey. To-day another courier made his appearance, bearing answers to the letter despatched to El Graoui, on the morning of the 12th, so that we had no reason to complain of remissness on his part. Probably for the purpose of shifting an unpopular measure from his own shoulders, the wary old Governor forwarded a letter from the Viceroy, expressly sanctioning our stay in the Aït Mesan valley, for as long a time as should be required for the objects of our journey. Along with this, El Graoui wrote to Hooker, expressing a hope that our stay would not be prolonged more than was really necessary, inasmuch as the villages of the valley were very poor. Further provision was made to meet Maw's desire to return to England, and an order sent that two soldiers should be detached from our escort to accompany him to Mogador.

It was curious to observe that whenever literary knowledge was in request, whether for reading and fully understanding letters addressed to us in Arabic, or for the composition of letters to be addressed by us in the same language, the member of the expedition always most

---

[1] For further particulars as to the mountain flora, see Appendix E.

relied upon was Hamed, Ball's personal attendant, one of
the poor fellows engaged at Mogador to act in a menial
capacity and accompany the expedition on foot. It is
true of the western dialect spoken in Marocco, as well as
of the purer Arabic of the east, that a familiar knowledge of the spoken tongue does not imply a full acquaintance with the written language, and Abraham was
evidently sometimes at fault. Education, in a literary
sense, is not among Moslems a privilege of rank or wealth,
and is quite as often found among the poorest as with
those above them. Our two Kaïds were both ostentatiously
illiterate, and the soldiers knew no more than their
officers ; and poor Hamed, alone of all our suite, seemed to
be worth taking into council on these occasions.

Once more the insatiable rapacity of our escort gave
us trouble, and proved to us that the objection to our
making a long stay among these poor mountaineers was
not an unreasonable one. We ascertained that the
demands of these shameless fellows on their own behalf,
apart from the rest of the expedition, rose to forty fowls a
day, with bread, tea, and sugar in proportion, while they
were constantly grumbling at the insufficiency of this
allowance for ten persons, and demanding money from the
natives in lieu of the other luxuries to which they thought
themselves entitled. We sincerely regretted our want of
power to put a stop to these abuses ; but it was impossible to sacrifice the main object of our journey, and we
merely resolved to acquaint El Graoui with the facts after
our return to Mogador.

We rose early on the morning of the 15th, and lost
no time in preparing for our departure. Just when all
seemed ready for a start a new and serious difficulty
arose. Hooker and Maw had both provided for the
journey several large tin cases painted green, and intended for the transport of living plants from Marocco to
England ; and, as a matter of course, some of these were
amongst the luggage packed up for the expedition to

Arround. When the sheik arrived about 7 A.M. he at once declared that he had undertaken to conduct us to Arround, but that to carry the luggage he saw prepared was utterly out of the question. A long and vehement controversy ensued; at first it was impossible to understand the real nature of the difficulty, and when this was gradually made clear, the objections seemed to us so incoherent and inconsistent that we suspected them to be mere pretexts to cover some unavowed obstacle in the sheik's mind.

It appeared that the tin cases were the real stumbling blocks. 'When the people of Arround see those cases,' said Si Hassan, 'nothing will persuade them that they are not filled with treasure—they will attack us in the night, and will kill you and me too, in order to get possession of them.' 'They will believe the boxes to be full of gunpowder'—so ran another version of the difficulty—'and think you have come to take possession of their valley, and will fight to resist your remaining there.' We suspected at the time that the unavowed cause of offence lay in the boxes being painted green—the colour of the Prophet and his descendants—but from the slight attention paid by the mountaineers to the observances of Moslem law, even in more important matters, we afterwards rejected this explanation, and were inclined to believe that the sheik knew his own people, and truly represented the strange fancies to which they are subject.

The tin boxes were reluctantly sacrificed, and with them the possibility of making any large collection of living plants in the upper region of the Great Atlas. The luggage was repacked, and after several hours' delay we started about 10.30 A.M. by the same track which had been the scene of our recent night adventure.

The weather during the last two days had been gradually changing from the condition of 'set fair' to which we had been used since our arrival in Marocco. The barometer had fallen progressively fully five millimetres, and the clouds had changed from their ordinary condition of light, separate fleecy masses gathered round the higher ridges during the day to a dense canopy stretching continuously over the visible portion of the sky. If choice had been left to us we should not have selected this morning for our excursion; but, after overcoming so many difficulties, there was no thought of letting weather stand in our way, and we could only resolve to make the best of it whatever it might be. On this occasion we had rather more time to spare than two days before, and we added a good deal to our previous collections in the valley, though less than we could have done if the precious morning hours had not been wasted in controversy.

We had many an occasion for admiring the sureness of foot of the mules and horses in this country; but we also noticed that, like their fellows elsewhere, they have some peculiarities of disposition that a traveller should take note of. Maw's mule, hitherto remarkably steady, had a trick of keeping to the outer verge of the path in steep places, and when his head was turned inward his hind foot would go over the edge. After recovering himself once or twice, he at last slipped completely. Maw saved himself in good time, while the animal rolled down the steep slope towards the torrent. In many places this is so precipitous that the beast must have been killed; as it was, he was stopped by some thorny bushes, and was with some trouble got back to the track, a sadder and a wiser mule for the rest of that day at least. It is well known that several fatal accidents in the Alps have occurred in places of the same character from interference with a mule, who should be left to take his own course. The now almost familiar road to Arround, with the ascent of the watercourse, seeming more objectionable each time that we passed it, was accomplished without further incident, and no other delays than those involved in plant-collecting. In six hours from Hasni we reached the village, whose inhabitants had all turned out of doors to gaze on the Christian strangers who, from some inscrutable motive, had come a second time to their secluded valley, and now seemed resolved to fix their abode there.

A house, the best in the village we were told, had been prepared for us by the simple process of turning out its inmates, and to this we were at once conducted. On the ground floor were two quite dark and low cellars or

HOUSES AT ARROUND.

dens, seemingly filthy, but which we were not inclined to explore. Ascending by some rickety steps, we reached the upper floor, the larger part of which was occupied by a rude open verandah, at each end of which was a little closet about seven feet square, one of which was occupied by Maw, while the other was used as a kitchen, the open verandah serving as our sitting-room, and as night quarters for Hooker and Ball. As usual in the Shelluh houses, the doors were only about four feet, and the rooms and verandah not over five feet high, making it inconvenient for us to move about. In most of the houses there are underground cellars to which the inhabitants retire in winter, as is the custom in Armenia and in some of the higher valleys of the French Alps.

We scarcely had been settled in the house when several applicants for medical advice presented themselves; but these were disposed of soon enough to leave some remaining daylight, which was devoted to a stroll up the left, or SE. branch of the upper valley. The flat ground was parcelled into small fields divided by stone dykes, and intersected by slender irrigation channels. The fields seemed to be carefully tilled, rye, barley, and beans being the only crop now above ground. Maize is sown in the latter part of May and ripens in the course of six weeks. We saw with surprise a few vines in this inclement spot, and also Madder (*Rubia tinctorum*) seemingly wild, but doubtless originally introduced for native use.

The mountains, as far as we could see them, looked forbidding, and the scene this evening was even more sombre than it had been two days before. Leaden clouds roofed the valley across, and completely hid the higher ridges; slight gusts of chilling wind blew at intervals, and all the tokens of impending bad weather warned us not to indulge in cheerful anticipations for the morrow. On this occasion we had limited our escort to two soldiers, whose presence showed that we were under the shadow of imperial protection, but who would doubtless have been utterly useless if the natives had harboured hostile designs. Of such, however, there was not the slightest indication. The demeanour of the people was respectful and friendly, rather than the reverse. Our every movement was watched, but from a distance, and there was none of the intrusive curiosity so often complained of by travellers among semi-barbarous people. The men all habitually wear the hooded cloak, of dark-coloured goat's hair, somewhat looser than the Moorish *jellabia*, which appears to be

peculiar to the tribes of the Great Atlas. Whether the
slight variations in the triangular patch of coloured stuff
with rude embroidery that is inserted at the back, serves
to distinguish the men of one tribe from another, we failed
to ascertain. The women, who make but a faint show
of concealing half the face when approaching strangers,
seemed to be rather better favoured than those we had
seen in the lower valley. They partly shave the head, and
twine the remaining hair into two broad plaits, bringing
these forward crosswise over the forehead.

With a view to possible difficulties arising with the
sheik, we had taken Abraham with us, leaving the camp
at Hasni for twenty-four hours in charge of Crump,
Hooker's English attendant; but it was arranged that the
former was to return on the following morning, much to
his own satisfaction, as the cold and discomfort of this
Alpine village seemed to make both him and the Marocco
soldiers perfectly miserable.

Our evening meal was enlivened by one of some
precious bottles of generous wine that Maw had added to
our stores, reserved for special occasions, such as the
present. After this we should gladly have gone to sleep,
if stern duty had not forbidden any such luxury. The
minimum of evening work for the travelling botanist is to
lay out between dry paper the contents of his boxes and
portfolios filled during the day. On this occasion the
operation was more troublesome than usual, as we struggled
to screen our single flickering candle from the night wind
in the least exposed corner of the verandah, and midnight
had come and gone before we stretched our mattresses on
the earthen floor, first duly dusted with insect powder,
and sought rest. In our exposed position the cold was
very sensible through the night, though the thermometer
did not fall below 45° Fahr.

The morning sky was so gloomy that no one awoke
so early as we had intended; and at sunrise on May 16,
when we loudly called for breakfast, the light was still so

imperfect that it seemed as though the day had but just
dawned.   There was less than the usual delay; but six
o'clock had passed, and we were not yet ready to start.   To
our great satisfaction we found that the sheik did not pro-
pose to accompany us to-day, but had appointed two men
of the village to act as guides.   With these, and our usual
personal attendants, whom we knew by experience to be
active pedestrians, we started about 6.30 A.M. to ascend
the main southern branch of the valley.   For rather more
than a mile the way over the filled-up bed of the old
moraine lake is quite flat, and for a considerable distance
beyond this the ascent along the bottom of the valley is
very gentle; but we were led by the aspect of the ground
to ascend the rather steep western slope at a part much
farther from the village than we had traversed three days
before.

Our inducement to leave the track was the wish to
examine certain solitary trees that we noticed scattered at
rather wide intervals on the slopes, nowhere descending
below the level of 8,000 feet above the sea, but extending
upwards, where they find a resting place, through a vertical
zone of about 1,500 feet.   The first we were able to reach,
which was similar in aspect to the rest, showed a trunk
more than two feet in diameter, and about thirty feet high,
but broken off and shattered at the top; the branches,
with their very dark foliage, diminishing in length upwards, give the whole a conical form.   We took it at the
time for *Juniperus phœnicea*, which is rather a common
tree in the lower valleys.   Subsequent examination showed
it, however, to differ from that species, and to be identical
with *Juniperus thurifera*, a tree hitherto known only in
Central Spain, Portugal, and Algeria, and apparently nowhere common.   From the number of dead stems seen, it
seems to have once girdled this mass of the Atlas with a
belt of forest, which has been gradually thinned, and is
doomed to ultimate destruction.   The existing trees are
probably of high antiquity, and their destruction is mainly

due to the practice of setting fire to the brushwood to
gain pasture for animals ; while the young plants, of which
not a single one was seen, would be cut off while yet seedlings by the tooth of the goat, the great enemy of tree
vegetation—an animal whose disastrous influence, acting
indirectly on the climate of wide regions, entitles it to
rank as one of the worst enemies of the human race.

Although the ground was to a great extent occupied
by the two dwarf bushes seen on our first visit, *Alyssum
spinosum* and *Bupleurum spinosum*, there was no lack
of new forms of plants to maintain our enthusiasm ; and, in
spite of the desire to push on, many a halt occurred as one
or other lighted on an object of fresh interest. As a
natural consequence of our having chosen to make our
way along the side of the glen, instead of following its
bed, we had to cross several projecting spurs, the last
rather steep, before descending to a spot where, at the
extreme head of the valley, our guides pointed out a Saint's
tomb, consisting of a rude stone hut with a space five or
six feet square in the centre. When we reached this, the
guides made it clear to us that we had arrived at the end
of our excursion. The hut stands at the junction of the
streams issuing from two rocky ravines. That on the
west side was apparently very steep and pathless ; the
other, mounting about due S., was nearly equally steep,
but we could see that a beaten track ascended along the
opposite bank of the slender torrent that tumbled over the
rocks at its entrance. The native guides confirmed the
statement before made to Hooker, that by that tract lay
the way to Sous ; but, by expressive pantomime, they explained that danger lay in that direction, and that the
people of the other side were addicted to the practice of
shooting at strangers. We were careful to avoid controversy, and set ourselves to collect plants in a patch of
boggy ground near the hut, where familiar northern species, such as *Stellaria uliginosa, Sagina Linnæi, Montia
fontana* and *Veronica Beccabunga*, grew in company

with a new species of *Nasturtium*, and others not before
seen by us.

So intent had we been on the surrounding vegetation, that we had scarcely cast a glance at the sky
overhead. This had continually assumed a more and
more gloomy aspect; and at length, after due notice
and preparation, the long-expected rain began, not in a
heavy downpour such as often occurs in southern countries, but in that fine steady drizzle which is known
to those whom the fates have led to the northern parts of
our island as a Scotch mist, hateful to the lover of the
picturesque and still more hateful to the botanist. On
this occasion, however, it seemed to us no unmixed evil,
as it furthered the execution of a stratagem that was
already in our minds. Our followers were scantily clad,
and felt more than we did the chilly temperature of the
day, and of course the rain increased their discomfort.
They were, therefore, in the right frame of mind to accept
at once the suggestion that they should light a fire within
the hut, therein following the example of preceding way-
farers. After muttering a few prayers, they proceeded to
gather some damp sticks, and presently were busy in the
attempt to make a fire out of them. Having continued
for a few minutes to loiter about, still gathering plants
near the hut, until the men appeared to be fully engrossed
in their occupation, we started together to ascend the track
leading to the summit ridge of the Atlas.

We had reached the Saint's tomb about 9 A.M., and
found its height above the sea-level to be 7,852 feet
(2,393·2 m.). Little more than half-an-hour had since
elapsed, so that, if no unexpected difficulty occurred, there
was ample time to reach the summit of the pass which,
as we thought, could scarcely be 3,000 feet above us. A
number of interesting plants soon rewarded our adventure,
and delayed us for a while on the rocky banks of the torrent near the bottom of the ravine, but out of view of the
Saint's tomb. On joining the track, we found it a well-

made mountain path, constructed with some skill, advantage being taken of the nature of the ground to make zigzags, evidently intended for the passage of beasts of burthen.

We had ascended several hundred feet, and were looking about for plants among rocks to the left of the path, when some faint sound made us look up, and we descried, amid the rain and mist, a party of men and laden mules descending towards us down the steep ravine. There was some obvious awkwardness in the impending encounter of three Englishmen, utterly ignorant of the native tongue, with a set of wild mountaineers of the Atlas, in a spot where no stranger had ever before been seen. In such cases, the less time that is left for deliberation the better. Suspicion or greed may prompt an attack where time is left for consultation; but if people are suddenly confronted by peaceable strangers, they will rarely, unless robbers by profession, think of molesting them. The shape of the ground happened to favour this obvious bit of policy, and some projecting rocks concealed the approaching train until we suddenly confronted them at a turn of the path, and passed within a few yards, with something approaching to a grave salute. The mules appeared to be laden with goat-skins, along with other articles that we could not distinguish. Whether these were people from the northern side of the chain returning from a trading expedition in the Sous country, or men of Sous carrying goods to the capital, we never certainly ascertained; but, from noticing pieces of orange-peel on the track, we inferred that they must have descended rather low in the Sous valley; while it is certain that people going from the lower part of that valley to the city of Marocco would not have followed this circuitous and difficult track, unless urged by special reasons.

A little farther on we found, on ledges of rock near the track, several of the most interesting plants seen during the day. Thenceforward all botanising became difficult.

The rain turned to sleet, and before long to snow; and, though the roughness of the ground still enabled us to discern the more conspicuous plants, it was almost impossible to secure satisfactory specimens.

Soon after the snow had set in, we heard, from below, yells and screams, and immediately guessed that the caravan from Sous had brought news to our guides at the Saint's tomb of our escapade towards the summit of the pass. The guess was correct; and though we pushed on rather faster than before, the foremost guide soon overtook us, and addressing himself especially to Maw, who led the way during the ascent, with vehement gestures and emphatic phrases, that seemed to combine threats and injunctions with supplication, urged an immediate return. Maw judiciously had recourse to an argument of universal efficacy, and, presenting the man with a piece of silver, pointed upwards and strove to explain, by signs, that we meant to go to the top and then return. Shortly afterwards, the second man appeared, panting from the pace at which he had run up the steep ascent. He addressed himself to Ball, who came next to Maw, but was answered by the same reasoning that had prevailed with his companion.

The upper part of the ravine was wider than it had been below, and the slope rather less steep. Here, as throughout the upper valley, porphyries and porphyritic tuffs of a prevailing red colour, form the mass of the ridge; but we observed at several points intrusive masses of diorite, sometimes much resembling granite in appearance. Higher up, near the summit of the ridge, Maw noticed white crystalline limestone, of no great extent, which appeared to be intercalated with the porphyrites.

To collect plants was now scarcely possible, for the snow covered the surface, and it was necessary to kick it away from the tufts of grass or dwarf bushes, in order to ascertain what might be growing beneath. The wind, which had hitherto spared us, now joined itself to the opposing forces, driving the snow with blinding force, and

making the cold, already severe, well nigh intolerable.
The poor fellows who had for some way followed us without further remonstrance, now renewed with redoubled
energy their appeal that we should return. Kissing the
hem of our coats at one moment, brandishing their arms
with passionate gestures, or actually pulling us back at
another, they really impressed us more by their pitiable
appearance, exposed with the slightest covering to the
bitter blast, their feet and bare legs cut and bleeding
from the rocks and thorny bushes of the way.

It was now apparent that the dread with which these
men were evidently impressed did not arise solely from apprehension of an encounter with human enemies. Firmly
believing that the heights of the Atlas are inhabited by
djinns, or demons, it was obvious to our companions that
the storm was caused by their anger at the intrusion of
strangers into their sanctuary. We had not before noticed
that one of the Shelluhs carried with him a live cock under
his arm. In a state of the utmost excitement, he now proceeded to cut the animal's throat, in order thus to appease
the wrath of our supernatural foes, then renewing the
appeal to us to forego further provocation.

In emphatic English, and such pantomime as we could
command, we explained that we were determined to reach
the top, but would then immediately return, and proceeded to face the last portion of the ascent. This lay
through a broad *couloir*, some twenty to thirty feet wide,
between steep walls of rock where, on narrow ledges giving
scarcely any hold for snow, the last plants were collected.
The storm, now almost a hurricane, raged with increasing
violence; it was scarcely possible to face it, and our hands
and feet gave scarcely any token of sensation. The thermometer, though carried in a pocket, marked 25° Fahr.
(or about − 4° Cent.) when last observed. Maw pushed on
with increasing vigour, and, in the driving snow, was soon
lost to sight. Presently, shouts were heard, and he re-
appeared, saying that he had reached the ridge where the

ground fell away on the southern side, that he could see
absolutely nothing in any direction, and, owing to the
severity of the cold, found it impossible to remain. He
estimated the height at rather more than one hundred
feet above the point reached by Ball, who in turn was
about sixty feet above Hooker. They descended through
that short space; and, after very brief deliberation, decided
that no more could be done, and that a speedy descent was
the only possible course. The appearance of the party
was singular, and not one could have been recognised by his
nearest friends. Faces of a livid purple tint were enclosed
by masses of hair thickly matted with ice, and the beards,
frozen in the direction of the wind, projected on one side,
giving a strangely distorted expression to each countenance.

After observing the aneroid barometer at the point
which we estimated at 200 feet below the summit, and
glancing at our watches, which marked about 2.30 P.M.,
we turned downwards, and set out as fast as our legs would
carry us, cutting across the zigzag track now deeply
covered with snow. Before long we got shelter from the
violence of the wind, and began to feel the tingling of
returning circulation in the hands and feet. In places the
ground was steep enough to require a little caution in
traversing the rocky slopes, partly grown over by tufted
bushes, all now veiled in fresh snow; but little delay
ensued, and in less than two hours we reached the bottom
of the ravine where the track passes close to the Saint's
tomb. The shouts of our Shelluh guides had announced
our approach, and we were met by the smiling faces of our
Mogador attendants, who had judiciously made themselves
as comfortable as circumstances permitted by keeping up
a fire in the hut.

In the valley little snow had fallen, and that was half
melted, and continued to fall in that intermediate condition between snow and rain that forms slush, a word of
odious import except for its associations with the Christ-

mas holidays. We learned that the sheik, Si Hassan, was
waiting for us some way lower down in the valley, and
without halting we pushed on to meet him. Long waiting in cold and wet does not mend any man's temper, and
the sheik, already much annoyed that his injunctions not
to let us go beyond the Saint's tomb had been ineffectual,
was doubtless in a savage humour when we at length appeared, after successfully breaking through all the re-
straints he had contrived. Yet he managed to put a good
face on the matter, offered his congratulations on our safe
return, and invited us to partake of some food that was
provided in a spot where an overhanging rock gave partial
shelter. This did not save the poor fellows who had done
their best to keep us within the intended limit from a
desperate ' blowing-up,' and many threats of future vengeance. Drenched and cold as we were, the invitation to
halt was anything but tempting ; but in this country the
obligations of hospitality are binding on the receiver as
well as the giver, and it was necessary to wait some time
and eat a few mouthfuls before proceeding on our homeward way to Arround.

As we approached the village, we witnessed a marvellous exhibition of colour that, even in our weather-beaten
condition, impressed us with admiration. The steep
ridges enclosing the, valley were now thickly powdered
with snow, but almost concealed from view by the clouds
that hung low over our heads. Towards sunset these
gradually rose up and melted into mist, and the whole
scene was transfused with a delicate sea-green hue that
seemed bright by contrast with the sombre tints in which
we had been enveloped during the day. It often happens
in bad weather that as the distant horizon is lit up towards
sunset the rays, travelling under the dense strata of cloud
that cover a mountain district, produce at that hour the
effect of sudden illlumination ; but whether the green tint
on this occasion was due to a similar colour in the distant

horizon, unseen by us, or was complementary to the prevailing red colour of the surrounding rocks, we were unable to decide.

About sunset we reached our house at Arround. The open verandah on this chilly evening, with the thermometer little over 40° Fahr., was not the most comfortable place for the evening toilet, nor for working in after supper. Under ordinary circumstances, two at least of the party would infallibly have been laid up with heavy colds or worse; but the last three weeks of open-air life in this fine climate had put us all into excellent condition. A moderate supper was despatched with general satisfaction, and no one suffered further inconvenience from the roughness of the day or the coldness of the night.

As might have been expected from the unsettled state of the weather, the observations taken to determine the altitude of Arround had not been quite satisfactory. The heights deduced from comparison with Mogador, where the weather was also unsettled, were discordant to the extent of about 80 feet. A comparison of observations taken here and at Hasni, on the 13th and 15th inst., with a few hours' interval, gave a much nearer agreement; and the mean of these, being 6,463 feet (1,970 m.), is that which we have adopted. A boiling-water observation at 8 P.M. in the evening, with the temperature of the air at 40° Fahr., gave a result higher by 20 feet; but it helps to show that the probable error is not large.

The answers to our inquiries as to the Pass reached during the day were, as we had reason to believe, designedly vague and indefinite. It appears to be known as Tagherot, and to serve for communication with Tifinout, which is the name of a mountain district with one or more large villages, whose drainage is carried to the Sous. The main valley of the Sous must, however, lie to the south of Tifinout, and extend much farther east. Our corrected results make the point at which our last observation was taken, 4,821 feet above Arround, giving for the Tagherot Pass 5,021 feet above Arround, or 11,484 feet (3,500·4 m.) above the sea.

In Appendix A the reasons which showed the necessity for a considerable correction to the original observations made since our arrival at Mogador are fully explained. The difference in the resulting heights throughout our journey is not of much moment as regards the lower stations; but it increases rapidly with increasing altitude, and in the case of the Tagherot Pass amounts to about 500 feet. As this correction was disclosed only after careful examination and comparison of all the observations, the first result, which was derived from the rough reduction made at the time, communicated in a letter from Hooker to the late Sir Roderick Murchison,[1] and which appeared in other published notices of our journey, is probably erroneous to the extent above mentioned.

Assuming our final results to be pretty nearly correct, and having been unable to hear of any other easy or frequented pass in this part of the range, we seem to be justified in concluding that this section of the Great Atlas chain, as compared with any of the mountain systems of Europe, maintains a remarkably high mean level. The height of the projecting summits in the adjoining portions of the chain was variously estimated by us at 1,500 or 2,000 feet above the Tagherot Pass. Taking the lower of these estimates, and assuming the other depressions to be no higher than Tagherot, we should have for the mean height of the main ridge at least 12,200 feet. Judging from all the distant views we were able to obtain, the portion of the Atlas chain near the head of the Aït Mesan valley is very similar in character to that extending eastward to the sources of the Oued Tessout, and does not reach a higher elevation. If this opinion be well founded, we have in this part of the Great

[1] Printed in the *Proceedings of the Royal Geographical Society*, for 1871.

Atlas a range, fully 80 miles in length, which in its mean elevation surpasses any other of equal length in Europe, or in the countries bordering on the Mediterranean. The chain of the Pennine Alps, from the Col de Bonhomme to the Simplon, alone approaches the same limit, as, excluding those limiting passes, the mean elevation of the dividing ridge for a distance of over 90 miles is about 11,800 English feet. That of the Mont Blanc range, from the Cime des Fours to the Pointe d'Orny, probably equals the mean height of the Great Atlas, being about 12,300 feet, but this is only about 25 miles in length. Excluding the mountains of Central Asia, and the Andes of Bolivia and Peru, neither of which can be spoken of as mountain chains in the ordinary sense, the only considerable range surpassing the Atlas in height is the higher part of the Caucasus, between the peaks of Elbruz and Kasbek, whose mean height, for a distance exceeding 110 miles, must reach, if it does not surpass, the limit of 13,000 feet.

Our usual evening occupation was pursued under greater difficulties than usual. There was not much wind; but the cold was severely felt in the open verandah, and the portion of our day's harvest that was gathered in a wet state had to be left till the morning's light should enable us to give our specimens the requisite treatment.

Our design, not disclosed to any of our native followers, had been to remain another day at Arround, and, if circumstances were favourable, to ascend some projecting point in the range that should command a panoramic view. We had, however, scarcely opened our eyes on the morning of May 17 when we clearly perceived that the fates had decided against our scheme. Snow had fallen steadily during the night, and both branches of the valley above the village were thickly covered. The sky overhead was of the same leaden complexion as that of the previous day, and flakes of snow falling slowly showed that the disposition of the weather continued unchanged.

The continuous fall of the barometer for three days before the rain set in had prepared us for a persistent fit of bad weather; so we were less disappointed than we should otherwise have been, and acquiesced as a matter of course in the preparations for our departure.

The natives still flocked to the entrance of our house, seeking medical advice from the Christian *hakim*. When these had been disposed of, and all seemed ready for our departure, an unexpected incident occurred. Eight or ten women, dragging with them a sheep, entered the house in a tumultuous way, crowding up the stairs and into the verandah, addressed vehement entreaties to Hooker, and suddenly cut the sheep's throat in his presence. Then followed more passionate entreaties, a document was thrust into his hand, and we were left at a loss to guess the meaning of the strange scene. At length, through Ambak's increasing skill as interpreter, the matter was made sufficiently clear. A number of men of the village, the husbands or fathers of our suppliants, had been carried off as prisoners to Marocco, for non-payment of taxes, and were there confined in the horrible subterranean dungeons that serve as prisons. The object of these poor women was to obtain from El Graoui an order for their release, through the intercession of Hooker. A promise to do what was possible on their behalf was readily given; but, although a courteous answer was afterwards sent through the consul at Mogador, it may be feared that little attention was paid by the powerful Governor of this region to the representations of Christian strangers.

The state of the prisons in Marocco is one of many scandals that disgrace the administration of this country, though an apologist might suggest that in this respect Marocco is only a century or two behind the most civilised States of Europe, and not thirty years behind the late kingdom of Naples. When in the city of Marocco, we were told that about 4,000 prisoners, of whom the large majority were unlucky peasants, unable or unwilling to pay taxes,

were confined in dungeons.   Criminals who have committed murders and robberies frequently escape by taking
refuge at some of the numerous sanctuaries scattered over
the whole territory, while lesser offenders and mere de-
faulters are caught wholesale.   No food is provided for
prisoners by the authorities; but the means of keeping
body and soul together are generally forthcoming, through
the kindness of relatives, or the charitable feeling which
is common here, as in other Mohammedan countries.

The survival among the Bereber tribes of the practice
of sacrificing an animal to propitiate the favour of a man
in authority, is a fact deserving the attention of ethnolo-
gists.   Another instance of a similar kind came to our
knowledge a few weeks later, and we had recently seen
that the same rite is observed to avert the displeasure of
evil spirits.

Our increased acquaintance with the flora of the Great
Atlas did not much modify our first impressions.   Making
due allowance for the earliness of the season, and for the
adverse conditions that may have concealed from us some
species inhabiting the higher zone, it was clear that the
vegetation here differs very much from that of all the lofty
mountain masses of Southern Europe and Western Asia,
and especially in the absence of those families that elsewhere form the chief ornaments of the higher mountain
zone, and which we are accustomed to associate with the
glories of the Alpine flora.   There was here to be seen no
gentian, no primrose or *Androsace*, no rhododendron, no
anemone, no potentilla, and none but lowland forms of
saxifrage and ranunculus.

Our first impression had been that the flora is absolutely very poor; but this was due mainly to the fact that
so large a proportion of the plants have inconspicuous
flowers.   Comparing the produce of our day's work with
that of high mountain excursions made elsewhere, the
species are not deficient in variety, but show a singularly
small proportion of showy flowers.   As regards novelty, we

had nothing to complain of; for, in the upper part of this valley, out of 151 species collected, 31 are described as new; and, so far as we know, are peculiar to the Great Atlas chain. This gives about the same proportion of endemic species as the Sierra Nevada of Granada, always regarded as a singularly rich botanical district.

The most remarkable feature of the flora of this region, is, undoubtedly, the very large proportion of common plants of the colder temperate region (Central and North-Western Europe), here found associated with species of very different type. Nearly one-half (70 out of 151) of the species found in the upper zone, belong to this category, and the proportion is here actually larger than it is in the higher mountains of Southern Spain. It was further remarkable that several of these northern species, such as the wild gooseberry, are plants that do not extend to the South of Spain, although climatal conditions must be at least equally favourable, and whose nearest known habitat is six or seven hundred miles distant. Especially to be noted was the fact that, with the doubtful exception of *Sagina Linnœi* (the *Spergula saginoides* of the older botanists), not one of the plants in question is characteristically an Alpine species, or typical of the Arctic or glacial flora. Combining this with the almost complete absence of rushes and sedges, we are forced to conclude that, whatever agencies may have contributed to make up the existing flora of the Great Atlas, transport by floating ice during the last glacial period cannot have been amongst them. If such ice-rafts were ever borne to what was then probably a long western peninsula of Northern Africa, they must either have foundered at sea with all their vegetable crew, or, if cast ashore, must have found an inhospitable region where the voyagers were starved, and left no descendants.

As was to be expected, from the habitual dryness of the climate, ferns were here deficient in number and variety. In the upper region we found very sparingly

six of the common species of Northern Europe; and lower
down, in the middle part of the valley, we were able to
add to our lists but two southern forms.

About one-third only of the species found in the upper
region could be described as properly belonging to the
Mediterranean flora; most of these being widely-spread
plants, while a few are exclusively confined to the nearest
neighbouring mountain regions—the Lesser Atlas of Algeria or the mountains of Southern Spain. But there
was little in the general aspect of the vegetation to suggest any special connection with either; and several of the
conspicuous plants have been hitherto known only in very
distant regions. A bright-flowered *Veronica* appeared to
be no more than a large variety of a species peculiar to
Asia Minor; *Medicago suffruticosa* had hitherto been
seen only in the Pyrenees; and *Evax Heldreichii* had
been detected nowhere nearer than the mountains of
Sicily and Eastern Algeria. Our original expectation of
finding some connecting links between the special flora of
the Canary Islands and that of North Africa was so far
completely negatived, and we saw nothing to suggest their
existence.

The most prominent characteristics of the mountain
flora of the Great Atlas were found to be of a negative
character. If asked to point out the positive features that
most struck us, we should in the first place note the
prevalence of *Cruciferæ* and *Caryophylleæ*, the former
reckoning one-ninth and the latter one-tenth of the whole
number of flowering plants. Of conspicuous genera we
had especially remarked *Chrysanthemum, Galium,* and
Linaria. Of the first of these we found two new species,
one of which, from its remarkable buff-coloured rays and
large scarious involucres, has been named *Chrysanthemum Catananche.*

The state of the weather and the earliness of the season
may partially account for the scarcity of animal life in
the upper part of the Aït Mesan valley; but, from all we

could learn, this appears to be a characteristic feature of the Great Atlas. The lion is said to exist in the lower valleys, and especially in Sous, but we were led to suspect that the animal so called by the natives is the leopard. The lion undoubtedly exists in the low country, but appears to be now rare. Those sent as presents from the Sultan to crowned heads have generally been taken in the valleys east of Fez. The only one of the Carnivora seen by us was a rather large creature, resembling a civet in form, but with no markings on the yellowish brown fur, once seen near our camp at Hasni. Birds were remarkably scarce, and the only conspicuous kind observed near enough to be identified much resembled the red-legged partridge, and was seen at a height of from 9,000 to 10,000 feet in the ascent to the Tagherot Pass. Instead of going in pairs, as that species is used to do, these formed a small covey. A single scorpion, of large size, seemingly of the species common in the low country, was found under a stone by Maw. Of the numerous reptiles that abound about the skirts of the mountain range, few, except lizards, seem to frequent the interior valleys; and the latter are wanting, or at least rare, in the higher region. Insects were also infrequent, and none were found under stones above the level of 9,000 feet.

Our brief stay among the Shelluhs in this valley helped to confirm our previous impression that they form the best element in the population of Marocco. How much of this superiority is due to race, and how much to the conditions of existence in a mountain country, where steady labour is indispensable, may be a question for discussion; but as the same is also apparent among the people long settled in the low country at the foot of the mountains, it may be inferred that the inherited qualities of the mountain tribes are not speedily lost when they are subjected to altered conditions. We are told by Rohlfs that on the northern skirts of the Marocco Sahara the Shelluhs have adopted a predatory life, and are the most dreaded of all the wandering robbers of that region. But when we learn that all the fertile oases of the Sahara have been monopolised by a small class of Arab descent, who rest their claims on religious authority, it is not apparent that there is any alternative for those who do not belong to the privileged class; and, under such an anomalous condition of society, the energy of the superior race will show itself in robbery, where that becomes the only means of obtaining a livelihood. To judge from what we saw of the country, the best thing that could happen for Marocco would be the substitution for the Moorish government of an authority strong enough to keep the Bereber tribes from intestine feuds, and intelligent enough to leave them a large measure of self-government, under a moderate and just fiscal system. Gradual extension of irrigation works would fit for cultivation large tracts now unproductive, and the superfluous population of the mountain valleys would spread into the plains, and develop the latent resources of the country. If it be said that the gradual diffusion of more intelligent ideas of government may gradually draw the Sultans of Marocco into the path of progress, and thus effect without violence as rapid an advance as is compatible with the ideas and character of the native population, the answer seems to be that this supposition is not probable in itself, and is not justified by experience.

For over a thousand years since the date of the Saracen conquest the two races that make up the population have remained perfectly distinct. The gradual extension of the central authority may have done something for the maintenance of external tranquillity, but it has been marked by a general and persistent decline in the prosperity of the country. It suffices to read the description given by Leo Africanus, himself a Moor, of the numerous large and thriving towns visited by him in the early years of the sixteenth century, just at the time when they were brought under the rule of the Moorish Sultans, to measure the vast falling off that has since followed.

The Moorish government is marked by two fatal defects, from which it seems unlikely to free itself. That
religious fanaticism should have taken deep root in a
country long exposed to the attacks of not less fanatical
enemies was quite inevitable; but for two centuries there
has been peace with Portugal, and the brief Spanish war in
1861 does not seem to have much altered the state of feeling as to Europeans; yet the hatred to Christians as such
seems to be quite as strong among the Moors at this day
as at any former period, and while it exists must continue
to be a serious barrier to industrial progress. Among the
Shelluhs fanaticism has evidently no deep hold on the
people. Some of the chiefs may share, or affect to share,
in what they doubtless consider the tone of good society
among the rulers of the country; but our experience of the
people agrees with that of Jackson, who lived for some
time near Agadir, and found there a positive desire among
the people of all classes that he should establish himself
permanently among them.

More serious even than fanaticism, as an obstacle to
good government, is the seeming incapacity of the Moor
to estimate any but immediate results, or to make any
effort of which the good effect will not be very speedily
visible. To prove to a Sultan of Marocco that such a
public work or other improvement would double his
revenue at the end of twenty or thirty years would be
sheer waste of breath. It would never occur to a Moor
that a benefit so distant, however great, was worth the
slightest present exertion. Hence the utter neglect of
public works, of mineral wealth, and of the forests which
should be an abundant source of national wealth.

We were somewhat surprised to find among the Aït
Mesan people, a decided taste for ornament, of which no
trace is perceptible among the Arabs of the low country.
We noticed that the lintels of the doors at Arround were
decorated with rude carving in geometrical patterns, diamonds, circles, and triangles. Saddlecloths had similar

chequered patterns in black and white, reminding Hooker of fabrics made by the Lepchas in Sikkim.

We had scarcely started to descend the valley when it began to sleet. As soon as we reached the village below, this turned to rain, which continued with little intermission throughout the day. The halt for luncheon offered little attraction to any of the party, and was curtailed as much as possible. It is a proof of the variety of the vegetation, that although we had already twice passed through the valley, and made considerable collections, we noticed on this occasion two or three plants growing close to the path which none of us had before seen.

A little anxiety was felt as to the ford in the river; and in truth it was a piece of good fortune that the weather had been so cold, and that snow instead of rain had fallen on the mountains. A slight rise of the stream must make the ford impassable. We might have made our way on foot along the E. side of the valley, though this appears pathless below the ford, and discovered some way for passing the stream lower down; but our baggage and precious collections could scarcely have followed that way, and practically we should have been prisoners for several days. As it was, the water at the ford was no higher than usual; the increased supply from the lower slopes was doubtless compensated by the cutting off of the drainage of the upper region, now deeply covered with snow.

Soon after crossing the stream, we were met by Abraham, dressed in his best, who in this valley assumed quite a lofty air. He informed us, with a tone of great elation, that the country people took him for a sherreeff, and, had come out to ask a Jew for his blessing. The path was in places very slippery, and it was impossible to travel fast, and so it happened that the daylight was beginning to fade when we returned to our camp. This displayed a doleful spectacle. After more than twenty-four hours' rain the piece of flat ground which we had selected was

turned into a large pool of slippery mud. Hooker's tent
had been left standing; but the wet had worked its way
inside, and there was scarcely a dry spot to be found,
while there was no choice but to pitch the smaller tents
in the midst of the general sludge. In such a case the
best plan is to make a floor of branches and leafy twigs,
and the carob trees that grew close at hand were at once
requisitioned for the purpose. The Alpine Club tent,
though the canvas was wet, when set over this rough
flooring, afforded very tolerable shelter and freedom from
the all-pervading mud.

The advantages of this tent were further proved during
the night. When pitched in open ground it depends, like
every other tent, on such support as tent pegs can give ;
but when trees are at hand the supporting rope can be
rigged fore and aft in such a way as to defy any storm to
upset it. During the night a furious gale arose, as it
seemed, quite suddenly, accompanied by torrents of rain ;
the tent pegs were drawn out of the muddy soil; and
twice Hooker's tent fell bodily to the ground, luckily with
no other bad result than to envelop the sleeper in the
clammy folds of the wet canvas. The soldiers contrived
to keep their tent standing ; but those who mounted guard,
and the natives sent by the sheik for the same duty, must
have passed a miserable night. The excitement caused
by the falling of the tent, when the whole camp turned
out with loud outcries, must have been welcome as a relief
from the dreariness of the time.

The storm of the night marked the end of the bad
weather, and the morning of the 18th brought back to us
the clear air and blue sky to which we had been hitherto
accustomed, and which rarely left us during the remainder
of our journey. The morning hours were fully occupied,
and the short time we could spare for examining the low
ground near our camp enabled us to add several species
to our lists. Having packed up our collections in tolerable
condition, notwithstanding all the difficulties of the last

few days, we made ready to start a little before mid-day, having arranged for a short day's journey.

The vegetation of this valley offers so many points of interest that it has seemed better to reserve a fuller account of the flora of this and the Amsmiz valley for the Appendix.

CHAPTER X.

IN departing from our camp at Hasni, on May 18, our
cavalcade was escorted by the friendly sheik, Si Hassan,
and two other native chiefs. Up to the last moment, sick
people had continued to arrive from distant villages, and
some of the late-comers were left unattended. As we
started, the population of the adjoining hamlet, who were
gathered round the camp, gave unequivocal tokens of
good will and kindly wishes towards the strange visitors,
doubtless due to the good effects of Hooker's medical
advice; and more friendly salutations reached us at the
villages as we passed. After descending the main valley
for a distance of three or four miles, we turned to the left,
and began to ascend in a westerly direction towards a
depression in the hills that enclosed us on that side. The
opuntia and palmetto here grew to a large size ; and among
many less familiar forms, the oleander was a conspicuous
ornament, growing freely up to about 4,500 feet above
the sea.

  As we gradually wound upwards, and the Aït Mesan

valley was finally lost to view, we found that, instead of
reaching a pass whence we should descend into an adjoining valley more or less parallel to that which we had left,
the country before us was an undulating plateau, extending
over a space of many miles, through which no stream runs
from the higher mountains towards the plain.   This
plateau does not subside gradually towards the low country
as might have been expected; for at almost every point
we found higher ground lying between us and the plain,
in the form of rounded eminences, rising some three or
four hundred feet above the plateau.   The soil was calcareous, and the underlying pale limestone cropped up here
and there; but the stratification appeared very irregular.
In some places we noticed bosses of intrusive igneous rock
of dark colour.   Though no villages were in sight, most of
the surface was under rude tillage; but the fields were gay
with a multitude of wild species in full flower.

After the excitement of the preceding days, the afternoon ride seemed uneventful in a botanical sense, as we
failed to find much that was altogether new.   The most
interesting forms were several fine *Orobanches*, which
might here be studied with profit by a traveller less
pressed for time than we were.   A great feast of colour
was presented to us as we approached Sektana, our camping place, by a magnificent new *Linaria*, of which we had
hitherto seen only stunted and starved specimens.   In some
fields of corn not yet in ear, the spikes of numerous dark
crimson flowers all but concealed the green, and gave to the
surface a tone of subdued splendour.   The plant has been
described by Hooker, in the *Botanical Magazine* (vol. 98,
No. 5983), as *Linaria Maroccana*.   The artist, who had
not seen the wild plant, has failed to attain the rich tint
of the native flowers.   In cultivation, the colour loses its
original depth, and in some gardens it has faded to a pale
purple or violet tint.   From this, and other differences
shown in cultivation, it seems possible that this may be
an extreme form of *Linaria heterophylla* of Desfontaines,

a plant so different in appearance that, at first sight, no one would suspect very near relationship between them.

Our camp this evening was fixed on open ground, near the village of Sektana. To the north, between us and the plain, a hill rose some 400 or 500 feet, crowned by a castellated building, somewhat similar to that at Tasseremout, of which all that we could learn was that it had been built by Christians or Romans, the same word, as before observed, bearing either interpretation. To the south, the plateau stretched away in rolling downs, unbroken by tree or house, save a few small plants—probably fruit trees—growing near the village, about half-a-mile from our camp. We were received, on our arrival, with some show of cordiality by three native sheiks; and a *mona*, on a scale sufficient to satisfy even our greedy soldiers, was forthcoming during the evening.

It appeared that Hooker's fame as a physician had already spread far and wide, to an extent that might, indeed, have been inconvenient if we had remained longer in this district. On this evening, and the following morning, troops of applicants for medical relief continued to arrive at our camp, and amongst them a *moullah* of reputed sanctity, from Moulaï Ibrahim, troubled with some painful affection of the eyes.

Between the ordinary work at our plants, writing up journals, and completing a letter from Hooker to the late Sir Roderick Murchison, with a brief account of our proceedings up to this point, the evening was fully occupied, and we enjoyed the change of climate that had accompanied the return of fine weather. The thermometer at 8 P.M. did not fall below 58°, and the mean of two closely accordant observations gave for the height of our camp, 4,523 feet (1,378·7 m.) above the sea level.

The morning of May 19 broke brilliantly. Although on the preceding day we had travelled under a blue sky, the higher mountains had been concealed by dense masses

R

of fleecy *cumuli*, and we were not prepared for the
grandeur of the panoramic view that was spread before
our eyes, as we sallied from our tents in the early morning.
A large portion of the range of the Great Atlas of
Marocco stood robed in glittering snow down to a height
of about 7,000 feet above the sea level, only the projecting ribs of rock appearing through the white vestment
along the higher and steeper ridges.   In the annexed
sketch is shown the part of the range nearest to our camp,
lying between the head of the Aït Mesan valley, and that
of the next adjoining (much shorter) valley that opened
nearly due south from our station.   The view, however,
extended in both directions far beyond the nearer part of
the range.   The high peaks of the Ourika district were
sharply cut against the sky, but so crowded together that
their relative position was not apparent.   Farther east
were other high peaks, probably belonging to the district
of Glaoui.   Turning westward, it was seen that to the
right of the high group shown in the sketch, the valley
which feeds the main branch of the Oued Nfys runs deep
into the main range, which here sinks to the comparatively
low level of about 7,500 feet.   Towards this valley the
high snow-clad mass before us fell with comparatively
easy slopes, nowhere difficult of access; and we indulged
in the hope that, by fixing our camp pretty high up, we
should be able to effect another ascent.

To the west of this great gap the main chain rises
again to a considerable height, but less by at least 2,000
feet than that of the central range which we were now
about to leave.   The western range also differs in being
less continuous; the peaks are comparatively isolated, and
of massive, conical form; and the intervening passes do
not seem to rise above the level of tree vegetation.   We
observed that, even allowing for its lesser elevation, the
western range showed much less snow, whether because
during the recent bad weather the precipitation was more
considerable on the eastern group, or because in the

J. B. delt.

WEST END OF THE MAROCCO ATLAS FROM SEKTANA

region nearer to the ocean this had fallen mainly in the form of rain.

After a rather late breakfast, the hour fixed for the departure of our travelling companion having arrived, we with much regret bade adieu to Maw, whose engagements in England hastened his return. He carried with him a considerable collection of living plants which, owing to his skill and experience in managing this difficult process, arrived in excellent condition, and have since thriven in his garden in Shropshire. The soldiers of our escort who accompanied him to Mogador, bore orders to the local authorities which ensured their respect and attention, and, as we afterwards learned, his journey was in every way successful.

Mid-day had passed before we started from Sektana, the morning hours having been employed in collecting and laying in paper a tolerably large mass of specimens. Our course lay over the plateau, whose undulations gradually subsided towards the Oued Nfys. At a distance of some seven miles from our camp we reached the brow of a range of low broken cliffs of white limestone, facing westward towards the broad valley of that stream. They are at a considerable distance from the present bed, but were doubtless formed by erosion at a distant period when the level was much higher than it now is. Among other plants, a variety of the wild caper (described as *Capparis ægyptia* by Lamarck) was here common, and was afterwards often seen in similar stations as we travelled westward. The flowers-buds are eaten raw by the natives, who call them *Pan*.

Below us, on the fertile tract extending for three or four miles from the foot of the cliffs towards the Oued Nfys, stood the village of Gurgouri, overlooked by two *kasbahs* belonging to the Governor of the district which we had now entered, also called Gurgouri. The older fortress-like building, standing on a projecting rock, was apparently uninhabited, and the Kaïd dwelt in a less imposing structure close to the village.

Our present design was to approach the high summits of the Atlas that we had viewed in the morning through one of the lateral valleys of the Oued Nfys, whence, as it appeared, the ascent might be effected without serious difficulty. The leaders of our escort had ridden on before to announce our arrival, and, after a short halt, we approached the village through a belt of gardens and olive groves. No Governor appeared to meet us, but only a messenger with some lame excuse for his non-appearance. It would seem that our dissatisfaction at this want of attention was speedily reported, and that the Kaïd's second thoughts were different from his first, for he presently appeared just at the entrance to his *kasbah*. He was a tall, handsome man, courteous, but no way cordial in his greeting. He invited us to stop at this place, offering at the same time a suitable *mona*. It was necessary, however, to bring the question as to our further progress to a speedy decision; and when the proposal to ascend the neighbouring mountains was met by a positive refusal, and an intimation that such an expedition might be effected from Amsmiz, the adjoining district to the W. of the Oued Nfys, Hooker at once decided to continue our journey, and to refuse the proffered entertainment.

It appeared that our refusal was felt by the Kaïd as a slight; so, by way of offering an irresistible attraction, a cow was led out and slaughtered on the spot, close to the *kasbah*. Fresh beef is a delicacy rarely found in Marocco; but even this failed to move us. Our greedy soldiers were furious at being baulked of the opportunity for feasting and idling, which they evidently considered the main object of their mission, and our departure from Gurgouri was accompanied by the surly faces and muttered grumblings of our escort.

It was only on the following day that, owing to the continuing feud between the two men, we learned, through our Mogador captain, that Kaïd el Hasbi had here once more been scheming to frustrate the objects of our journey. In announcing our arrival, he at once prejudiced the Governor of Gurgouri against the intrusive Christians, who had come to visit his district, and directly advised him not to let us enter the mountain valleys. It is likely that this conduct was as much prompted by a keen recollection of the discomfort of his recent five days' stay in the Aït Mesan valley, and the poorer fare there available, as by mere fanatic dislike to Christians and strangers; but we all know hów readily fanaticism allies itself with the baser passions of human nature, and neither were wanting in Kaïd el Hasbi. In any case, it was only natural that a local Governor should take his cue from the man who seemed to be the personal representative of the authorities in Marocco.

If our soldiers were disappointed at missing a feast, we were in no better humour at being foiled in what appeared a hopeful project. We silently rode for nearly an hour amidst well-cultivated fields and gardens before we finally reached the banks of the Oued Nfys, at a village called Nurzam. The channel was some 300 yards in width, cut out from the soft limestone strata that rose on either side in steep banks about thirty feet high ; while, in spite of the recent rains, the stream was only about twenty yards in width, and everywhere shallow. The day was so far advanced that we could not linger here—a fact the more to be regretted as we found, on the dry gravelly bed of the stream, several plants not before seen during our journey. Among these were *Salvia ægyptiaca*, and a curious *Antirrhinum*, nearly allied to, but different from, the Algerian *A. ramosissimum*.

On the west side of the Oued Nfys, the ground rises gradually, but not nearly to so high a level as the plateau of Sektana. The underlying rock throughout the space between this and the next valley descending from the Atlas appears to be covered with a thick red earthy deposit, sometimes of the consistence of clay, sometimes of a more friable character, doubtless formed at the expense of the portion of the Atlas at the head of the valley of the Oued Nfys.

No indications of glacial action were observed in this region, or in the Amsmiz valley which we were about to visit. On the way to Amsmiz we crossed a ravine fully 200 feet in depth, cut by a streamlet through the clay beds, without reaching the underlying rock. The country was in great part under tillage, and, although we passed no villages, must maintain a considerable population. A few interesting plants were seen; but time was pressing, and we could not afford to halt. The sun set, casting a brilliant red glow over the heaving plain that lingered for a short space longer on the flanks of the mountains which here rise more abruptly than in the district near Marocco. Our course was directed towards the narrow opening of a valley, cleft through the outer range of the Atlas, which we had already descried from a distance; and, after a gradual descent, we arrived, about 8.15 P.M., some time after dark, at Amsmiz.[1]

This is the most considerable place on the northern declivity of the Great Atlas, and, from the number of inhabitants, may deserve to rank as a town. It stands on a shelf of flat rocky ground, somewhat above the level of the adjoining plain, and nearly 200 feet above the stream issuing from the mountains close at hand, which, for want of any other name, we have called the Amsmiz torrent. The Governor of this district was, as we learned, a man of some consequence, being a nephew of El Graoui, and brother-in-law of the Governor of Mogador. We considered it a favourable omen that, as we approached our camping ground close to the town, this functionary, with a motley train of torchbearers, came out to meet us, and, with much show of cordiality, welcomed us to his district.

[1] The final letter is nearly or quite mute, and the name would by an ordinary Englishman be written Amsmee.

He was almost quite black, and of nearly pure Negro type, with the sensual, but apparently good-humoured expression that is common among that race. It was too late to discuss business on this evening; and the less necessary to do so, as we knew that our large and precious collections made in the Aït Mesan valley, and put up in indifferent condition, would require a full day's work, before we could undertake a new excursion of any importance. An abundant *mona* was provided; and general satisfaction appeared to prevail in the camp at having reached this Capua of the Great Atlas.

The 20th of May was a day of rest for the men and animals of our party; but of rather hard work for the two botanists, who were for eight or nine hours busily engaged in putting their collections into order, and transferring the specimens from damp to dry paper. The system of ventilating gratings which we adopted, works admirably in a dry climate, and especially when it is possible to expose the parcels to sunshine; and in such conditions most plants may be dried without a single change of paper. The case is different when, owing to rain, or the dampness of the climate, the paper cannot be well dried, and the plants have to be laid in in a more or less moist condition. Artificial heat may sometimes be applied; but this is rarely available for travellers in such a country as Marocco.

It was necessary to interrupt the work during the forenoon, in order to pay a visit to the Governor. This was no matter of mere ceremony, as it was essential to obtain his consent and assistance towards carrying out the design— on which we had fixed our hearts—of penetrating to the head of the Amsmiz valley, and climbing some one of the higher adjoining peaks. We knew, indeed, that in this part of the range, the Great Atlas does not attain so high a level as it does farther east; but as the summits must reach a height of fully 11,000 feet, they could not fail to exhibit the characteristic vegetation of the higher zone, and at the same time, unless we were again pursued by

bad weather, command a wide view over the unknown
country, on the south side of the chain.

The Governor was courteous and even friendly in manner, and in general terms expressed his readiness to forward
the objects of our journey. He seemed pleased with the
articles which Hooker presented to him — a musical box,
an opera-glass, and a long sheath-knife; but when a thermometer was added, and an attempt made to explain the
use of the instrument, he at once returned it, saying that
it would be of no service, and that he would much prefer
a brace of pistols. The pistols were promised, and an unsuccessful attempt was actually made to forward them a
year later. All had gone smoothly so far; but we were
much disappointed when, the practical question of our
intended exploration of the Amsmiz valley being brought
forward, our friendly Governor expressed himself distinctly
opposed to it, the only ground assigned being some
doubt as to our safety. As the misbehaviour of Kaïd el
Hasbi on the previous day had come to our knowledge
during the interval, we at once came to the conclusion that
the real obstacle was due to his machinations. On returning to our camp, Hooker summoned El Hasbi, and administered a 'blowing-up,' which produced the most
salutary results. He was told that we were thoroughly
aware of his treacherous conduct, and duly warned that if
any further difficulty were thrown in the way of our
reaching the high mountains, as we were fully authorised
to do, a report should be sent to the Viceroy and to El
Graoui, with a request that another officer should be sent
to take charge of our escort. This drew forth a multitude
of excuses, and profuse promises to do all that was possible
to carry out our wishes. The effect was soon apparent; for
we learned in the course of the afternoon that the Governor had summoned the sheik of the valley, in order to
arrange for our visit, while at the same time we received
an invitation to sleep that night in the *kasbah*, which we
thought it judicious to accept.

In the afternoon we went out for a stroll, and were able to form a better idea than we had hitherto done of the character of the scenery. The position of Amsmiz somewhat reminds one of that of villages in Piedmont, that stand at the opening of some of the interior valleys of the Alps, and still more of similar places in the Apennines of Central and Southern Italy. The lofty hills that form the outer extremity of the spurs diverging from the Great Atlas slope rather steeply towards the plain, while the torrent issues from them through a cleft so narrow that no path is carried along it into the valley. Trees, that naturally clothe the outer ranges of the Alps, are here very scarce, and the upper declivity, as commonly in the Apennine, is covered with brushwood and low shrubs; while the lower slopes are partly under tillage, or else planted with olive and fig trees. We descended from the plateau, where our camp stood close to the town of Amsmiz at 3,382 feet (1,030·7 m.) above the sea, by steeply sloping banks to the level of the torrent; and followed this for some distance, collecting plants by the way; and then made a circuit among fields, enclosed by high hedges, in which grew a profusion of climbing plants. The chief prize of our excursion was a curious new species of *Marrubium*, whose spherical heads of flowers are beset with long stiff bristles hooked at the end, formed by the elongated lower teeth of the calyx.

It was not without misgivings that we quitted our tents in the evening to repair to the Governor's *kasbah*. We had hitherto been very successful in escaping the varied noxious insects that prey on the human body, and which the walls of the first house we had seen in Marocco and the concurrent testimony of all who know the country declare to abound throughout the empire; but it now seemed as if we must confront these enemies under circumstances where we could not, without giving offence, resort to energetic measures of precaution. We were, therefore, agreeably surprised, when our host con-

ducted us to a room which, at least to the eye, seemed
scrupulously clean. Two beds were arranged, nearly in
European fashion, on low bedsteads; of other furniture
there was none, excepting a low carpet-covered divan. On
the sill of a window we found four or five Arabic books,
the only ones which we saw in the possession of a native
in this country. Our report the next morning agreed,
that we had not been attacked by any enemy more formidable than a few intrusive fleas.

Our slumbers during the night were made more agreeable by the satisfactory information that all the requisite
arrangements had been made for our excursion in the
Amsmiz valley, on the following day. . The sheik, as we
were told, would provide a house for our occupation in the
highest village; and though nothing definite was said as
to the precise limits of our expedition, we relied on luck
and good guidance to turn our opportunities to account.

We rose early on the morning of the 21st, and employed some spare time in looking about us. The *kasbah*
was not nearly so large as many that we had seen, but
was distinguished by a certain air of neatness, and there
were sundry indications that its possessor was superior in
general intelligence and appreciation of civilised life to
those we had hitherto dealt with in Marocco. In the
court there was a small garden, wherein grew some large
bushes of a curious variety of the common myrtle, having
the young branches and leaves covered with a fine downy
pubescence, the leaves were of large size, and much
crowded together, giving the plant a peculiar aspect. We
did not observe the myrtle in a wild state anywhere in
South Marocco; and these plants probably came from
some gardens in the city of Marocco.

The sheik of the valley made his appearance in due
time, and we at once perceived that he was reluctantly
pressed into the service. He presented an example, unusual among the Shelluhs, of genuine religious fanaticism,
never relaxing, during the three days which we passed in

his company, from an attitude of undisguised aversion to the Christians, whom he was forced to treat with a faint show of outward civility. We afterwards learned that it was by his express order, that his people were prevented from applying for medical advice, and kept aloof from us during our stay in the valley, not even replying to the ordinary courteous salutation. The latter is quite a remarkable incident, and without example in our pretty wide experience of Oriental people.

Our party was as far as possible reduced in number, most of our followers and all the escort, except two soldiers, being left behind. After the usual delay, we were under way soon after 10 P.M. Instead of directing our course towards the cleft by which the torrent issues from the mountains, we left the little town by the side farthest from the stream, and rode across the strip of plain lying between it and the outer range of steep hills. After riding about a mile we came to a place where, according to the usage of the country, the weekly market was being held. A considerable crowd of wild-looking people, most of them apparently mountaineers, formed a busy throng, wherein, under different dress and aspect, human nature showed itself much the same as it does everywhere else. One half of the crowd was intent on business, and hard bargains were driven where the difference between the seller and the buyer may not have exceeded the tenth part of a farthing. The rest were mere idlers, come to while away the time in gossip, or in listening to professional story-tellers, or in beholding the feats of serpent-charmers, who make a precarious living by frequenting these gatherings.

We soon reached the hills, and began to mount by a well-beaten, but rather steep path. The vegetation on the dry stony slopes, mainly covered with brushwood, was already much parched, and we noticed nothing of especial interest till we reached the top of the ridge overlooking the valley of Amsmiz. Here stood two lonely poplars, and an old weather-beaten trunk measuring about five and a half feet in circumference, seemingly of high antiquity. The tree appeared to be in no way different from the *Juniperus Oxycedrus* of Southern Europe, except that this rarely exceeds the dimensions of a bush five or six feet high.

We now began to obtain a clearer view of the portion of the Atlas chain which it was our present object to explore. The valley before us was evidently different in character from those which we had hitherto seen. Especially throughout its lower part, it is a mere trench, whose sides slope with increasing steepness towards the bed of the stream; while the flanks, throughout a zone of from one to two thousand feet above the water, are but slightly inclined, and afford space for numerous villages and for cultivation. It is much shorter than the valleys in the Ourika and Reraya districts; and, instead of being enclosed at its head by a continuous ridge of great height, we here saw a single lofty snow-streaked peak at the head of the valley, apparently separated from the next eminences on either side by comparatively low passes, over which an easy passage to the Sous valley must be found. The peak which was the obvious aim of our expedition is known to the Moors as Djebel Tezah, or Tezi, and its summit, which had already attracted our notice from Sektana, can scarcely be more than fifteen or sixteen miles, as the bird flies, from the point where the torrent enters the plain near Amsmiz; whereas, in the part of the chain first visited by us, the watershed must be everywhere more than twenty miles from the northern foot of the mountains.

Our track descended slightly from the top of the ridge above Amsmiz, and then continued nearly at a level for a considerable distance, the torrent, which ran at a great depth below us on the right, being usually concealed from view by the convexity of the slope. We soon observed that several villages lying on the upper slopes were mere piles of ruin. Some, as we learned, had been destroyed by hostile tribes, and others had been abandoned by their inhabitants who had migrated elsewhere. Perhaps owing to the scantiness of the present population, timber was more abundant here than usual in the Great Atlas. Besides larger trees at intervals, the slopes along which we rode were to a great extent covered with oak scrub to such an extent as to leave little space for herbaceous plants. The prevailing form of oak in this part of the valley tends to confirm the opinion of those botanists who, with Visiani, consider the common evergreen oak (*Quercus Ilex*), and the cork oak (*Quercus suber*), to be forms of the same plant. In general appearance, the oak here quite resembled *Q. suber*, and in the older trees the tendency to form a corky outer bark was apparent; but the anthers all showed traces of the sharp points which are supposed to distinguish those of the common species.

At a point where a slender rivulet from the dry flanks of the hills on our left enters the valley, we passed close to a small village, with a belt of cultivation surrounding it, and soon after began to descend steeply to the bed of the torrent, no longer so distant from us as it had been in the lower part of the valley.   Here we found several fine plants not before seen, and then, somewhat to our surprise, began to ascend, by a narrow and difficult path, the steep rocky slope above the left bank of the torrent.   Our object was to pass the night at the uppermost end of the valley, as near as possible to the foot of Djebel Tezah, and the sheik had undertaken to conduct us to the highest village.   We were now informed that this stands on the west side, and some 600 or 700 feet above the level of the torrent.   This was evidently inconvenient, as our route on the following day must clearly keep to the eastern bank; but there seemed to be no help for it, and, as the rocks on this side were different in appearance, we perceived a fair prospect of adding some novelties to our collections.

About 2 P.M. we reached Iminteli, the poor village which served as our head-quarters for the next two days.

The weather being now settled, our observations for altitude above the sea give nearly the same results when compared with Mogador or with Amsmiz, and that adopted
was 4,418 feet (1,346·5 m.). The house of a Jew had been
cleared out for our reception. It was of rather more solid
construction than that in which we lodged at Arround, in
the Aït Mesan valley, but similar in plan. It appeared to
be tolerably clean, though on closer scrutiny a few bugs
were detected, but in far less numbers than we have seen
in Sicily and some other places in Europe. In pursuance
of the sheik's order, the Shelluh inhabitants of this place
kept carefully aloof from all communication with us; but
there were several Jew families who were clearly well
disposed towards the Christian strangers. When we sallied
forth, soon after our arrival, to examine the vegetation of
the rocky slopes above the village, one of these Jews
volunteered to accompany us. Conversation either in
Hebrew or Shelluh being unfortunately out of the question, our intercourse was necessarily of the most limited
character; but we could not fail to be struck with the
man's air of intelligence and friendliness.

The character of the rocks throughout this valley is
altogether different from that of the more lofty range
which we had previously visited. The red sandstones,
there so prevalent, are here absent, and the strata are,
without exception, schistose, though seemingly varying
much in mineral composition. In some places, and especially on the ridge above Amsmiz, mica is present to an
appreciable extent; in others, and notably in the mass of
Djebel Tezah, the rock would pass under the old designation of clayslate; while in this part of the valley calcareous schists prevail. Intrusive dykes and bosses of porphyry and other igneous rocks were seen in many places,
but not to such an extent as to affect much the general
aspect of the surface. The stratification appeared to be
very irregular, but in general the beds are inclined at a
high angle to the horizon.

We were led by the difference in geological structure to anticipate a considerable change in the flora, as compared with that with which we had already made acquaintance; and this expectation was confirmed by what we had seen during our morning's ride, and still more by our afternoon herborisation on the rocks and slopes above Iminteli.

We had already perceived that the vegetation of this part of the Atlas is much less varied than that of Aït Mesan; and this impression became stronger with subsequent experience. Something may be due to the fact that we had there more time and opportunity for close examination; but this could only slightly alter the general result This may be partly due to the more permeable character of the rocks in the Amsmiz valley preventing moisture from resting on the surface; and in part to the fact that, although nearer to the Atlantic, the range here is less lofty, and the precipitation of vapour, in the shape of rain or snow, is less frequent and abundant during the summer months. The subject is more fully discussed in Appendix E; but it will be seen by reference to our lists, that only seventy-six species seen by us were confined to the Amsmiz valley, that nearly twice as many were found common to both valleys, while more than three times that number were seen in Aït Mesan, but not in Amsmiz. If a similar comparison be made as to the endemic species which, as far as we know, are peculiar to the flora of the Great Atlas, the proportions are nearly identical. A further peculiarity of the vegetation here, is the greater resemblance which it bears to that of the mountains of Southern Spain than does that of the Aït Mesan valley. Several plants not before seen by us were identical with Spanish species; and of the undescribed forms here collected, at least four have their nearest allies among the endemic plants of the Peninsula. Two or three characteristic species of the Algerian flora were also seen here for the first time; but these will doubtless be found to extend through the intervening region when this has become accessible to scientific travellers.

At two or three spots, during the afternoon, we had noticed fragments of pine cones, which the more excited our curiosity, as up to this we had not seen a true conifer in Marocco, and there was no reason to expect the appearance in this region of any other species than the Atlantic Cedar. When asked on the subject, the sheik declared that they had been dropped by strangers who had come from a great distance. The only explanation for a lie, apparently so objectless, is to be found in the deep-rooted suspicion which the mountain tribes feel as to all strangers, but which in the case of the sheik was intensified by religious bigotry. It was satisfactory to find a few trees about the highest point which we reached—700 or 800 feet above the village—and to ascertain that the cones belonged to the *Pinus halepensis*. It is remarkable that, when questioned, our Jew guide gave for the tree the name *Tœda*. There is much doubt as to the species which Pliny knew by the name *Pinus Tœda*, but it is pretty clear that the Romans applied the name *tœda* to various species of pine used for torches ; and to this day, in the Italian Tyrol, where brands of *Pinus Mughus* are commonly adopted for that purpose, they go by the name of *tea*. The preservation of the exact Latin term in this remote region, along with that of the *furbiune*, already mentioned, appears worthy of note.

Well content with our collections, we returned before sunset, but our good-humour was soon troubled by the recalcitrant sheik, who seemed resolved to frustrate the main object of our expedition. When our interpreter was charged to arrange with him for our departure early next morning, he returned an answer to the effect that we could go with safety no farther than the village of Iminteli ; that the great mountain was frequented by the Sous people, and could not safely be approached by us; and this was wound up by a flat refusal to let us proceed any far-

J. D. H. delt.

DJEBEL TEZAH FROM IMINTELI

ther. When it was found that he stuck doggedly to this
resolution, Hooker judiciously resolved to assume a more
resolute tone. 'Tell him,' he said to Abraham, 'that the
Sultan has issued his order that we should go to the snow
—El Graoui has ordered that we shall go the snow
—the Kaïd of Amsmiz has ordered that we shall go to the
snow—if he refuses to carry out their orders, we shall
return to Amsmiz, and send a courier to the Viceroy and
to El Graoui, and we shall see what will be the consequence.' After some delay, the answer came, that if we
went to the mountain, it would be necessary to take an
escort of fifty armed men. 'Tell him,' was the reply,
'that he may take as many men as he likes—five men, or
fifty men, or none at all—we do not care as to that; but
the Sultan's command must be executed!' The sheik sul-
lenly gave way, and promised that all should be ready for
the next morning. After achieving this victory, we proceeded to dust the floor profusely with insect powder, and
to spread our mattresses : our insect enemies either held
aloof, or performed their operations so deftly that we were
unconscious of their assaults, and slept till the first gleam
of dawn showed in the eastern sky.

When we sallied forth at 5 A.M., on the morning of
May 22, the air was cool, but a light mist hung between
us and the mountains, the usual precursor of a hot day.
When our preparations for starting were complete, the
sheik was not to be seen; but presently a message came
to say that he had gone on ahead, and would await our
arrival on the banks of the torrent below the village. It
seemed as if it involved an unnecessary detour to return
by the path which we had ascended the day before, instead
of aiming at a point higher up the valley; but a native
who was left to act as guide, insisted on keeping to the
steep rocky path with which we were already acquainted.
At the appointed place, by the bank of the torrent, we
found the sheik, with four or five ragged fellows, of whom
but two were armed with long guns. Anticipating any

remark as to this sorry substitute for the promised escort of fifty armed men, the sheik announced that more men would join the party farther on. As we firmly disbelieved the stories of danger from the terrible men of Sous, who are the bugbears of the population on the northern side of the mountains, we never cared to call attention to the fact that the promised reinforcements did not make their appearance.

In the upper part of the valley, the trench which the torrent has cut for itself is less deeply excavated than through its lower course, and leaves space for a path, and a few straggling olive and walnut trees; and in some spots for small patches of cultivation. For about six miles we kept to the torrent bank, our horses sometimes preferring its stony bed, till we reached the junction of the two streams that feed the Amsmiz torrent. Between them rises the peak of Djebel Tezah, and here the ascent of the mountain begins. As the slope was still very gentle, we rode on a short distance farther, after hurriedly collecting some interesting plants, but soon came to a halt at a clump of fine walnut trees, standing by our observations at 5,604 feet (1,708 m.) above the sea. We had seen no village by the way, but only a few men engaged in fashioning gunstocks from walnut wood. It appeared, nevertheless, that there was a small village near at hand, and this place would be the proper starting point for travellers intending to make the ascent. There would be no difficulty in conveying small tents hither from Amsmiz.

Much to our satisfaction, the sheik now withdrew, committing us to the charge of an active, but unarmed young Shelluh, with strict injunctions to lead as far as the snow, but not to allow us to proceed farther. It is hard to say whether the sheik and his people felt any real uneasiness as to the possibility of a casual encounter with natives of the Sous valley; but it was pretty clear that they had succeeded in frightening our attendants, as our Mogador men, usually so active and attentive, soon dropped behind, and were not again seen till our return in the afternoon. We took the most direct course in the ascent, following a slight gully down which flowed a mere trickling rivulet, fed by the snows on the upper slope of the mountain, and pushed on rather fast with a view to get as high on the mountain as possible before the sun reached the meridian.

Bearing in mind the great diversity in the vegetable population which is seen in Southern Spain (the high mountain region nearest to the Great Atlas), where neighbouring peaks of different mineral structure exhibit numerous quite distinct species, and very few identical features, and having found the flora of the lower valley to a great extent different from that of Aït Mesan, we confidently reckoned on obtaining still greater evidence of distinctness in that of the upper region. It was therefore with some surprise that, as we continued the ascent, we met, one after another, many of the peculiar species that we had first seen in the ascent from Arround to the Tagherot Pass, and comparatively few not already familiar to us. For once, however, it must be owned that during part of this day, our emotions as botanists yielded to the interest that we felt in the near prospect of a peep into *terra incognita*.

If but little had been hitherto known of the northern slopes of the Great Atlas from the reports of the few travellers who had viewed the range from the low country, or attained its outer slopes, the southern side of the main chain remained a sealed book to geographers, whose reliance on the vague reports of native informants has led them, like the chartographers of the middle ages, to fill up the blank space on their maps by representations utterly discordant and contradictory. Ever since we had been in South Marocco, we had heard of the Sous valley, as the proper home of everything strange and marvellous to be found in the empire. It is there, our informants assured us, that lions and other savage beasts roam at leisure,

there pythons twenty or even thirty feet long lie in wait
for the traveller, mines of the precious metals abound in
Sous, and in Sous the soil is so fertile that all the products of nature are obtained without labour.   But of the
physical features of the country we could learn nothing.
Whether it were enclosed on the southern side by a second
lofty range, or Anti-Atlas, parallel to that we already knew,
or merely by secondary branches diverging from the main
chain, and from how far eastward the sources of the Sous
might flow, were all matters quite unknown to us.   One
European, indeed, had traversed some part of the valley,
and should have been able to throw some little light on
these obscure points ; but unfortunately the few lines in
which Gerhard Rohlfs recounts his adventurous journey to
Tarudant, and thence eastward to the northern skirts of
the Sahara, give scarcely any information.   He speaks of
high mountains lying south of the Sous valley, but says
nothing to show what relation these bear to the main
chain.   It appears from his account, that no considerable
ascent is necessary in order to pass from the southern
branch of the Sous to the streams that flow southward
towards the Great Desert ; but whether the Great Atlas
and the Anti-Atlas are throughout their length separated
by a broad trough, in the same way as Lebanon and Anti-
Lebanon, or Anti-Atlas be a diverging range over which
Rohlfs made his way by a deep pass or depression, it is
impossible to infer from his narrative

By the time we reached the lower skirts of a long snow
slope that stretched upwards towards the summit of the
mountain, the sun, which had now ascended nearly to the
zenith, beat down upon us with intense rays, that drove
two of the party to seek some temporary shelter.   The
Shelluh guide probably considered that he had done his
day's work ; and, finding a narrow rim of shadow under
an overhanging rock, lay down, with his head screened
from the blazing heat.   Ball, who was suffering from a
violent head-ache, also found a spot that gave partial

shade. Hooker took advantage of the halt to push on at
a steady pace that soon carried him beyond the reach of
interference from the guide. When Ball felt able to
resume the ascent, the guide sprung to his feet, and for
the first time became aware that one of the party was
already too far ahead to be easily overtaken. He proceeded by a series of unearthly yells and frantic gesticulations, to attempt to attract Hooker's attention, and urge
his return. When these demonstrations were found to be
useless, and he perceived that Ball was also about to follow
in the ascent, he commenced a fresh series of exclamations
and pantomimic gestures, of which the burden seemed to
be that if we went to the top, we were certain to be shot ;
but the same argument that was used with effect on the
Tagherot Pass—the gift of a silver coin—was so far successful that no attempt was made to arrest Ball's progress,
and, after ascending a few hundred feet higher, the unwilling guide gave up the attempt, and rested comfortably until he had an opportunity of rejoining Hooker in
his descent.

It was perhaps fortunate for our object of reaching the
summit of the mountain as early as possible, that the slope
by which we ascended is extremely dry and barren. A few
species, already seen on the Tagherot Pass, were gathered
near the snow, but the upper ridges showed only a few
perennial species in flower, of which the most conspicuous
was a variety of *Alyssum montanum*. Most of the others
were stunted bushes, one of them being a dwarf form of
the common gooseberry, with stems about a foot long,
lying flat on the surface of the rocky soil. Throughout
the ascent the rock was of schistose structure, seemingly
argillaceous, but in some places containing a notable proportion of lime, and here and there showing traces of mica.
Intrusive dykes and bosses of reddish porphyry appear in
places, but do not play a conspicuous part in the aspect
of the mountain.

Hooker reached the summit about 2 P.M., and was re-

joined by Ball nearly half an hour later. Excepting some
light fleecy cumuli floating over the low country to the
north, at a lower level than the eye, the sky was cloudless; but in some directions a thin haze obscured the
details of the vast panorama. Our first glance was inevitably directed towards the unknown region to the south,
and there, at a distance of fifty or sixty miles, rose the
range of Anti-Atlas, showing a wavy outline, with rounded
summits, and no apparent deep depression, rising, as we
estimated, to a height of from 9,000 to 10,000 feet above
the sea. The highest portion within our range of view,
and the only part with a somewhat rugged outline, bore
a few degrees west of due south, and corresponded in
position with the Djebel Aoulouse of the French map.
A somewhat darker shade traceable at some places on
the flanks of this dimly seen range, possibly indicated
the existence of forests, or at least of shrubs covering the
slopes.

When the first impulse of curiosity was partially satisfied, we began to take more careful note of our position,
and to study in detail a view which had been so long
denied to us. The first fact that struck us, was that the
peak on which we stood lies a considerable way north
of the watershed. The axis of the main chain, which
here subsides into undulating masses from 2,000 to 3,000
feet lower than Djebel Tezah, lay between us and the central portion of the Sous valley, and, even if the prevailing
haze over the lower districts had not veiled the details,
would probably have cut off the course of the stream and
the rich tracts that are said to fringe its banks. The
higher strata of the atmosphere, above the level of about
7,000 feet, were, however, delightfully clear towards the
east and west, and every feature of whatever portion of the
main chain lay within our range was easily traced even at
distances of thirty or forty miles. An extraordinary
change had occurred during the three days since we had
viewed the chain from Sektana, covered in deep snow

down to the level of about 7,000 feet, and showing only
a few crests of precipitous rock here and there protruding.
The white mantle had now completely disappeared, and
only long streaks of snow filling the depressions of the
surface now seamed the flanks of the higher mountains,
leaving the summit ridges everywhere bare. During the
ascent of the northern face of the mountain, we had kept
close to one of these long and comparatively narrow snow-
slopes that extended through a vertical zone of over 2,000
feet, with a breadth of some 300 to 400 feet, and we now
saw a still longer and wider strip of the same character,
filling a shallow trough below us, on the east face of the
peak.   Near to the summit, and on the ridges leading to
it, not a trace of snow was to be seen, even in the crevices
of the rocks, where it would find partial shelter from
the sun.

    We now proceeded to survey the field of view, in order,
if possible, to fix the positions of any conspicuous summits.   Looking due west, nothing approaching our level
lay between us and the dim horizon.   A succession of
projecting spurs of the Atlas, dividing as many successive
valleys, subsided into the plain ; the most prominent, and
that extending farthest from the main chain, being the
mountain above Seksaoua.   Turning the eye a little to
the left, about west by south, we saw crowded together
many of the higher summits of the western portion of the
main range, which was here seen foreshortened, so that it
was impossible to judge of their true relative position.
The highest of these, seamed with snow, we judged to be
about twenty-five miles distant, and higher than Djebel
Tezah by 600 or 800 feet.   In nearly the same direction,
but only about ten miles distant, was a rugged projecting
peak, rising some 300 feet above our level, and very many
more of somewhat lower elevation were discernible in the
space between us and the more distant points.   Between
SW. and SSE., the range of Anti-Atlas, rising behind the
broad Sous valley, bounded the horizon.

At our feet, and cutting off from view the course of the river Sous, the mountain mass that here forms the axis of the main chain presented the appearance of a troubled sea of a light ferruginous colour, declining gradually in elevation from W. to E. At a distance of about eight miles ESE. of Djebel Tezah it sinks to an estimated height of little over 7,000 feet, at the head of the main branch of the Oued Nfys, and offers the only apparently easy pass over the main chain which we had yet seen.[1] The rocky sunburnt flanks of the mountains were dotted with trees of dark foliage, doubtless some form of the evergreen oak, up to a height of about 8,000 feet above the sea, for the most part solitary, sometimes in clumps, but nowhere forming a continuous forest. The numerous feeders of the Oued Nfys had cut deep ravines in the flanks of the mountains, and were lost to sight, except where gleams of silver light shot upwards from the deeper valleys amid the walnut trees that fringed their banks. Numerous hamlets were seen, some perched upon projecting ridges, some lying in hollows and girdled with a belt of emerald-green crops.

It was impossible not to speculate on the condition of these primitive mountaineers, who have since the dawn of history preserved their independence. Leo Africanus, speaking of the very district now overlooked by us, which he calls Guzula, says that the people were in his day molested by the predatory Arabs and by ' the lord of Marocco ;' but they successfully resisted all encroachments, and no attempt is now made to assert the Sultan's authority among them, or to enforce tribute. Something they have doubtless gained in material, and still more in moral, welfare by stubborn resistance to alien rule ; but the prosperity that is sometimes attained by tribes subject to the

---

[1] This is apparently the pass spoken of by Leo Africanus as leading from near Imizmizi (Amsmiz ?) to the region of Guzula (the northern branch of the Sous valley). He says it is called Burris, that word meaning downy, because snow frequently falls there.—See *Ramusio*, vol. i. p. 17, B.

semi-feudal rule of chiefs, and among whom intestine feuds
are rooted in immemorial tradition, is usually short-lived.

Our hope of getting further knowledge as to the
eastern extremity of the Sous valley, and the orographic
relations between the Atlas and Anti-Atlas ranges was
not to be satisfied. Djebel Tezah, as we found, stands
some way north of the axis of the chain, while the great
mass that rose over against us between ENE. and ESE.,
extending to the head of the Aït Mesan valley, sends out
massive buttresses to the south, and by these our view
of Anti-Atlas was cut off to the SE. On one of these
western projecting buttresses, we could distinguish a large
village belonging to the district of Tifinout, and standing
at an elevation of nearly 7,000 feet. Turning our eyes to
the north of true east, many of the higher summits of the

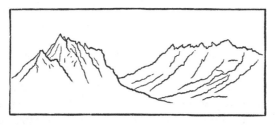

chain were seen rising above the intervening ranges, the
most distant probably belonging to the Glaoui group, east
of Tasseremout. Once more we came to the conclusion,
that throughout the portion of the Great Atlas chain visible from the city of Marocco, between the easternmost
feeders of the Oued Tensift and those of the Oued Nfys,
there are no prominent peaks notably surpassing the
average level. Many of them must surpass the limit of
13,000 feet above the sea, but it is not likely that any one
attains the level of 13,500 feet. The last object that attracted our attention in the panorama, in a direction
about east by north, was an isolated mass, forming a bold
promontory on the northern side of the chain, of which a
rough outline is here given.

When the engrossing interest of the distant view had so far subsided as to let us pay attention to nearer objects, we were struck by the unexpected appearance of considerable remains of dwellings on a platform of level ground, only a few feet below the actual summit of the mountain. About a dozen rude stone dwellings, all in a ruinous condition, with chambers sunk a couple of feet below the level of the ground, and the roofs fallen in, had at some former period been here erected; but we saw no traces of recent occupation. It seemed most probable that they were intended as shelter for herdsmen, who had driven their flocks in summer to this lofty station.

As we lingered on the topmost point of the mountain, the intense silence of the scene was broken by the distant scream of a large grey eagle that soared over our heads, and then sailed away southward over the Sous valley, making the deep stillness still more sensible than before.

The interval allowed for musing was not long; there was still much to be done, and we started to our feet to make ready for the descent. The observations for altitude had still to be recorded, and the usual difficulty of ascertaining the temperature of the air was here experienced. With a hot sun falling on an exposed rocky ridge, it is impossible to isolate the instrument from the effects of radiation. The result is usually to register too high a temperature; but the effect of placing the thermometer in a cleft where the rock is much below the temperature of the air leads to error in the opposite direction. The temperature here adopted was 60° Fahr., and the result of a comparison with Mogador gives a height a few feet above 11,000 feet, while the comparison with Iminteli falls a little below that level, the mean adopted being exactly 10,992 feet (3,350·1 m.)

After bottling a few beetles that were brought to light by turning over some flat stones, we gave a last glance at the ridge of Anti-Atlas, and at a quarter-past three

turned to the descent. It was clearly desirable to take
a different line from the straight course followed in the
ascent, and we speedily agreed on the plan of action most
likely to add to the botanical results of the ascent, which
hitherto had fallen somewhat short of our expectations.
The round-backed ridge sloping westward from the summit
throws out a massive spur, projecting nearly at right
angles or somewhat E. of N., so as to enclose a recess in
the mountain into which a large part of the drainage of
the northern slope is collected; there was reason, therefore,
to count on finding there a more varied vegetation than
on the bare slopes enclosing it. On the projecting spur
above it, we were struck by the appearance of trees,
evidently not coniferous, scattered at intervals along the
slopes, while the greener tint of the surface gave some
promise to the botanist. It was, therefore, desirable that
this ridge should not escape examination. A rapid descent
soon brought us to a point overlooking the hollow recess
of the mountain where we were rejoined by our Shelluh
guide, who had now assumed a crest-fallen air, and we at
once determined to separate, Hooker with the guide descending into the hollow, Ball making a circuit by the
ridge to the left. The time at our disposal being so short,
it was impossible to examine the ground carefully, and
many species were doubtless overlooked, but we were both
rewarded by finding several plants not seen elsewhere
during our journey. Among others Hooker secured a
dwarf, very spiny barberry, with blueish-black berries,
seemingly not different from the Spanish variety of *Ber-
beris cretica*; and lower down, near the base of the mountain, a fine white-flowered columbine, fully four feet high,
probably a variety of the common *Aquilegia vulgaris*,
widely spread throughout the mountain regions of Europe
and Asia, but not, as we believe, before found in the African continent. Ball, who reached the rendezvous half an
hour after Hooker, brought down with him a curious little
succulent plant, forming a new species of the genus *Mon-*

anthes, hitherto known only in the islands of the Canary
and Cape de Verde groups, along with three species of
the lily tribe, all of them found in Southern Europe,
but not before seen in Marocco. The tree was found to
be the belloot oak (*Qu:rcus Ballota* of Desfontaines),
a variety of the evergreen oak, which is spread through
North Africa and Spain, where the sweet acorns are commonly roasted and eaten, as chestnuts are elsewhere.
Many of the trees are of great age and have thick trunks,
and weather-beaten stunted branches, and are apparently
the remains of extensive forests that once clothed the
flanks of this part of the Atlas up to a height of about
8,500 feet above the sea.

Hooker found the sheik in a state of thorough exas-
peration at our success in defeating his orders, probably
aggravated by the tedium of waiting for our return. He
discharged volleys of fierce abuse at the guide who had
failed to keep us within the prescribed limits, but was not
openly disrespectful in his manner towards the Christian
hakim who had come to his country under the immediate
shadow of imperial protection. Foreseeing future trouble
in returning through the valley after dark, he was
evidently much annoyed at the necessity for awaiting
Ball's arrival. To calm his impatience, Hooker lent him
a field telescope, and the novel experience so much amused
him that his ill-humour appeared to vanish for the time.
Uncivilised men are like children, rarely remaining long
under the same impression; and even when seemingly quite
possessed by some strong feeling, are led away from it by
the veriest trifle.

As required by the inevitable rule of hospitality, a
mona was offered by the people of the adjoining hamlet
in the shape of a dish of *keskossou*, barely tasted by us,
but speedily despatched by our followers, and at 6.15 P.M.
we started on horseback to return to our night-quarters
at Iminteli. The sun set before 7, and a brief interval
of twilight soon gave place to a dark, though star-lit,

night. In the open there was no difficulty in following
the track along the torrent; but at one place, in riding
through a walnut grove, we were reminded of our night-
adventure in the Aït Mesan valley. The thicker branches
of the walnut do not, however, lie low, as do those of
the olive, nor are they beset with the stiff jagged leafless branchlets that made the latter so dangerous in the
dark.

The grove was traversed without trouble; but another
unlooked-for experience was in store for us. We had
complained in the morning of what seemed a round-
about way taken in descending from Iminteli to the
bottom of the valley; and perhaps the sheik now took
a malicious pleasure in showing us the advantages of a
short cut. Leading the way, he rode across the torrent,
which barely reached the horse's knees, and began to
ascend the slope above the left bank. Before long he
struck into the bed of a brawling streamlet that came
tumbling over loose boulders down the declivity. As we
advanced, the way became steeper, and shut in on either
side by tall bushes and straggling climbers, all, as it
seemed, beset with hooks and spines. There was nothing
for it but to rely on our riding animals to carry us through
as best they could, and wonderfully they demeaned themselves. Though patches of sky showed overhead, to our
eyes the ground below was absolutely invisible; the
boulders were evidently very large and slippery, and it
was only by the most desperate struggles that the poor
beasts succeeded in clambering up the slope, pausing
frequently, with muscles quivering all over from the
violence of their renewed struggles. The only thing for
the riders to do was to hold on at all hazards, and keep
their heads bent low, so as to save their faces from
the spiny branches, that made havoc of their nether
garments.

The time seemed very long before we finally emerged
on the shelf of more level ground which lies along this

side of the valley, and soon after reached our quarters at Iminteli, at about half-past eight. As we knew that we should have time on the following morning, and the day's work had been rather fatiguing, we yielded to the claims of nature, let our collections rest in their boxes and portfolios for the night, and soon after supper lay down to sleep.

CHAPTER XI.

Return to Amsmiz—Arround villagers in trouble—Pains and pleasures
   of a botanist—Ride across the plain—Mzouda—Experiences of a
   Governor in Marocco—Hospitable chief of Keira—A village in
   excitement—Arrival at Seksaoua—Fresh difficulties as to our route
   —A faithful black soldier—Rock vegetation at Seksaoua—Ascent
   of a neighbouring mountain—View of the Great Atlas—Absence of
   perpetual snow—Return of our envoy from Mtouga—Pass leading
   to Tarudant—Native names for the mountains—Milhaïn—Botanising in the rocks.

THE morning hours of May 23 were devoted to the
necessary work at our collections of the preceding day ;
but before our departure we once more took a short
ramble through the ground surrounding the village.
With a single exception, all the plants seen were species
common to the Spanish peninsula, two or three being
characteristic of Central Spain. Apart from the style of
building and the dress of the inhabitants, a stranger
transported to the spot might easily suppose himself
somewhere in Southern Europe, though closer examination
would suggest differences to the naturalist. At noon we
started on the way to Amsmiz, halting at the torrent in
the bottom of the valley to secure specimens of two very
fine and undescribed plants, both very troublesome to
the collectors. One was a spiny *Genista*, with very
numerous, stiff, intricate branches ; the other a fine
thistle, five or six feet high, whose long woolly leaves
were beset with sharp, slender, golden spines, fully an
inch in length.

Without much further delay, we retraced the track
that we had followed on the morning of the 21st. We
were once more struck by the remarkable coolness of

the climate of this region as compared with somewhat similar positions in higher latitudes on the north and east sides of the Mediterranean. Although the sun at noon now approached within less than 15 degrees of the zenith, the temperature in the shade was pleasantly cool, scarcely rising above 70° F. At the same season, and at a greater height, on the Lebanon we have seen the thermometer stand above 80° in the shade by day, and scarcely fall to 70° at night. This is mainly due to the cool winds that prevail along the coast, and extend some way inland, though not much felt in summer in the city of Marocco. As we rode along the eastern flank of the valley, and down the slopes above Amsmiz, we were pleasantly fanned by a NW. breeze that often lasts throughout the day, but subsides at night.

On the brow of the declivity overlooking Amsmiz, we met a messenger from Arround, our stopping-place at the head of the Aït Mesan valley, come to implore our protection for the unfortunate inhabitants, whose appeal to us had only brought them into fresh trouble. The story had of course been reported to the Vice-Governor (El Graoui's deputy) with the circumstance of the sacrifice of the sheep. He had resented this attempt to escape from his authority, had had some of the suppliants severely beaten, and sent two more men of the village to prison. It seemed very doubtful whether any interference on our part might not merely aggravate the condition of these unfortunate people. We promised, however, to do what we could for them; and before we left Mogador it was reported, whether truly or falsely, that we had been successful in our intercession.

On returning to our camp at Amsmiz we found work in abundance ready to hand. Our precious collections from the Aït Mesan valley, including, as they did, the most interesting results of our expedition, had been lying for three days untouched; and it was necessary to go through them all again, putting into separate parcels those that were dry and those still requiring pressure between dry paper. With the exception of half an hour given to another interview with the friendly Kaïd of Amsmiz, we were thus occupied until long after midnight. Although our store of drying-paper was large, the demand often exceeded the supply, and many a friendly contest arose as to respective rights of property in parcels of soiled paper, here priceless, which elsewhere would have seemed of no value. Those who have had experience in this line know that the labour of a botanical collector is not light, and in truth it would be almost intolerable if it were not for its compensating pleasures. It often happened that the solitary candle was in use throughout the entire night, Ball working till two o'clock or later, when Hooker would rise, more or less refreshed, and keep up work till daylight.

But in the pursuits of a naturalist there are abundant sources of satisfaction not suspected by the uninitiated. These are not merely derived from the objects themselves, suggesting as they often do interesting trains of thought and speculation ; there are further springs of keen enjoyment in the countless impressions with which they are linked by the subtle influence of association. Much of the pleasure that an artist, however unskilled, derives from travel, arises from the power of each sketch to bring back again to the mind the original scene of which it is but the imperfect transcript. If he be active and industrious, he may preserve a dozen such keys to the impressions of each day's journey. But to the botanist almost every specimen is indissolubly linked in the memory with the spot where it was collected ; and as he goes through the produce of his day's work, every minute detail is vividly presented to the mind, along with the wider background that lay behind the original picture. The wonder and awe that dwell around the mountain fastnesses, the consolation of the forest glade, the indefinable grandeur of the desert plain, nay, even the bleak

T

solitariness of northern moorland and morass—these
dominant impressions suggested by the aspects of nature
are varied and enriched for the naturalist by the myriad
phases of beauty that are disclosed to the eye of the
observer. The glory of colour in the gentian and
saxifrage and golden *Alyssum*, and the other bright
creatures that haunt the mountain tops; the tender
grace of the delicate ferns that dwell in the rocky clefts;
the teeming life of the warm woodland; the strange
beauty of the unaccustomed forms that spring up in the
desert solitudes; the purple glow of the heath relieving
the sombreness of the leaden sky, and the delicate
structures of the *Drosera* and *Menyanthes*, and bog-
asphodel, and many another inhabitant of our northern
bogs—these and countless other images are instantaneously
revived by contact with the specimen that grew beside
them. Strangest of all is, perhaps, the enduring nature
of this connection. Often does it happen, as many a
botanist can testify, that after a lapse of a quarter, nay,
even half, a century, the sight of a specimen will bring
back the picture, seemingly effaced long ago, of its
original home.

We were on foot again at 5.30 A.M., May 24, and the
order for departure went forth. But, as usual, there were
unexpected causes for delay. Many sick came to invoke
Hooker's medical skill, some trifling presents were to
be distributed, and finally word was brought that the
Kaïd meant to accompany us for some distance on our
day's journey, and it was necessary to await his appearance.
Among the articles provided for presents we had included
scissors and needles; but such things, especially the
needles, were everywhere disregarded by the natives,
whether Moor or Shelluh; and it appeared that the art of
sewing, as well as every other occupation requiring the
slightest manual dexterity, is—at least in country places
— exclusively practised by the Jews; to them, accordingly,
such gifts were very welcome. In the larger towns there

are, of course, many handicrafts, and notably the making of slippers and boots, practised by the Moors ; but such trades are for the most part hereditary in certain families, and the ordinary Moor affects to despise all occupations of the kind.

At half-past eight the Kaïd appeared, mounted on a strong serviceable horse; and, everything being ready, we rode down the steep bank above which stands the town of Amsmiz, and, after following the torrent for a short distance, reascended to about the same level above the left bank. We now found ourselves on the verge of a wide open plain, sloping gently from south to north, and our course to Mzouda—the next stage in our journey—lay a little north of due west, while the outer range of the mountains trended away to S.W. We had been led to suppose that Mzouda lay, like Amsmiz, at the foot of the Great Atlas, and might therefore serve as starting-point for another excursion into its recesses; but it was now clear that it must stand far out in the plain, many miles from the nearest range of hills. We were somewhat comforted, however, by the positive assurance that Seksaoua, the next stopping-place beyond Mzouda, stands close to the mountains at the opening of a considerable valley, and was therefore a promising spot for our purposes.

The difficulty of getting correct information in such a country as this, as daily experience proved to us, is one of the most serious difficulties of the traveller, and depends quite as much on the incapacity of the natives as on the habitual suspicion with which all strangers are regarded. One day when Kaïd el Hasbi appeared to be in unusual good-humour we were endeavouring to obtain from him information as to some place on our route, and the interpreter being told to ask if he could make the matter clearer by reference to the cardinal points, he answered in a tone of contempt, ' Does he take me for the captain of a steamer ? '

After riding with us for two or three miles the Kaïd of Amsmiz bade us a friendly farewell ; and we continued in our course across the plain, with occasional halts, in order more closely to examine the vegetation, which was here less varied and interesting than usual. Most of the surface was under grain crops— chiefly wheat and barley —now ripe, and in great part cut and carried away. In the drier waste tracts we once more came upon the characteristic vegetation of the plains, *Acacia*, *Zizyphus Lotus*, *Rhus pentaphylla*, and *Withania frutescens* being the prevailing shrubs. Of herbaceous plants *Elæoselinum meoides*, and other large *Umbelliferæ*, with *Compositæ* of the thistle tribe, were most conspicuous.

In Beaudouin's map the chief branch of the Oued Nfys is shown as flowing parallel to the Atlas range from the south-west of Amsmiz, and receiving as tributaries the Amsmiz torrent and the broad stream that we had crossed near Gurgouri. On this day we satisfied ourselves that this representation is erroneous. The unanimous statements of the natives, confirmed by our own observations, proved that all the waters flowing northward from the mountains between Amsmiz and the borders of Mtouga are united in the stream that we had traversed at Sheshaoua on its way to the main river Tensift. The practice of intercepting the streams from the Atlas, and carrying them across the plain through irrigation channels below the surface, makes it extremely difficult to unravel the hydrography of this region.

About noon we reached the boundary that separates the districts of Amsmiz and Mzouda, and agreed to the suggestion of a mid-day halt under trees near a large village, of which, as we learned, many of the inhabitants are Jews. From early morning the clouds had been gathering along the mountain range, and by this time had quite covered the sky. The temperature was unusually low, not rising above 62° F. in the shade, and our hard work during most of the preceding night supplies the only excuse for the fact that, after a light luncheon, we both fell fast asleep, until aroused by the information that it was two o'clock, and high time to continue our journey. The flora being somewhat monotonous, we did not, perhaps, lose much by this unusual neglect of duty ; but we remembered with regret that we had not ascertained to what species the tamarisk tree belonged under which we had taken our rest.

The boundary between Amsmiz and Mzouda is here formed by a torrent bed, now nearly dry, called by our escort Asif el Mel. This stream, as was agreed on all hands, joins those farther west that run by Sheshaoua to the Oued Tensift. As we rode onward across the plain several heavy showers passed over, which thoroughly drenched the scantily clothed men of our party, without at all quenching their habitual good-humour, but the soldiers were well provided with woollen coverings that kept them tolerably dry. There was little attempt at collecting plants during the afternoon, as it requires a strong inducement to make a horseman whose outer clothing is thoroughly wet set foot to the ground. We found the village of Mzouda rather different in appearance from those we had hitherto seen. The houses—small cubical blocks built of clay dried in the sun—were less solid than the rough stone dwellings of the Atlas mountaineers, but much superior to the miserable huts of the Arab tribes in the plain of Marocco ; and instead of the unsightly piles of thorny branches commonly used by the latter, these were enclosed within massive hedges of *Opuntia* whose dimensions showed that they were of considerable age.

As usual Kaïd el Hasbi had ridden forward with one of the soldiers to present the letter to the Governor, and to announce our arrival; and when, about 6 P.M., we reached the *kasbah*, quite a mile from the village, we received a message inviting us to take up our quarters within the building. As the ground outside was already wet, and the evening sky threatened more rain, we at once accepted the offer, and were conducted to two small but clean-looking rooms in a square tower that formed one of the angles of the building. Between the care of the

small collections made during the day, and writing up our
notes, and a frugal supper, the time was fully occupied
until 10 P.M., when by previous arrangement we paid our
visit to the Governor.

We found a spare-looking man of serious mien, quite
devoid of the coarse, overfed, sensual aspect common among
the men in authority in Marocco. The usual conversation as
to the objects of our journey, led to an assurance that the
district under his jurisdiction did not extend to the higher
peaks of the Atlas, or, as it was expressed, 'did not go to
the snow.' This may not improbably have been quite
true, but our experience of El Hasbi's machinations made
us now very incredulous as to such statements. It was, however, obvious that Mzouda was not a convenient centre
for mountain excursions, and we made no objection to
the proposal that we should on the following day proceed to
Seksaoua, which stands close to the foot of the mountains.

When we came to know more of his history, we found
no cause to wonder at the grave and depressed demeanour of our host. He had succeeded to the government of his native district in early life, and had held it
for many years when he was invited by the Sultan to Fez.
On his arrival he was thrown into a dungeon where he had
remained ten years, frequently subjected to torture, until
so much of the wealth he was supposed to have amassed
during his administration had been disgorged as satisfied
the demands of the sovereign or some ruling favourite ; and
then, being released, he was sent back again to govern his
district with the agreeable prospect of renewing the same
experience after some uncertain interval. If actual fact
in this country did not supply frequent proof, it would
seem scarcely credible that the attractions of power and
comparative wealth should induce men to face such a terrible, yet almost inevitable, future.

The sky had cleared during the night, but the morning
of May 25 was unpromising. At 8 A.M., shortly before
we started, the thermometer marked only 65° F., although

our observations showed that in our yesterday's ride of rather more than 20 miles we had descended fully 1,000 feet, the height of the *kasbah* above the sea being calculated at 2,367 feet. The night had not been altogether pleasant, for, in spite of insect powder, the bugs had made a vigorous and successful attack, and we should have preferred to start at an earlier hour. But as usual Kaïd el Hasbi stood in the way. He was quite determined not to let the unbelieving strangers put him to the slightest inconvenience that could be avoided.

Before we started the Governor sent Hooker a present of 20 dollars, which was of course immediately returned. The poor man doubtless thought it well to lose no chance of propitiating any influence that could possibly be of avail in the hour of future need. With the same object, he took the opportunity of sending through Abraham a dog to Mr. Carstensen at Mogador, and doubtless made presents to the officers of our escort.

After leaving the *kasbah* we rode through a narrow belt of tilled land, and soon reached the verge of a tract of open country remaining in a state of nature, with but few and scattered traces of population or cultivation. In some parts the soil was stony, and the presence of *Arthrocnemum* and other Salsolaceous bushes indicated the presence of soluble salts, but in others the absence of cultivation was probably due only to the want of irrigation. There can be little doubt that by a more skilful distribution of the drainage from the northern slopes of the Great Atlas, the area of land producing human food might be largely increased.

Our course lay between WSW. and SW., and we observed as we advanced that in that direction the outer ranges of hills did not rise so nearly parallel to the axis of the main chain as they do in the districts lying between Tasseremout and Amsmiz. A very considerable mass, extending northward as a promontory from the main range, became gradually more conspicuous as we advanced

towards it, while a minor mass lying much nearer to us
was seen on our left.   About noon we approached the latter range in which the stratification appeared very irregular
with a prevailing southward dip, and the strike NE. to
SW.   At its western extremity this range showed a line
of steep cliffs, reminding us of those near Tasseremout,
with the difference that the strata were here crumpled or
contorted in a remarkably uniform manner, the same cur-
vature of the folds being repeated nine or ten times.   The
compressing force must here have operated nearly in the
direction of the axis of the main chain, and in a distance
of some two miles the beds whose exposed edges we
viewed must have originally covered a space of nearly
twice that length.

As often happens when the air is nearly saturated with
moisture, the horizon was to-day remarkably clear, and we
made out the position of the city of Marocco, more than
40 miles distant, and bearing nearly due NE.   About
due north, and not quite so distant, rose the hills near
Sheshaoua, and about midway between them a remarkable
conical hill seen from near Misra ben Kara.

Before 2 P.M. we approached a large *kasbah* at a place
called Douerani.   When we afterwards learned that this
belonged to the same chief who hospitably received M.
Balansa, and assisted him in exploring the neighbourhood
until orders from Marocco cut his stay short, we had some
doubt whether this was not the place described by him as
Keira.   An examination of his map and the account of
his expedition leads us, however, to the conclusion that
Keira must be the name of another habitation belonging
to the same chief, lying a few miles farther north, and
that the mountain called Djebel Aït Ougurt, ascended by
M. Balansa, must be some eminence in the range near at
hand which we had just before been scrutinising.   We now
perceived that there is a considerable valley or depression
lying between this outer range and the main mass of the
Atlas, which is, indeed, indicated in M. Balansa's sketch map.

Before long we received a courteous message inviting us to stop at the *kasbah*; but as it seemed clear that Seksaoua promised more easy access to the higher mountains, we had no hesitation in adhering to the plan already fixed, and declining the proffered hospitality.   It was not without regret that we adhered to our resolution, when the chief came out with a numerous suite to visit us at our halting-place close to the *kasbah*.   The friendly air of the worthy old man, which evidently made a deep impression on M. Balansa, was not without effect upon us.   Failing to induce us to stop on our way, he sent an ample *mona*, including, besides tea and sugar, a parcel of candles of French manufacture, the more acceptable as our supply threatened to run short before we could reach Mogador.

Our halting-place was in a pleasant spot overlooking the broad bed of the Oued Usbi, which appears to unite the torrent from a considerable valley south of Seksaoua with several minor streams from the Atlas, and to be the main affluent of the river of Sheshaoua.   The weather had improved, and the thermometer stood at about 70° F. in the shade, our height above the sea being 2,671 feet (814·3 m.).   Spiny *Compositæ* belonging to the genera *Scolymus*, *Echinops*, *Cnicus*, and *Onopordum*, were the most conspicuous plants; but, as no species not already gathered were seen here, we dispensed ourselves from collecting and drying these troublesome inmates of the herbarium.

It was near 4 P.M. when we started for Seksaoua, and, after crossing the Oued Usbi, held on in a SW. direction nearly parallel to its course.   In little more than an hour we came to a large village, which was the scene of unexpected commotion. As our cavalcade was seen to approach, some natives ran on to announce the fact to the villagers, and by the time we reached the first houses the whole population turned out, and a scene ensued of which no description can give an idea.   The men who lined the way on either side shouted with emulous vehemence and fury guttural sentences, illustrated by frantic gesticulations, while the women and children kept up a deafening accompaniment of shrieking, wailing, and howling, and the whole formed a scene worthy of Pandemonium. It seemed sufficiently clear that no hostile intentions against us were expressed, but amidst the horrible din and confusion it was some minutes before we were able to learn from Abraham the meaning of this wild excitement. It appeared that, as constantly happens among the mountain people, there was a feud between this and a neighbouring tribe ; the village had been attacked, or at least approached by the enemy, and one of the villagers had been shot.

It was evident from the first that our brave escort felt extremely uneasy; but when it became clear that the object of the people was to invoke the protection of the soldiers of the Sultan against further molestation, our two Kaïds for once thoroughly agreed on a policy of strict neutrality, and in desiring to get as soon as possible out of harm's way. As for us, it may be feared that we failed to maintain the gravity which, to the Oriental mind, befits persons of distinction. Just when the confusion was at its worst, and before we well understood what it portended, we happened to look up to where on the top of the nearest house two or three storks, each poised on one leg, were looking down on the frantic crowd. There was something irresistibly ludicrous in the contrast between the air of solemnity that characterises these birds and the insane excitement of the human crowd below that set us off in a peal of laughter, which we found it hard to tune down to decent seriousness.

The uppermost anxiety of our escort being to get away from any chance of being mixed up in the local troubles, they proposed to push on as far as possible towards the mouth of the valley, and we were all the better pleased to find ourselves as near as possible to the mountains in which we still hoped to effect another excursion. It was not, however, practicable to go far. About two miles above the village a rocky spur projects from the mass of the Atlas towards the plain, and is backed by a mountain mass rising some 2,500 or 3,000 feet above the valley. At the eastern base of this rocky promontory, in a stony field planted with young olive trees, we pitched our tents on very rough ground, where it was not easy to find a level spot to sleep upon, but where we promised ourselves good botanising in the immediate neighbourhood, even if unable to penetrate far into the mountains.

Some unusual precautions were taken this evening to guard against a night attack upon our camp, and the Kaïds assumed an air of importance befitting men who felt that the time had at length arrived for a display of their professional skill and prowess; but, as we fully expected, the night passed without the slightest molestation. A few musket shots discharged at a distance were heard, exchanged between the hostile parties, or more probably fired *in terrorem* to show that the defenders were ready for action. As we heard no more on the subject, it is probable that no further disturbance ensued during our stay in the neighbourhood.

On the morning of May 26 our first anxiety was to ascertain what might be our prospect of reaching from this point the head of the valley, and making another ascent of the main range. We had already heard rumours of disturbances among the native tribes in the upper part of the valley, so that our expectations of success did not run high; and when the sheik of the valley was forthcoming we were not much surprised to hear him declare that an excursion in that direction was utterly impracticable. We at once suspected Kaïd el Hasbi of practising his usual machinations to defeat our intentions; but with the difference that on this occasion there was probably some foundation in fact for the tales that were told us of conflicts between the neighbouring tribes, and of possible danger for travellers. With an escort furnished by the orders of the Sultan, and quite numerous enough to inspire respect among the rude mountaineers, there would have been no real risk in proceeding along the valley—or anywhere else in this part of the country—provided we could have reckoned on our men ; but in the face of their refusal, there was no use in further pressing the point.

The next thing to be done was to make an arrangement for enabling us to see something of the outer range of mountains immediately surrounding our camp, and after some debating it was agreed that on the following day we should ascend to the higher ridge of the considerable mass already referred to as rising to the west of our camp. Much nearer at hand, extending from behind our tents towards the opening of the main valley, a steep rocky ridge, only from 400 to 500 feet in height, promised to show us what we had hitherto seen little of, the rock vegetation of the lower region of the Atlas, and we readily made up our minds to devote the remainder of this day to its careful examination.

There was, however, another matter of a practical nature requiring immediate attention. A glance at the map shows that in travelling along the skirts of the Atlas from Seksaoua to Mogador our route must lie through the district of Imintanout, and thence through the adjoining provinces of Mtouga and Haha. We had informed Mr. Carstensen of our intention to follow this line of route, and fixed the probable date of our return to Mogador at the 2nd or 3rd of June. During the last two days we had heard vague reports of disturbances going on in the provinces of Mtouga and Haha, and these were now confirmed and aggravated by the sheik of Seksaoua. War was actually raging, we were told, and the Governors had summoned all their people to arms. As was to be expected, the men of our escort, who clearly had no stomach for fighting of any kind, were becoming very uneasy at the idea of coming near to the seat of operations, and we apprehended that they might make an attempt to force us to diverge from our intended route and travel northward across the plain so as to rejoin the beaten road from Marocco to Mogador. Having ascertained that the distance from Seksaoua to the *kasbah* of the Governor of Mtouga is no more than an easy day's ride for men travelling without luggage, Kaïd el Hadj of Mogador with two of his men was despatched on a mission to Mtouga. He was to ascertain the truth as to the stories that had reached us, and to require the Governor, in case he considered extra protection necessary, to send additional soldiers to escort us through his territory, thus, as we hoped, committing us to keep to our intended route as far as Mtouga.

About this time we became a good deal interested in one of the soldiers of our escort who had travelled with us throughout our journey. He was a large man, with black skin, but with hair and lips of less pronounced Negro type than we see among the natives of western equatorial Africa. When leaving Mogador he had an ulcerous sore on one hand, which was much swollen and almost useless. The sore, under Hooker's treatment, was quite healed, and he was genuinely grateful for the benefit. Alone among the soldiers of our escort he did what he could to forward our desire to explore the mountain valleys; and of late, on more than one occasion, he had given useful information that helped us to defeat the petty intrigues of Kaïd el Hasbi. By our direction Abraham made some inquiry as to his previous history, and he quite readily told his story. He belonged, as it appeared, to one of the tribes that inhabit the skirts of the Great Desert on the south side of the Great Atlas. They led a predatory life, gaining an uncertain living by robbing travellers, and killing those who made resistance. After some years passed in this way, our friend seemed to have taken a dislike to the mode of life, and enlisted as a soldier in the service of the Sultan of Marocco. In his new position he had gained or developed some elementary notions of religion and morality, and he now expressed a strong opinion as to the impropriety of robbery and murder.

Here was a case such as is often cited by superficial travellers to show the absence of a moral sense among savage people. This man had no doubt robbed and murdered in his youth without the slightest compunction; but, given the conditions under which the ethical sense could be developed, the result was to produce an individual morally superior to the majority of those around him. The analogy, so well drawn by Reid, between the moral nature of man and the development of the plant from the seed holds good. External conditions are necessary; but they do not create the germ, without which no evolution can follow. The conditions vary from one individual to another. One requires to be fostered by many favourable influences; another, with stronger vitality, will bud forth under the least auspicious conditions. The assertion that there are human beings in whom it is impossible to awaken any sense of difference between right and wrong must be, at least, premature, until the world shall have reached a social condition in which each individual may be tried under appropriate conditions.

Our day's botanising on the rocks near Seksaoua was successful beyond our expectations. Many conspicuous plants peculiar to Marocco were here seen for the first time. Several of these had been gathered by M. Balansa during the four days which he passed in the adjoining district of Keira, but were known to us only by name. That active and successful botanical traveller was able to collect so few specimens that in several cases no duplicates were available for distribution, and the specimens exist only in the rich herbarium of M. Cosson. Among other novelties we here saw for the first time *Trachelium angustifolium* of Schousboe, utterly unlike any other species of that ornamental genus; *Teucrium rupestre* and *T. bullatum*, both described by M. Cosson from Balansa's specimens; and a single specimen of *Elæoselinum exinvolucratum* of Cosson, a fine umbelliferous plant, apparently very rare even in its native

district. A very fine *Brassica*, standing five or six feet
high, with a straight upright stem, set with candelabrum-
like branches, was the most remarkable new plant found
by us which had not already come in the way of M.
Balansa.

The morning had been cool ; the thermometer at 8 A.M.
did not rise above 64° F., and the sky was overclouded ;
but as the day went on the sun blazed out with great
power, and this was one of the hottest days we experienced.
The heat was, of course, especially felt on bare rocks which
became so hot that the hand could not bear them ; and
the soldier who had gone out by way of protecting us
judiciously retired to the shade of a fig-tree at the foot of
the hill. After some time, we separated and returned to
the camp by different routes. A portion of the slope not
far above our camp was altogether covered with broken
blocks of moderate size obviously derived from the steeper
crags above. This ground abounded in reptiles of various
kinds, which were, however, so shy that it was not easy to
get a favourable view of them. By sitting perfectly still
for some minutes, Ball was partly successful in getting
them to approach him. The most remarkable creature
much resembled a miniature Iguanodon in form, being
about eighteen inches long, with a row of thick conical
processes projecting upwards along the back, and gradually diminishing towards the tail from about two inches
in height between the shoulders. Numerous ·lizards were
also seen ; but no snakes, except a small black viperine
species, seen gliding between the stones, actually under
one foot, which fortunately did not touch or injure the
animal.

By this time we were beginning to feel the effects of
the unsatisfactory dietary to which we had been reduced
during the four weeks since we left Mogador. It may
seem unreasonable for men in health, plentifully supplied
with fowls, sheep, and eggs, to complain of their food ; but
those who have experienced the difference between the

meat of well-fed animals and the stringy tasteless fibre which is produced in such a country as this, will duly appreciate our longing for some variety. As the season advanced, and the herbage in the lower country became more and more parched, the sheep, always miserably thin, approached nearer and nearer to the condition of skeletons, covered with skin and ragged wool, and for some time back we had given up the attempt to eat any part, except the liver and kidneys broiled on short sticks; while the fowls had become equally distasteful. The *keskossou,* daily presented with the *mona,* was prepared with large quantities of rancid butter, to which, in spite of many experiments, we never could reconcile ourselves. Our attempts at obtaining any variety of diet were quite unsuccessful. Ducks and geese, being by Mohammedans considered unclean, were out of the question; and the turkey and guinea-fowl appear to be unknown to the domestic economy of the Moors. Our chief desideratum was fresh vegetables or fruit, but these were not to be obtained. Except in the neighbourhood of the coast towns, where they have been introduced by Europeans, none of our European vegetables are cultivated in South Marocco, except the cucumber and the pumpkin, and, owing to the want of the most elementary skill in horti-culture, these seem to remain in season for a very short time; while the cultivation of fruit, at least in the dis-tricts we traversed, seems to be generally neglected. In this respect Marocco presents a striking contrast to most places with a somewhat similar climate in the Mediter-ranean region. Egypt, Palestine, and Syria, however low they may have fallen owing to corrupt and oppressive government, have retained some share in the inheritance of an ancient civilisation. We had carried with us sundry tins of preserved vegetables, of which green peas were by far the most acceptable; but our stores were now nearly exhausted, and our chief remaining luxury was portable soup, made with compressed vegetables and biscuits, which was now served out very sparingly. Tea without milk was often pleasant in the evening; but cocoa, prepared with milk in small tins, was much preferred for the morning meal.

The evening air was cool and pleasant, and, in spite of the advancing season, the night almost cold, though the height of our station, by the mean of two observations, did not exceed 2,867 feet (874 m.). Even the horrible howling of the dogs in a neighbouring village failed to keep us from a good night's rest.

The morning of May 27 broke brilliantly, and, though the sun's rays were already hot, the thermometer in the shade at 6 A.M. did not rise above 60° F. Another attempt was made to induce the sheik to take us for an excursion up the main valley; but he held fast to his declaration that the country in that direction was too dangerous, and repeated his offer of the previous day to lead us up the nearer mountain. Failing anything better, we resolved to accept this.

The declivity of the hill immediately west of our camp being much too steep for horses, we followed a circuitous track, at first NW. and then SSW., chiefly along steep slopes, on which, among other novelties, we first gathered *Erodium atlanticum*, discovered in this district by M. Balansa. After an ascent of some 1,500 or 1,600 feet the track turned again nearly due west, and we found ourselves on the southern slope of the mountain, which we now saw to be almost completely detached from the main range of the Atlas. The slightly convex ridge on which we stood inclined gently to the south, forming the watershed between the Seksaoua valley and that of Imintanout which adjoins it on the west. The slopes of the mountains enclosing both those valleys are better wooded than usual in the Atlas, some variety of evergreen oak being apparently the prevailing tree. Behind us, as we stood facing the great range, the mountain rose some 1,200 feet above our present level, and as the sun was hot we did not imme-

U

diately dismount, but continued to ride some part of the
way, only the final ascent being made on foot.

The view was in many respects very interesting, as it
showed us a great part of the main range from an entirely
new point of view, and the air on this day was unusually
clear. Looking westward, where the horizon, at a distance
of at least eighty miles, must have been rather near the
Atlantic coast, we were able to assure ourselves that the
hills that extend through most of the great province of
Haha are all of moderate height, none of them approaching
that on which we stood.   In this respect Beaudouin's map
is much more correct than that of Gerhard Rohlfs, which
seems to show that the main chain at its western end is
broken up into lofty, diverging branches, some of which
extend far through Haha.   No prominent object caught
the eye to the northward, except the familiar flat-topped
hills near Sheshaoua.   For the last time we were able to
distinguish the site of the city of Marocco, bearing about
NE. by E., and over sixty miles distant.   About due east
the high range at the head of the Aït Mesan valley showed
much more snow than when we viewed it five days before
from the summit of Djebel Tezah, while the latter mountain seemed pretty much in the same condition in which
we had found it.   About due south a rugged peak towards
the head of the Imintanout valley had snow in rifts and
depressions; and another of somewhat similar aspect,
rising farther east and above the head of the Seksaoua
valley, seemed to be the highest point in the whole range
west of the sources of the Oued Nfys.   From their
position, and the ruggedness of their aspect, either of
these peaks promised well for a naturalist who could suc-
ceed in gaining access to them, but we felt that such good
fortune was not now in store for us.

The fresh sheet of snow which had fallen on the Aït
Mesan range within the last few days led us to what
seemed an explanation of the inconsistent accounts given
as well by travellers as by natives as to the existence of

perpetual snow on this part of the Great Atlas. From its position between the Great Desert to the south, the Atlantic Ocean to the west, and the low country to the north, it is obvious that a range of mountains from 11,000 to 13,000 feet in height must frequently be the seat of violent atmospheric disturbances. Whenever these draw from the ocean currents of heated air, nearly saturated with moisture, into the upper region, the cooling effect consequent on rapid expansion must produce copious precipitation, and it is most probable that on the higher part of the range this, even in the hottest season, takes the form of snow. But, as we had seen, the snow melts with extreme rapidity under the almost vertical sun during the summer months; and hence one traveller may have seen the range thickly snowed even in the hottest season, while another, with equal truth, may describe it as almost completely bare. The state of things is such that a very moderate change in the physical conditions might easily lead to the accumulation of an annual surplus of unmelted snow, which is the first condition for the formation of glaciers. A mere increase in the amount of precipitation, with little change in the general conditions of temperature of this region, might produce glaciers reaching as low down as that whose moraine we saw at the head of the Aït Mesan valley.

Many early-flowering plants were already withered, but we collected on the mountain several interesting species. Of two tall and very distinct *Resedas* found here, one is also a native of Spain; the other, *R. elata*, of Cosson, was first gathered by M. Balansa, and seems to be confined to this district. Of another curious plant discovered by the same active naturalist we now first saw satisfactory specimens. It is at first sight scarcely to be distinguished from a species characteristic of the hot and dry region of North Africa—the *Cynara acaulis* of Linnæus. The latter was discovered by Tilli, a Florentine physician, afterwards professor of botany at Pisa, who was

called to Constantinople early in the last century to cure
the favourite daughter of the Sultan. Being successful
in his treatment, he received many tokens of favour, and
seems to have made use of his opportunities to visit several
parts of the Turkish Empire, and certainly travelled in
the Regency of Tunis. The same plant was next seen by
the English traveller, Thomas Shaw, who mentions it in
the Appendix to his Travels published in 1738; and it was
at last more fully described and well figured by Desfontaines in his excellent work, the *Flora Atlantica*. Decandolle, in attempting to reduce to order the vast mass
of plants that belong to the natural order of *Compositœ*,
clearly saw that this differed essentially from the genus
Cynara (of which the type is the common artichoke), and
referred it first to *Serratula*, and finally to *Rhaponticum*;
and it has hence been generally known as *Rhaponticum
acaule*. Many botanists were somewhat startled to find
in the *Genera Plantarum* of Bentham and Hooker that
the authors had united all the plants hitherto ranked
under the generic name *Rhaponticum* with *Centaurea*, a
vast genus, containing species of the most varied aspect,
of which nearly 300 are already known in the Mediterranean region. It was interesting to us to find that the
new species discovered by Balansa, of which the foliage is
quite undistinguishable from the old *Rhaponticum
acaule*, is, as regards the flowering heads, intermediate in
structure between that and recognised species of *Centaurea*,
though nearer to the latter. If we had remembered
Shaw's statement, that the roots of his *Cynara acaulis*
have an agreeable flavour, and are eaten by the Arabs in
some parts of Africa, we should certainly have tried
whether the species are also similar in this respect.

During the ascent of the mountain we had passed near
a little hamlet, containing eight or ten houses of the poorest class; but the laws of native hospitality required that
refreshments should be offered to the strangers, and on
the way back a halt was called. The *mona* consisted of

eggs, wheaten cakes, butter, and milk, which were speedily despatched; and we added to our collections a curious biennial variety of *Rumex vesicarius*, having the membranous wings enclosing the fruit of a bright rose red.

By 4 P.M. we had got back to our camp, and the remainder of the day was devoted to the care of our collections. Before nightfall Kaïd el Hadj returned from his mission to Mtouga, bringing confirmation of the reports as to the outbreak of hostilities between the people of Mtouga and their neighbours of Haha, with an addition to our escort in the shape of six ragged-looking soldiers sent by the Governor of Mtouga.

On the morning of May 28 our numbers were further increased by the return of the two soldiers who had left us at Sektana for the purpose of escorting Maw to Mogador. They were welcome, for they brought letters from England, together with a good account of our travelling companion. He had reached Mogador early on the fifth day from Sektana, and happened to arrive a few hours before the departure of a small British steamer bound from the Canary Islands to London.

Before departing, we gave a last look at the neighbourhood of our camp, and reluctantly abstained from attempting a close examination of the ancient castle, or fort, which stood at the opposite side of the stream commanding the entrance to the main valley. We were well aware that any curiosity shown in that direction would have been set down to designs on buried treasure, and would have aggravated the suspicion with which all our proceedings were viewed by the native authorities.

We did not start until 10 A.M., and with an unusually long cavalcade followed a faintly marked track that winds round the northern base of the mountain which we had ascended on the preceding day, gradually attaining to a height of several hundred feet above the plain. Before long we crossed the borders of Imintanout, a district including several villages under a sheik who is dependent on the Governor of Haha. Through the valley, which here opens out, lies the main road from Marocco to Tarudant, the chief town of Sous. Jackson, who seems to have gained the especial favour of the reigning Emperor, received, about the beginning of this century, permission to accompany a military force despatched from Marocco to Tarudant, and no other European is known to have traversed this part of the Atlas.[1] Unfortunately his account of the expedition is limited to the statement that the way lies through a narrow defile, where the path cut in the rock is only 15 inches wide, with the mountain rising almost perpendicularly on one side, and on the other a precipice 'as steep as Dover cliff, but more than ten times the height.' It would have been a matter of great interest to us to make a short excursion up the valley, and to penetrate this defile, but once more we were doomed to disappointment. The sheik, having notice of our approach, met us near to what seemed the chief village. His language and manner were quite friendly, but he declared that it was quite impossible for us to enter the valley. Fighting, as he declared, was actually going on between the mountain tribes, those of Ida Mahmoud, to the east of the valley, taking part with Mtouga, and those of Ida Ziki, on the west side, holding with Haha. It was impossible to get any reliable information as to the nature of the country along the mountain road. According to one informant the distance to Tarudant may be traversed in two days, while another declared that time to be necessary to reach the summit of the pass. It seems certain that the main chain in approaching its western termination has a less regular structure than in the part nearer Marocco. It throws out numerous diverging ridges; the

---

[1] Mr. Lempriere, an army surgeon, who went in 1791 by the invitation of the reigning Sultan to treat his son Mouley Absolon, Governor of Sous, probably travelled this way on his road from Tarudant to Marocco; but his narrative is too imperfect to establish the conclusion. See Appendix C.

peaks, while inferior in absolute height, are more isolated ; and the valleys, or at least that of Imintanout, now opening in front of us, seem to be more deeply excavated. We certainly heard the names of the two mountains mentioned above, which appear on Beaudouin's map ; but no name at all resembling Djebel Aithadius, which M. Balansa gives for one of the higher snow-seamed peaks in this part of the range.

We were here again struck by the difficulty of catching the sounds from native lips, a feat to be achieved only by repeated trials. At a first essay two Europeans will often write down a name in ways so utterly different that they cannot be recognised as intended to represent the same sound. Though some of the Shelluhs understand and use the word *Gebel* (or *Djebel*) for a mountain, the native word, at least in this district, seems to be certainly *Ida*, probably connected with *Idrarn*, the plural form of *Adrar*, a mountain. Idrarn Drann is the name given by the Shelluhs to the whole, or some considerable portion, of the Atlas range; and etymologists, when they come to know more of the Bereber dialects, may consider whether the name Dyris, by which this part of the Atlas was known to the Romans, is connected with the same root. Captain Beaudouin, the author of the French map, seems to have been misled by natives of this region, who would sometimes call a great mountain well known to them by the generic name *Ida*, and sometimes by a special local name, and was thus led to consider these as alternative names. Thus he writes the names of three mountains, Ida *ou* Ziki, Ida *ou* Mahmoud, and Ida *ou* Mahmed.

When it was clear that nothing was here to be effected in the way of mountain exploration, and it was seen that the day was too far gone to reach Mtouga, we decided on proceeding to Milhaïn, a place, as we were told, standing close to the foot of the mountains. Only a slender stream which we crossed, issues from the valley above Imintanout, and conflicting statements were made as to the course of

this, as well as of the other stream which we saw somewhat later at Milhaïn. In ordinary weather both are
probably absorbed into irrigation channels before they
traverse the plain ; but it is most likely that their natural
course, which they must follow in rainy weather, joins
that of the Oued Usbi flowing from Seksaoua, and reaches
the Oued Tensift by the way of Sheshaoua.

An easy ride of two hours took us to Milhaïn. The
outer skirts of the Atlas here had an unexpectedly bare and
sterile aspect. We had supposed that in the portion of the
range approaching the Atlantic coast a more copious rainfall would produce more luxuriant vegetation. We were
now within about 70 miles of the ocean, but, as compared
with the valleys south of Marocco, the change had been
in the opposite sense. It may well be that owing to the diminished height of the mountains the cooling of the aërial
currents from the W. and SW. is here insufficient to cause
much rain, except in winter, or possibly this part of the
range is more exposed to hot and dry winds from the desert.
It may also be true that the difference in the vegetation
is largely due to the mineral structure of the rocks in this
district. They chiefly consist of hard brittle semi-crystalline
limestone, with softer beds intercalated, and the rainfall
must be very rapidly absorbed in crevices and fissures.
No trees were to be seen, except olives planted near the
villages, and a few white poplars near the banks of the
stream beside which we pitched our camp.

We were fully prepared for the assurance of the village
sheik that the condition of the country made it impossible
for him to conduct us into the valley which here issues
from the Great Atlas, and sends down a stream rather
more considerable than that of Imintanout ; and, as it was
clearly useless to press the point, we contented ourselves
by expressing a wish to take a short walk into a recess of
the mountain enclosed between steep rocky declivities
that opened within sight of our camp. A jocund young
Shelluh was appointed as a guide, though none in reality

was required; and he somewhat interested us by singing
lustily at the top of his voice songs of a lively character.
Hitherto all the mountaineers we had met were marked by
a serious and somewhat saddened demeanour, as of people
on whom the burden of life pressed heavily, the only
exceptions being among the men we had brought from
Mogador, of whom Ambak was especially noticeable for
his cheerful and lively humour.

The outer slopes of the hills about Milhaïn were
scantily clad with a meagre vegetation, in which a woolly
variety of *Ononis Natrix*, *Helianthemum virgatum*, some
variety of the ubiquitous *Teucrium Polium* and *Macrochloa tenacissima* were the prevailing species; and
the attempts at tillage seemed to produce only miserable crops of barley. We expected to find more
variety on the rocks which were before us, and were not
altogether disappointed ; but the season was already far
advanced, and the spring vegetation partially dried up.
Along the dry bed of the streamlet, that is probably filled
only after heavy rain, we gathered *Euphorbia pinea*,
not before seen in Marocco. On the dry rocks we found a
curious form of *Coronilla viminalis*, reduced to a stunted
bush, scarcely two feet high, with its curious jointed pods,
four or five inches in length. A range of quite vertical
crags was almost covered with two peculiar plants of this
region—*Euphorbia rimarum*, of Cosson, and *Andrachne
maroccana*, of Ball. The latter, though abundantly different in structure, has much the habit of *A. telephioides*,
the wide-spread Mediterranean species. In a crevice of
these rocks a single small specimen of the rare fern,
Asplenium Petrarchœ, was also found.

The hill opposite that which we ascended was crowned
by a fort, similar in character to those which we had seen
elsewhere on the skirts of the Atlas, to which our Shelluh
guide gave the name Taganagurt. Our involuntary
change of route prevented us from ascertaining whether
these extend westward along the northern base of the

mountains in the direction of Agadir, but this is probable.
Future travellers may be able to ascertain more about
them than we were able to do. To whatever date their
construction be referred, it is clear that they were erected
either by the people inhabiting the low country to restrain
the incursions of mountain tribes, or by the latter to repel
attacks on their independence, the former being, in our
opinion, the more probable opinion.

We returned to camp between 5 and 6 P.M., and
found that a courier from Mogador had arrived with letters
from Mr. Carstensen. The whole province of Haha was,
he assured us, in a most disturbed state; and besides, the
war with Mtouga was complicated by the insurrection of
some of the tribes in Haha against the authority of their
own Governor. He strongly urged that we should abandon
the intention of travelling along the skirts of the Atlas
through Haha, and make up our minds to return from
Mtouga by way of Shedma, telling us that he proposed
to meet us at the *kasbah* of the Governor of that province
on the last day of May. We were very loth to forego the
promise of seeing a district new to travellers, and far more
interesting than any lying on the direct way from Mtouga
to Mogador; but we felt it impossible to persevere in the
face of Mr. Carstensen's strong opposition. It was, indeed, open to question whether, under the ægis of the
Sultan's protection, we might not without serious risk have
carried out our original intention. Whatever might be
the intestine troubles of the country, it could not suit any
of the contending parties to provoke encounter with the
paramount authority of the Sultan; but we felt that we
had no right to take a course directly opposed to the
advice of the official representative of our Government,
and especially of one to whom we felt under so many obligations.

CHAPTER XII.

Departure from Milhaïn—Defile of Aïn Tarsil—Dwellings of the troglodytes—Arrival at Mtouga—Gloomy evening—Governor's return from the fight—Prisoners of war—Their fate—Ride to Mskala—A venerable Moor—Return to the *kasbah* of Shedma—Poisoned guests — Ride to Aïn el-Hadjar — The Iron mountain — Ancient mining work—Eccentric soldier—Ascent of Djebel Hadid—Ruins of Akermout—Ride to Mogador—A *kasbah* in ruins—Powder play on the beach—Return to Mogador.

THE partial failure of our plans doubtless had a depressing effect on the morning of May 29, and this was increased by the aspect of the weather, which was misty, and before long turned to fine, drizzling rain. At 8 A.M. the thermometer stood at 58° F., and we found the height above the sea of our camp at Milhaïn to be 3,397 feet (1,035·3 metres). We were in no hurry to start; but, as the rain grew lighter and finally ceased, we got under way about 11 A.M. The sheik, who had provided for our wants and those of our escort on a liberal scale, escorted us for a short distance, and we parted with friendly expressions on both sides.

Our course lay somewhat west of due north, over a bare and sterile country. Small hamlets, surrounded by a narrow belt of cultivation, were seen at rather wide intervals; and, save a few olive trees near the houses, we did not pass a single tree during the day. *Artemisia Herba-alba,* and *Chenopodiaceæ* of the *Salsola* tribe, were the prevailing plants, indicating the presence of gypsum and of soluble salts in the soil. About three o'clock we approached a large village, with a massive square *kasbah,* and soon after, following a dry watercourse,

entered the singular defile which, as well as the village,
is known by the name Aïn Tarsil. It has evidently been
formed by erosion from the limestone strata which dip
slightly towards the south. The surrounding country
here shows a hilly undulating surface, unbroken by any
marked inequalities; but the stream, though dry in
ordinary weather, has cut a trench from two to three
miles in length, and from thirty to fifty feet in width,
between steep walls of rock about equal in height to the
width of the trench.

M. Balansa, the only traveller who is known to
have passed through this defile, must have been more
hurried than we were, as he does not speak of the numerous rare and local plants which grow upon the rocks,
most of them, indeed, the same that we had found on the
rocks about Seksaoua. But he could not avoid being
struck by the singular excavations in the rock, evidently
used at some remote period for human habitation, which
extend at intervals along both sides of the defile. In some
cases there may have been a natural recess in the rock,
afterwards artificially enlarged; but the majority appear
to be altogether the work of human hands; and in most
of them, where the entrance had become difficult owing to
the breaking away of pieces of rock from the edge, this
was afterwards made good by building up a bit of loose
wall of irregular blocks of stone. The height of the
entrance does not exceed four feet, and is often less. The
most singular point about these dwellings is the fact that
they are all near the top of the cliff, where the rock is
nearly vertical, in positions that cannot now be reached
without a ladder, or other artificial assistance. It might
be suggested that since these prehistoric dwellings were
abandoned, the work of erosion has deepened the trench,
and thus increased the difficulty of access; but unless we
suppose that during the same period the climatal conditions have been profoundly modified, this seems a highly
improbable explanation. As far as we could afterwards

judge, the watercourse running through the defile receives the drainage of only a small tract of hill country, and the marks of water action do not extend to the rocks on either side.   It is impossible to see these remains without being reminded of the notions current in antiquity as to troglodytes who dwelt in the neighbourhood of the Atlas mountains, and who could run faster than horses ; [1] but until the dwellings can be carefully examined, all speculation as to their date and origin must be vague and unreliable.   It seems most probable that the rude savages who fashioned them for their own use deliberately chose positions offering the best security against attack, either from human enemies or wild beasts.   Whether to facilitate entrance they used a rude ladder, such as the notched trunk of a tree, or relied on the superior climbing power which the freer use of the foot confers on most savage people, must remain uncertain.

As usual in this country the Moors refer these, as well as all other antique remains, to the 'Christians,' and stories of concealed treasure connected with all such monuments, of whatever date, make it almost impossible to attempt to explore or examine them.   The work can be undertaken only by a traveller authorised by a special order of the Sultan, who should also be prepared by handsome presents to secure the goodwill of the local authorities.

[1] Τούτων δὲ καθύπερθεν Αἰθίοπες ᾤκουν
ἄξενοι γῆν νεμόμενοι θηριώδη, διειλημμένην
ί ρεσι μεγάλοις, ἐξ ὧν ῥεῖν φασι τὸν
Λίξον, περὶ δὲ τὰ ὄρη κατοικεῖν ἀνθρώπους
ἀλλοιομόρφους, τρωγλοδύτας · οὓς
ταχυτέρους ἵππων ἐν δρόμοις
ἔφραζον οἱ Λιξίται.

Hanno 'Periplus,' 7.   Text of C. Müller.

Contrary to the opinion of most commentators, we are disposed to think that the river Lixus of Hanno is the Sous.   The description, 'a large river of Libya, said to flow from the great mountains,' must refer either to the Sous or the Draha ; but it is not likely that at any period there can have been a considerable population about the mouth of the Draha, as there evidently was at the place commemorated by Hanno.

We halted for luncheon in a convenient spot, and gave some time to botanising on the rocks, where, along with other plants, we found a beautiful variety of the *Stachys saxicola* of Cosson, densely covered with very long, white, silky hairs. It was near to five o'clock before we were again under way. For some time the defile continued, the cliff-like walls still showing at intervals excavated rock dwellings, and at one point it receives a tributary stream, with a bed now dry, which had cut a similar trench, and whose cliffs also showed the traces of rock dwellings. As we advanced, always ascending, we gradually emerged from the defile, and found ourselves on the slope of the hills that extend northward from the base of the Great Atlas for a distance of thirty or forty miles, and are probably continuous with the low range that we crossed between Shedma and Aïn Oumast. There must be some change hereabouts in the mineral composition of the limestone rock, if not in its geological age; as from about this point the surface was much less barren, and the vegetation more varied. Among other fine *Cynaraceæ* we saw here *Atractylis macrophylla* of Desfontaines, only once before met in our journey.

The sky was overcast, and evening coming on, when we reached the summit-level, which by our observations is 3,905 feet (1,190·2 m.) above the sea-level.[1] It did not appear to us that the surrounding hills anywhere rise more than 200 or 300 feet above the point where we passed, and, as we afterwards assured ourselves, they gradually diminish in height as they stretch northward from the main range of the Atlas. Although the matter is not free from doubt, we incline to agree with M. Balansa in believing that the hills we had now crossed form the watershed between the affluents of the stream running northward by Sheshaoua, and those of the Oued Kseb, which reaches the sea close to Mogador. Beaudouin's

---

[1] M. Balansa gives for the height in round numbers 1,100 metres, or 3,609 English feet.

map, usually correct as regards the accessible parts of the
country, represents things quite otherwise; according to
him all the drainage of this district is carried NNW. by a
stream which passes east of the Djebel Hadid, and reaches
the sea between Mogador and the mouth of the Oued
Tensift; while the Oued Kseb, or Oued el Ghored of
Beaudouin, has a course of only some twenty miles, and
drains but a small tract of country near the coast. In
most countries such a question would admit of no doubt;
but, between absorption by irrigation and loss by evaporation, the streams in South Marocco dwindle away on
their course so fast, that only at certain seasons it is possible to trace their course. As now advised, we believe
that many considerable streams unite in the Oued Kseb,
although this, at its mouth, is a mere trickling rivulet,
unable to keep a definite course through the Mogador
sands to the sea; while the stream laid down east of the
Djebel Hadid has not, so far as we could ascertain, any
real existence.

A large flowered form of *Nigella arvensis*, with a few
other plants which we had not seen since we approached
the mountains, indicated a change in the soil, now much
less barren as we descended the north-west slope of the hills;
and as twilight had set in we reached the *kasbah* of the
Governor of Mtouga, a large pile surrounded by lofty stone
walls. The soldiers, who had ridden forward to announce
our approach, found for some time no response to their
summons at the gate; and it was after some delay that
two or three slaves presented themselves, and we then
learned that the Governor and all his men had gone forth
to fight against their enemies from Haha. No orders had
been given for our entertainment; but we were told that
within the enclosure of the *kasbah* there were rooms
which were at our disposal. Our brave escort at once
grasped at the prospect of shelter and safety within the
walls, and were urgent that we should at once decide on
accepting the offer. On inspecting the room to which he

was conducted, Hooker had no difficulty, however, in at once refusing the proffered accommodation, and the alternative course of pitching our tents within the enclosure, on ground constantly trodden by cattle, was equally uninviting. Much to the trouble of our followers, the order finally went forth that the tents should be pitched on some moderately level ground outside the *kasbah*.

Slowly and sulkily the order was obeyed, we meanwhile sitting on our horses, while the night fell gloomily around us. There was no real ground for the uneasiness which our people undoubtedly felt, as night attacks are quite foreign to the usages of the country; but there was a genuine feeling that the Mtouga people were greatly overmatched in their struggle with the Haha tribes, three or four times more numerous than themselves. In the absence of supper every ear was on the alert for the approach of some one with tidings of the fray. At last, about 9 o'clock two men appeared; whether they had taken part in the fight, or judiciously taken flight, did not clearly appear. They claimed a victory for Mtouga, declaring that eighteen men of Haha had been killed, and that many prisoners had been taken, while acknowledging that the victors had also suffered losses. Half an hour later the main body approached, but it was soon evident that the return was anything rather than a triumph. The night was too dark to take account of the whole number that passed by our camp, or to observe their countenances; but the Governor with a good many mounted soldiers, and a file of from twenty to thirty prisoners tied together to bring up the rear, passed close before our tent, and the dark outline of each figure against the sky passed in long succession before us.

There was something weird and uncanny in the deep silence of the nocturnal procession. The Governor, wrapped up in a white *haik*, did not turn his head, or seem to notice the strangers, and his followers copied his demeanour. Not a sound was uttered until the file of prisoners passed by, when one man made a sudden rush towards us, imploring our protection. Of course the attempt was vain, for they were all securely tied together, and each end of the rope was held by a mounted soldier. With many a blow and curse the wretched man was driven along to share the fate of his companions in captivity.

A little later came a message from the Governor, excusing himself for not coming in person to see us. He owned to having lost many of his men during the day's encounter, and said he was too anxious and disturbed to be able to entertain us. An ample *mona* was at the same time sent, and this helped to restore comparative cheerfulness among our followers.

As was natural under the circumstances we were anxious for information, but Abraham was not able to learn any reliable particulars as to the proceedings of the day. He professed, however, to be well acquainted with the method of warfare carried on between these turbulent tribes. The fighting consists in irregular skirmishes by men who keep as far as possible under cover, in which large quantities of powder are consumed with comparatively insignificant results. But, whatever be the result, it is a point of honour with each party to bring back prisoners. It is not often that these are made among the fighting men. Harmless peasants are seized—if of the enemy's tribe so much the better; but, if these are not to be had, those of their own tribe are made the victims. We were assured that the same thing happens with the Sultan's troops on the rather frequent occasions when they are despatched against some refractory mountain tribe. The mountaineers commonly make good their retreat to some spot not easily reached by horsemen ; but, in order to be able to announce a victory, the detachment seizes any hapless people that come in their way on their return to the capital. When we inquired as to the destiny of the captives it was horrible to be told that some of them would certainly be butchered during the night.

x

It is strange, although similar anomalies are found in all ages and countries, to learn that, along with an utter absence of rudimentary feelings of humanity, these people show indications of the sentiment of chivalry. However strangely understood, the point of honour has a recognised place in their ethical system. The feud between Haha and Mtouga had been smouldering for many weeks, and hostilities were to have commenced soon after our arrival in the country. But the brother of the Kaïd of Haha was about that time seized with small-pox, and it was thought proper to await his recovery before commencing the war.

The night was overclouded, dark, and almost cold, and we were on foot at an early hour on the morning of May 30. The *kasbah*, standing on an elevated plateau, 3,085 feet (940·3 metres) above the sea-level, does not, owing to the undulating character of the surface, command any extensive view. Close beside it, a stream from the Atlas has excavated a broad trench, similar in structure to Aïn Tarsil, but very different in aspect. A streamlet here meanders along the flat bottom between walls of rock some 40 feet in height, and the more constant supply of moisture suffices to cover the floor of the miniature valley with a carpet of vegetation, and to support a fringe of tall water plants along the banks. We might probably have added several species to our lists if we could have devoted a day to botanising along the course of the stream ; but in the existing state of the country that could not be thought of, and we contented ourselves with a morning stroll over the ground surrounding the *kasbah*, and along the neighbouring banks of the stream. As we wandered separately, Hooker was assailed with extraordinary vehemence by a negro woman. Not a word could, of course, be understood ; but the objurgations of a virago are to some extent intelligible in every language. It was not possible to guess what induced this outpouring of threats and abuse, but it seemed probable that a botanist might unconsciously have done something to clash with the superstitious feelings of the natives. It is clear that the religion of Mohammed reaches but skin deep with the Bereber population, while traditional observances, derived from far more ancient religious systems, are still deeply rooted among them.

Much to the relief of our escort, whose first anxiety was to get away from the troubled district, we started about 10 A.M. The Kaïd sent another message, again excusing himself from a personal interview. We afterwards learned that he had made a present of four dollars to each of our escort, of course with the object of procuring some degree of support and countenance at head-quarters.

Our course during the day lay between NW. and NNW., keeping for a short distance along the course of the stream. At two or three points we saw traces of rock dwellings, not nearly so well preserved as those of Aïn Tarsil. Before long the walls of rock subsided on each side of the stream, which bent in a westerly direction, while our track ascended gently amid undulating downs, till we reached a point commanding a wide unbroken view of the northern horizon. To the eye the country before us seemed almost a dead level, but there is a decided general slope from south to north, as we more fully ascertained in the evening, when we found that we had descended fully 1,500 feet during the day's ride. The outline of the Iron Mountain (Djebel Hadid) was now clearly traceable in the distance, the highest part bearing about due NW. As we advanced there was a manifest improvement in the fertility of the soil, and for a space of five or six miles we rode amidst cultivated fields, apparently the most productive that we had seen in South Marocco. The same conditions necessarily affected the wild plants of the country, and made it difficult to secure specimens of manageable dimensions of several interesting species. A great *Daucus* (*D. maximus* of Desfontaines), probably a luxuriant form, or sub-species, of the common

carrot, grew to a height of four or five feet, and the
flowering umbels were often more than a foot in diameter.
More ornamental was a splendid *Centaurea*, three or four
feet high, with very large heads of deep orange flowers,
often tinged with purple, which we took to be altogether
new. Subsequent examination made it doubtful whether
it should be separated from a very variable North African
species, *C. incana* of Desfontaines, though much larger
in all its parts, and differing in the colour of the flowers.

We halted for luncheon near a village called Hazarar
Assa, standing, as we were assured, at the frontier, between
the provinces of Mtouga and Shedma. In Beaudouin's
map those provinces are separated by an intervening strip
of territory, apparently belonging to the Ouled Bou Sba
tribe. As political divisions in this country are subject
to frequent alteration, the map may have been correct at
the time it was made.

Maize was grown near the village, although no means
for irrigation were apparent. As the growth of this plant is
in general absolutely dependent on a frequent supply of
water to the soil, we could only infer that we had come
within the limits of the coast climate, and that as a
general rule rain cannot there be very unfrequent during
the months of April, May, and June.

During the halt the villagers reported that fighting
was still going on in the neighbourhood, and a few gunshots
were heard in the distance; but, as our course led us away
from the scene of action, the result, if any, was never
known to us.

In the afternoon our track bore more westward than
before, keeping about due NW. in general direction. We
soon left the fertile ground behind us, and for many miles
rode over slightly undulating stony downs, where the
prevailing slope is always to the N. or NW. The vegetation in these barren tracts is mostly of the social kind,
two or three species, or sometimes one only, prevailing
over a wide area, and then being suddenly supplanted by

others. *Artemisia Herba-alba*, and *Retama monosperma*, are in this part of the country the dominant species. The monotony of the way was pleasantly broken by crossing a little valley traversed by a mere rivulet, which, however, sufficed to support a more varied and cheerful vegetation. From the rising ground beyond it, we gained a view to the south, and saw at a distance of eight or ten miles a massive pile of building crowning a low, thickly-planted hill, with a large enclosed space occupying part of the slope. The *kasbah* belonged to the Kaïd of Haha, and was on a scale proportioned to the importance of the Governor of one of the largest and richest provinces of the empire. We found additional reason for regretting the ill-timed local war, which prevented us from paying him a visit. Although guilty of frequent acts of atrocious cruelty, he was said to receive strangers with the utmost courtesy and hospitality.

As evening was fast approaching, we rode hurriedly through a little defile among the hills, where the rocks promised plants of interest, and soon after came upon the first Argan trees that we had seen since we quitted the province of Shedma, on April 30. Associated as the tree was in our minds with Mogador, where we had bidden farewell to European life, the sight awakened feelings of regret at the approaching termination of our tour, tempered by satisfaction at the prospect of returning to the usages and intercourse of civilised life.

About seven o'clock we reached Mskala, a rather large village on a stony slope, and saw close beside it an extensive camp where the old Governor of Shedma, with a considerable force, was, by the Sultan's order, watching the progress of events in the contest between the adjoining provinces. We established ourselves in a stony field a few hundred yards away from the Governor's camp, and before long an ample *mona*, consisting of two sheep, twenty-four fowls, tea, sugar, butter, and other luxuries, followed a little later by seven large dishes of cooked food, satisfied

the cravings of our ever greedy soldiers, and the greater
part of the night was devoted to general feasting throughout our camp.

On the morning of May 31 our men were in no haste
to bestir themselves after the orgy of the preceding night,
and we indulged in a longer rest than usual. The weather
was fine and clear, but remarkably cool considering the
moderate elevation of this district. By our observations
our camp stood at 1,562 feet (476·3 m.) above the sea;
yet at 8 A.M., when the sun was already high above the
horizon, the thermometer marked only 65° F. It was
suggested that politeness required a visit to the old
Governor, and Hooker, with our interpreter and some of
the escort, devoted himself to that duty, while Ball set off
alone for a short botanical ramble over the bare, stony
hills surrounding our camp. The excursion was not very
fruitful, except in the way of illustrating the effects of barren
soil and exposure, without the slightest shade or cover, on
the growth of many species that here assumed a dwarfed
and stunted condition.

Hooker, who had seen too much of the people of
barbarous countries to be open to the illusions that many
travellers, new to their manners, readily fall into, was, for
once, very favourably impressed by his interview with the
Governor of Shedma. He had found an old man of venerable aspect, with remarkably fine features, whose
conversation displayed a happy union of dignity and
frankness. He was engaged in superintending the distribution of pay to his soldiers, and the subject that naturally
arose for discussion was the part which he and his forces
were destined to play in the intestine troubles of their
neighbours. His instructions from the Sultan were, as it
seemed, of an indefinite kind. He had, in the first
instance, endeavoured to play the part of mediator and
avert the outbreak of hostilities His present duty was to
hold himself in readiness to carry out such further orders
as he might receive. Considering the jealousies that

always exist between the people of neighbouring provinces, usually inhabited by tribes of different race and origin, it may be doubted whether the troubles and losses of their neighbours are in Marocco viewed as matters of deep concern ; or whether, as sometimes happens among the statesmen of more important countries, the mediator may not feel some secret satisfaction at the failure of his own proposals.

Shortly before 10 A.M. we started for an easy ride of three hours over the undulating country that lies between Mskala and the *kasbah* of Shedma, where we were to meet Mr. Carstensen. It was an agreeable change from the bare hills, with which we had of late been familiar, to enter on a comparatively well-wooded country. The Argan trees were nowhere so near together as to form what could be called a forest, but scattered in small clumps or single trees over the surface, so that nothing but a carpet of green turf was wanting to complete the resemblance to an English park. On reaching the *kasbah,* we found that our arrival was already expected. Our former host, the Governor's corpulent son, had two rooms within the castle walls prepared for our reception, and before one o'clock we were installed in clean quarters, with iron bedsteads of European make, and cushions covered with Rabat carpets to complete the furniture. Soon after two o'clock notice of Mr. Carstensen's approach reached the castle ; the Governor's son, with several armed men, went forth to meet him, and before long we had the pleasure of again greeting a gentleman to whose activity and thoughtful care we felt so much indebted.

Much of the afternoon was naturally employed in giving an account of our doings. When we reached the close of the story, and Hooker spoke of his morning's interview with the aged Governor, in whose stronghold we were lodged, and the favourable impression made by his appearance and demeanour, the reply was somewhat startling even to men who had learned something of the

manners of the country. ' Yes,' said Mr. Carstensen, ' he
is a fine-looking fellow, but he is not much better than
other men of his class.   Last year he poisoned two friends
of mine under very discreditable circumstances.'   The
victims were men of consequence, near kinsmen of the
Governor, and supposed to have much influence among
the Shedma people who resided in Mogador.   Early
in the preceding year they were induced by hospitable
messages to pay a visit to their powerful relative.   Familiar
with the ways of Marocco, and feeling sure that his friends
were objects of jealousy and suspicion to the great man,
Mr. Carstensen at once wrote an urgent letter, in which he
expressed his strong anxiety for the safety of the visitors.
He soon received a reply written in the most reassuring
terms : ' Far be it from me,' wrote the Governor, ' to harm
these men ; I shall take every care of them, and cherish
them as if they were my own children.'   A few weeks
later another letter reached Mogador : ' Nothing could
exceed the Governor's grief at having to announce that
one of his guests had been taken suddenly ill, and soon
after died.   Such, however, was the decree of Allah, and
we must all be resigned to his will.'   Mr. Carstensen was
not surprised when, a little later, another letter reached
him, conveying in nearly the same terms an account of the
death of the second guest.   He had no doubt of foul play
having been used ; but some months later received further
assurance, when, on taxing the Governor's son (our fat
friend) with his suspicions, the latter answered : ' Well,
the fact is that my papa did not know what to do with
them, so he had them poisoned.'

It seems strange at the present day to find so near to
Europe a condition of society in some respects so like
that of Italy in the fifteenth and sixteenth centuries,
wherein no deed of atrocity committed by men in authority awakens the slightest feeling of moral reprobation.
In the present instance local ideas had so far prevailed
that Mr. Carstensen did not consider it expedient to allow

what had occurred to interrupt his amicable relations with the Governor.

To lessen our regret for having failed to see something of the western extremity of the Great Atlas, Mr. Carstensen proposed that we should visit the Djebel Hadid, or Iron Mountain, a range of hills, about 2,000 feet in height, that approaches the sea about fourteen miles NE. of Mogador, and extends inland in a NE. direction for a distance of some five-and-twenty miles. So far as we knew, this had not been visited by any European naturalist except M. Balansa, and we willingly accepted the suggestion, and, following Mr. Carstensen's advice, arranged to fix our camp next day at Aïn el Hadjar, a spot where some copious springs burst forth at the SW. extremity of the range.

The forenoon of June 1 was spent in the *kasbah*, and in a short ramble on the adjoining slopes. The appearance of the country was much altered since we had passed here at the end of April. The spring vegetation was then far advanced, and many annuals had ripened their seeds; but, thanks to the rain which had fallen at intervals during May, a new crop of young plants had sprung up; and during this and the following days we were able to gather several species in flower that we had before seen only in fruit. The fields where, at our last visit, the corn was being cut, had been ploughed up, and pumpkins had been extensively sown round the *kasbah*. There was more appearance of a taste for ornamental plants than we had seen anywhere among the Moors. The Governor had transactions with many of the foreign merchants at Mogador, and in that way had no difficulty in obtaining seeds or cuttings of many garden flowers. Amongst these we noted roses, pinks, garden-stock, geraniums, dahlias, *Tagetes*, and *Coreopsis*. Oranges and bananas were also cultivated; but it seemed doubtful whether in ordinary seasons the climate is suitable. The spring rains had recently extended over a wide tract of country, but they seem to be more often limited to the zone surrounding the higher mountains.

Among other articles intended for presents, Ball had picked up in London a large Highland brooch, with a yellow cairngorm crystal set in silver. This, with an opera-glass, was given to the Governor's son as we took leave of him. The use of gold or silver and jewels for personal adornment is forbidden by law or custom to Moorish men; and the gift, which was sure to be transferred to a favourite wife, did not seem to be much appreciated.

About mid-day we started for our short day's ride, forming, with Mr. Carstensen and his suite, a numerous cavalcade. Our course lay about WNW., over low undulating hills, dotted with Argan trees. Most of the surface was under cultivation, and appeared to be moderately fertile. At 2.30 P.M. we reached an olive grove near to a *zaouia* or sanctuary, called El Masaats. Close to this was a dwelling, on a larger scale than is common in this country, belonging to a man of some substance, with whom Mr. Carstensen had friendly relations. It would have been impossible to pass his home without a visit, and equally impossible, according to local ideas, for him to neglect the rites of hospitality. Luncheon for the entire party was speedily provided, and, while assisting as spectators at the lively conversation, we once more had to admire Mr. Carstensen's perfect command of the native dialect.

As we sat under the trees several parties of natives, dressed in their best, passed by on their way to the adjoining *zaouia*. This was the anniversary of the death of the local saint buried at the sanctuary; and on such occasions the people of this country, whether Moor or Shelluh, do not fail to resort to the sacred spot. For the great majority the occasion seems to be no more than a welcome opportunity for breaking the monotony of their daily life. Excepting our Mogador Kaïd, who was most exact in the performance of his devotions, we saw little indication throughout our journey of regular compliance with the injunctions for daily prayer, so strictly observed in most Mohammedan countries.

After resuming our route we soon found evidence that we were entering upon a new botanical region—that of the Atlantic coast. Besides numerous species not seen since we had left Mogador at the end of April, we here found for the first time several conspicuous plants characteristic of this region. In hedges, and among bushes, a tall *Bupleurum* (*B. canescens* of Schousboe) grew to a height of eight or ten feet, and in similar situations *Periploca lævigata* was just forming its fruit.

The facts known as to the distribution of the last-named plant, and the allied species, *Periploca græca*, suggest speculations as to their past history that deserve some passing notice. The genus *Periploca*, which takes its name from the twining stems of the species first known to botanists, has its centre in the sub-tropical zone of the Asiatic continent. The single mainly western species is *Periploca lævigata*. This appears to be common in the Canary Islands, and grows freely in the tract now visited by us to the north-west of Mogador. It has been found in abundance on some rocky islands near the coast of Sicily; but, in spite of the silky hairs attached to the seeds, it has not spread itself to neighbouring islands, nor to the Sicilian coast. It has been detected in two or three places in the south-east of Spain, and here and there in rocky places on the skirts of the desert in the interior of Algeria and Tunis. Finally, it was long ago found by Labillardière in one place on the coast of Syria. All this points to the former wide diffusion of a plant which no longer finds favourable conditions of existence, unless, perhaps, in the Canary Islands. Its presence in the interior of North Africa may possibly date from the period when it grew near the coast of a great gulf opening to the Atlantic; but it is not easy to understand how it has held its ground in a climate so different from that of its natural home. This plant has inherited from a remote ancestor a habit which is now of no service to it. The young branches near the root twine round any adjacent support; but as they grow older they become stiff and straight, and the taller specimens derive no adventitious support from this source.

The history of *Periploca græca,* the only species known to the older botanists, is somewhat different. It is rather common in Georgia, and in parts of Persia and Asia Minor. Less common in Greece, it becomes extremely rare to the west of that limit, being found only in Montenegro, at one place in Dalmatia, at another in South-eastern Italy near Otranto, and, finally, in the pine-woods on the Tuscan coast near Pisa. These facts indicate the former wider extension of the species towards the west, and its gradual retreat towards its primitive home in Asia. But we have more direct evidence to that effect. The prints of leaves unmistakably belonging to this species are not uncommon in the quaternary deposits of the valley of the Arno. It may probably have flourished in thickets on the Monte Pisano, and on the Monte Nero near Leghorn, when these were islands in a tertiary sea, and gradually descended towards the Mediterranean as the coast line was advanced by a change of level, and by the formation of the deltas of the Arno and the Serchio.

Another conspicuous plant, now seen for the first time, was *Odontospermum odorum* (*Asteriscus* of De Candolle), forming a dense dwarf bush, about two feet high. The whole plant gives out an agreeable scent; but, except in this respect, and in having the leaves covered with white silky hairs, it differs very little from *O. graveolens,* a characteristic species of the desert region, remarkable for its offensive smell. The sweet-smelling species had been hitherto found only in the Mogador district, and in the Canary Islands; but it was afterwards gathered by us near Saffi.

We reached our destination at about 5.30 P.M., and

were agreeably surprised at the verdure and freshness of
the spot. Our camp was pitched among large olive trees,
near to the stream flowing from the principal spring. The
position somewhat resembles that of the so-called fountain
of Elias near Jericho, well known to travellers in Palestine ;
but the contrast offered by the vegetation was remarkable.
If a few plants close to the stream appear to thrive about
the waters of Elias, the surrounding vegetation is meagre,
and amid the straggling bushes of exotic aspect that sur-
round the spot the traveller seeks in vain for effectual
protection from the sun.   Here, besides the gigantic olive
trees that must have been planted at a remote period, the
white poplar grows to a great size, and wild herbaceous
plants were still green, many of them in flower as well as
fruit, at this advanced season.   At a time when the summer heat has become intolerable at most places in North
Africa the thermometer in our tents stood at about 70° F.
an hour before sunset, and the nights were even cooler
than some might have wished.   Something was no
doubt due to the unusual amount of rain that had fallen
during the month of May; but if the climate of the
coast region of South Marocco were altered so as to resemble that of other places in the same latitude, much of
the existing vegetation would soon disappear.   On dry
sandy slopes above our camp the effects of the late rains
were plainly seen, and before nightfall we collected a considerable number of annual species in flower, sprung from
seeds borne by the first crop, and ripened two or three
months before.
    We did not visit the remains of ancient miners' work
that are visible at several places about the base of the
hill ; but we found scoriæ in abundance, and some fragments of the ironstone from which the mountain takes its
name. It is not easy to conjecture the date at which
these iron mines can have been worked.   There is no reason to believe that any Moorish ruler ever attempted to
turn them to account ; and although the Portuguese once

built a small fort near Mogador, it does not seem probable that they ever held control over the adjoining country. As to the long interval between the establishment of Roman power in North Marocco and the disappearance of Roman civilisation after the Saracen conquest history is silent, and it would be as unsafe to assert as to deny that the workings of Djebel Hadid are to be referred to that epoch. The only apparent alternative is to attribute them to a still more remote period, when Carthaginian colonies flourished on the coast.

In connection with this subject it is curious to remark that Leo Africanus, in his account of the hilly range of the Djebel Hadid, makes no allusion to the working of the mines there, although his work contains frequent reference to the extraction of metals, not excepting iron, from mines in the Great Atlas. In his day the Djebel Hadid seems to have had a rather numerous population of Bereber stock. He describes them as of gentle and inoffensive manners, who expelled from among them men guilty of robbery and violence. They had been much molested by the Arabs of the neighbouring plains, and had agreed to purchase tranquillity by the payment of black mail in the form of tribute, when the reigning Sultan, whose policy it was to protect and favour the Bereber population, despatched a military force (which Leo Africanus himself accompanied), brought the Arabs to order, and relieved the Berebers from tribute. At the present day the Shelluh stock has apparently disappeared from this part of the country, being either driven away, or absorbed by inter-marriage into the surrounding Arab population.

A young man, the son of one of the wealthiest Jews in Mogador, had been invited by Mr. Carstensen to accompany him in this excursion. He was absolutely ignorant of the country beyond what he may have learned in a daily canter over the sands at Mogador, and was far less fitted for rough life than the majority of English young ladies of the upper class. Everything in tent life seemed to him strange

and rather terrible. In the course of conversation over the
evening cigar it came out that he had never seen a scorpion; whereupon, by order from Mr. Carstensen, a corner
of the carpet within the tent was turned up, and a scor-
pion-hole speedily found.   When the ugly creature was
dug out of his hole and produced to the company, the
genuine consternation and disgust of our young friend
were irresistibly ludicrous. We afterwards heard that he
passed a miserable night, in constant terror of encountering
the enemy, and on the next day returned to the paternal
home, whence he will not again be easily lured.

The natives show no especial dislike for reptiles, excepting poisonous snakes, which, in spite of reports to the
contrary, must be rare. We heard so much of them, and especially of the *Cerastes* (*El Efah* of the Moors)—popularly
called the ‘two minutes’ snake,’ because a person bitten is
supposed to survive so long—that at first we always carried
about us a bottle of liquid ammonia, as the best, though
very uncertain, antidote. But when we failed to see a
single specimen, and were assured that they are found only
on the coast, we gradually laid all precautions aside, and
thought no more of serpents than we should have done in
Europe.

About this time we discovered that one of our escort
had a decided taste for reptiles, which we might have
turned to account, if we had known of it, by getting him
to collect specimens. He was a tall, lanky man, with a
prominent nose, whom we had nicknamed, from his peculiar personal appearance, ‘Don Quixote,’ but whose real
name was Sherrif Mouley Mohammed.   He had captured
several toads and lizards, which he carried about with him,
and showed another trait of originality in being the only
one of our native followers who willingly drank coffee.

The morning of June 2 was brilliantly fine, and the
sun remained unclouded throughout the day, although the
heat was at no time oppressive.   At 9.30 A.M. we started
for the ascent of the Djebel Hadid, directing our course to-

wards a hollow in the face of the hill, for the most part
thickly clothed with bushes, but showing here and there
outcropping escarpments of rock that promised a more
varied vegetation than the otherwise uniform stony slopes.
We at once found that, in comparison with the outer slopes
of the Atlas, we had entered into a region botanically
new to us. The evergreen oak had disappeared, and the
Arbutus, though seen near Aïn el Hadjar, was evidently
rare. The *Callitris*, which is abundant near the base of
the hill, does not ascend on its flanks, and *Juniperus
phœnicea* was either altogether absent or very rare. In
the place of all these there was an extraordinary abundance and variety of spiny bushes, such as made the day's
excursion severely remembered by the destruction of our
garments and the multitude of pricks and scratches with
which our bodies were covered. *Rhus oxyacantha* and
R. pentaphylla, *Celastrus senegalensis* and the wild
olive, with *Genista ferox* and *G. tridens*, were our chief
tormentors, all, except the olive, characteristic North
African species, though two or three of them have been detected in Southern Spain or in Sicily. Leaving our horses to
be led up the slope, we had hot work in climbing the hill
under a sun only a few degrees from the zenith, contending
the while with the various thorns and hooks and prickles
that molested us on every side. Every forward movement
would be resisted by a dozen spines running deep into our
legs or arms, and each attempt to draw back by the strong
hooks with which some part of our dress was sure to be
held fast. When we reached the top of the acclivity we
found ourselves on the verge of a very extensive plateau,
in some parts nearly dead level, in others undulating, and
rising into knolls of tolerably uniform height. Before
long we reached a point commanding a wide view over the
country on the north side of the range of hills. The
slopes below us appeared to be under cultivation, and
suggested the presence of a numerous population; but the
distant plain of Akermout, lying somewhat east of due

north, did not to our eyes afford any sign of cultivation. Jackson gives a view[1] of the Djebel Hadid as seen from the plain of Akermout, with a ruined town in the foreground, which he declares to have been utterly destroyed by the plague about the middle of the last century. The ruins, which have been seen by other travellers, being about 30 miles distant from Mogador and fully 15 from the point where we stood, were naturally not perceived by us.

Along the range of these hills are many saints' tombs, usually standing on some prominent point above the general level of the plateau. One of these was on a slight eminence somewhat higher than that first reached by us. Our escort, on this occasion limited to three soldiers, displayed great anxiety lest we should attempt to enter the *zaouia*, doubtless believing that our presence would profane the sanctity of the spot.

We strained our eyes to make out as much as possible of the Great Atlas range from the vantage ground we had now attained ; but the air was hazy towards the southern horizon. A faint outline was, indeed, distinctly traceable, and was sketched by Ball, but no details of any kind could be distinguished.

We estimated the height of the hilly range, where we ascended it, at about 1,500 feet above our camp at Aïn el Hadjar, which we had found to be 504 feet above the sea-level. It is not likely that any part of the Djebel Hadid much exceeds the limit of 2,000 feet above the sea, but, in a range fully 25 miles in length, it is not possible to compare altitudes accurately by the eye.

The *Cistus* tribe was the chief ornament of the vegetation here ; and it was interesting to observe that the species were to a great extent different from those that

---

[1] Like the other illustrations in Jackson's work, this must be derived from a very imperfect sketch, or else much altered by the fancy of the draftsman. In this the great city shown in the background is a mere fiction of the imagination.

abound in the inland districts and on the lower slopes of the Atlas. *Cistus salviæfolius* and *C. polymorphus*, both variable species, are common to this and the Atlas, although, strange to say, the latter widely-spread Mediterranean species has not been found in North Marocco. *Helianthemum virgatum*, hitherto seen everywhere on dry stony ground, was here wanting, as were *H. niloticum* and the less common *H. glaucum* and *H. rubellum*. In their places the top of the hill was in some places quite covered by large bushes of *H. halimifolium* and *H. lavandulæfolium*, both laden with masses of bright yellow flowers. On the slopes we also found *H. canariense*, one of the very few species, not strictly confined to the coast, that are exclusively limited to the Canary Islands and South Marocco. Another rare species of the same genus, now first seen by us, was *H. Lippii*. This seems to have been originally a desert plant, the sole representative of the genus in the arid regions of Beloochistan, South Persia, and the Arabian desert, whence it has spread westward through Egypt to the skirts of the Sahara. Beyond its natural home, it has been found here and there, but rarely in Syria, Asia Minor, Sicily, and South-western Marocco, and may not improbably be detected in Southeastern Spain.

Growing among the bushes on the upper part of the hill we found, in some abundance, the wild spiny form of the cultivated artichoke; whether truly indigenous, or carried hither by former inhabitants, it was impossible to decide.

On returning to our camp, some time before sunset, we found that Mr. Carstensen had received letters from Europe that bore intelligence of the terrible scenes enacted in Paris during the last days of the Commune and the final suppression of the insurrection. As was but natural, this completely engrossed our thoughts and our conversation during the evening. It was depressing to think that in the midst of the so-called advanced civilisation of Europe, to which we were now returning, ferocious passions, surpassing in their destructiveness those of the barbarian or the mere savage, may lie concealed until some unexpected shock causes their explosion.

The night was even cooler than the preceding one, and to our surprise the thermometer, about a quarter of an hour after sunrise on June 3, marked only 56° F. We employed a couple of hours in the morning in rambling about the gardens and irrigated ground near the springs, without adding much of interest to our collections, and at about 8 P.M. started for Mogador.

For some distance the country was well wooded. Orchards and olive groves did not extend much beyond the bounds of the irrigated tract ; the *Callitris* then became predominant, intermixed, here and there, with scattered Argan trees. In open spots the two showy species of *Helianthemum* seen the day before, *H. halimifolium* and *H. lavandulæfolium*, were still in full flower, and we gathered, for the first time, a charming little *Eryngium* (*E. tenue*) with extremely delicate spiny leaves and involucre. Mediterranean shrubs, such as the *Arbutus* and *Phillyrea*, growing along with such local forms as *Rhus oxyacantha, Statice mucronata,* and *Bupleurum canescens*, would have sufficiently informed a botanist that he was approaching the Atlantic coast of North Africa.

We soon after crossed a belt of land showing marks of former cultivation, where no dwellings were in sight, but where we passed close to a considerable group of earthy mounds, partly overgrown by vegetation, and showing here and there the remains of massive walls of *tapia* that had partially resisted the process of destruction. These ruins marked the site of the large *kasbah* of a former Governor. According to the custom of the country this had been pillaged and destroyed some thirty years before, when the owner fell from power. The traces of man's former presence were speedily lost as we entered on a tract of rocky ground, where the tertiary calcareous rock lay in horizontal beds, slightly excavated in places by watercourses, and cut into irregular steps. The increasing prevalence of blown sand now gave warning of a nearer approach to the shore; the distant roar of the ceaseless Atlantic breakers fell distinctly on the ear; amid the increasing masses of sand vegetation became more and more sparse; we rode on amidst undulating dunes of sand until, at length, on reaching the summit of one of the ridges, the blue Atlantic lay before us.

With mingled feelings we cast our eyes on the waters that were so soon to carry us back within the accustomed round of civilised existence. If the prospect before us were in many ways most welcome, there was yet some inevitable regret at the termination of a journey so full of interest, and, in spite of trifling drawbacks, so full of enjoyment. We felt that the time at our disposal had been too limited, and that what we had accomplished in the way of exploration fell far short of what we had expected; but enough had been done to reward us amply for the labour expended, and we indulged, as almost all genuine travellers are wont to do, in the hope of returning again to the country we were now about to leave.

We had reached the shore at a point about five miles north of Mogador, which, however, was concealed from view by the lofty sand dunes that have accumulated on the reef of rocks that stretches out seaward on the north side of the town. For more than half the distance we rode along the flat beach, where the sand gave somewhat firmer footing than it did above high water mark. Our soldiers took the opportunity for celebrating the prosperous termination of our journey by an exhibition of 'powder-play,' for which the ground was admirably adapted. Starting together, but not attempting to keep line, they urge their horses to their fastest gallop; while at full speed they discharge their long guns at an imaginary foe, fling the gun up in the air, and catch it again; and finally the horse is stopped short, and thrown upon his haunches, by the sudden pressure of the severe bit used in this country.

To avoid a long detour our course to Mogador lay over the high sand dunes that encompass the town on the land side. The forms into which the sand is fashioned by the wind here attracted our attention. In many places the appearances were exactly those that are found in the higher region of the Alps immediately after a fall of fresh snow, and in truth the phenomena are nearly identical. At a temperature considerably below freezing point snow commonly falls in the condition of fine grains that do not cohere when they meet, and, in a mechanical sense, differ from those of sand only by being lighter. Our observations on the relation between the form of the larger ridges and the smaller ripple marks, and the direction of the wind, quite agree with those published by Maw.[1]

When at length we escaped from the maze of ridges and hollows, and stood upon the brow of the last sandy eminence, rather before 2 P.M., we found ourselves unexpectedly near to our journey's end. The town of Mogador, backed by the island, with a few small coasting vessels lying in the channel between them, presented to our unaccustomed eyes an almost imposing aspect. As usual, one of the soldiers had ridden ahead to announce our approach ; and when, after passing by the Christian burialground, we drew near to the walls, crowds of people came out to meet us, and to gaze upon the strangers, of whose adventures in the Great Atlas fanciful reports had gone abroad. At the gate several mounted soldiers, sent by the Governor as a guard of honour, joined the procession ; and thus heralded, with all due state, we made our solemn entry into Mogador, and, along with our kind host, rode directly to the British Consulate.

[1] See *Quarterly Journal of the Geological Society*, vol. xxviii. p. 88.

CHAPTER XIII.

HOWEVER pleasant were our recollections of the rough life
we had been leading for the past five weeks, we could not
fail to appreciate the physical comfort involved in a return
to the habits of civilised life at Mogador. A chair to sit
upon, a table at which to eat one's meals, a hundred other
things to which daily use makes us quite insensible and
indifferent, become luxuries to one who has for some time
been deprived of them.

We had no lack of occupation during the four days of
our stay at Mogador. The large harvest of dried plants
collected during our journey had to be put in order and
safely packed for conveyance to England. It had been
our prime object to obtain as far as possible a complete
representation of the Flora of the territory which we had
been able to explore; and for that purpose we had made
it a point to carry away a specimen, or at least a fragment,
of every species, even the commonest, from each district
that we traversed. In this way our collections represent,
not merely the constituents of the South Marocco flora,
but, to a great extent, the distribution of the several
species. Besides attending to this, our main object, we
did not fail to collect duplicates of most of the new and
rare species seen during our journey for distribution to the

chief public herbaria, and to the botanists who have illustrated the flora of the Mediterranean region.   To avoid chances of future error, it was necessary to give much care to the labelling and packing of our collections, which we could not expect to see again until some time after our return to England.   Fortunately our specimens were nearly all quite dry, and in excellent condition ; and we had not to complain of the moist condition of the air which had given us so much trouble during our first stay on the coast.   The cool breezes from the N. and NE., which make the climate of this region so agreeable and healthful in summer, now steadily prevailed.   The air at this season is relatively dry and free from haze, and, as a consequence, the daily range of the thermometer is greater than at any other season.   Yet, as compared with any other place we know, the extremes are singularly moderate, and never exceed the limits conducive to full health and enjoyment.   The thermometer, observed pretty frequently by night as well as by day, only twice rose during our stay to 77° F.   At about 3 A.M. it usually fell to 63° F., and on one occasion to 61°.

On our return we found awaiting us a small addition to our Marocco herbarium.   With the kind assistance of Mr. Carstensen, we had arranged during our first visit that two natives, who had received a first lesson in the art of drying plants, should start for Agadir early in May, and should bring back whatever plants they could then find in flower.   The collection, which we now shared between us, was not of much importance, including, as it did, but a single species not found by us.   We doubted, at the time whether the men had, as they solemnly asserted, really reached the neighbourhood of Agadir ; but we have now reason to believe that the vegetation of the coast region from Mogador southward to about the twenty-ninth parallel of latitude is very uniform.

Since our return from Marocco our friend M. Cosson, with the active assistance of the late M. Beaumier, has succeeded in engaging the services of two native collectors, one of them a very intelligent Jew, the other a Shelluh mountaineer. From the former large collections from the country as far south as the borders of Oued Noun, and as far east as the oasis of Akka, have been sent to Paris; and the latter has contributed a few additions to the flora of the Great Atlas, along with many of the species collected by us. With his accustomed liberality, M. Cosson has sent us duplicates of these fresh contributions to the flora of Marocco.

In the afternoon of June 4, we went to pay a farewell visit to the banks of the Oued Kseb, which had been the scene of our first botanical excursion in South Marocco. During the interval of six weeks a great change had passed over the vegetation; most of the annuals were completely dried up and had disappeared; but the excursion was not altogether unproductive, and we were able to add a few plants to our collections.

There remained a point of some botanical interest, which it was very desirable to clear up before our departure. The curious cactoïd *Euphorbia*, producing the Gum Euphorbium,[1] written of by Dioscorides and Pliny, grows in the interior provinces of South Marocco. The only modern writer who has given an account of it is Jackson,[2] who, though no botanist, was a careful and conscientious observer. In his account of the plant, and the accompanying plate, we had been struck by some apparent discrepancies. The gum, as he says, is obtained from the plant growing on the lower region of the Atlas; but the same plant is, according to him, abundant about Agadir, and is carried thence to Mogador for the use of the tanners. The Agadir plant, however, he declares to produce no gum. Further than this, in the plate annexed to his description, the left-hand figure, giving a view of the

---

[1] See Appendix D.

[2] Jackson, *Account of the Empire of Marocco*, p. 134, 3rd ed. London: 1814.

whole plant on a reduced scale, shows the thick fleshy branches, with four angles, as we had seen them in the specimen given to Hooker by the Kaïd at Mesfioua; while the right-hand figure, showing a fragment of the natural size, represents the end of a branch, with numerous (about ten) projecting fleshy ribs, beset with spines. Hooker came to the conclusion that there were possibly two quite distinct plants known to Jackson, and on returning to Mogador he proceeded to make inquiry on the subject.

Before long a native was brought to us who appeared to be well acquainted with the Agadir plant, and who declared that it grows in abundance about half-way between that place and Mogador. Upon this Hooker became anxious to start at once for the purpose of personally examining the suspected plant, and securing live specimens for Kew. To this Mr. Carstensen felt it necessary to object. Matters, as he informed us, had been getting from bad to worse in the great province of Haha, which includes the whole sea-board between Mogador and Agadir. Between the discontent caused by repeated acts of unprovoked cruelty on the part of the Governor, and the results of the war still proceeding between him and his neighbours in Mtouga, the province had lapsed into a complete state of anarchy, and a European attempting to travel at such a time would be exposed to serious risk. It was reluctantly agreed by Hooker that our native informant, with a companion, should depart for the spot, which he professed to know, charged with the commission to bring back a donkey-load of specimens of the living plant.

June 5 was a day of some anxiety in Mogador, and the news which we received at breakfast brought full confirmation of Mr. Carstensen's apprehensions as to the disturbed state of the country. We had already noticed that the camel-drivers arriving from Marocco, or other places in the interior, are used, after discharging their goods in the town, to litter their camels outside the wall at a place close to the eastern gate. We now heard that during the preceding night a party of marauders from Haha had pushed their audacity so far as to attack and kill a camel-driver sleeping at the foot of the city wall, and drive off seven camels that were in his charge. Whether the guard at the gate close at hand slept soundly through the scene, or had their own reasons for non-interference, we failed to learn.

This was one of a series of incidents that was not completed until after our return to England; but as we were so directly concerned in the results of the disturbances in Haha, and as we learned the particulars in an authentic way from Mr. Carstensen, who had daily intercourse with eye-witnesses and actors in the drama, it is as well here to give the story as we learned it.

The Governor of Haha, the largest and most important province in the empire, which long maintained its independence of the Sultan, had hereditary claims to the government of the twelve cognate Shelluh tribes who make up the population. Although miserably fallen away from its ancient prosperity—in the time of Leo Africanus there were six or seven populous towns and several fortified places, where there is now nothing better than a village— the province still furnishes much agricultural produce and live stock, and sends hides, grain, oil, and other merchandise for exportation to the port of Mogador. The Governor, at the time of our visit, had long held his office ; by liberal contributions to the Imperial treasury, he had kept himself in the favour of the Sultan, while amassing for himself vast wealth ; and, according to the testimony of the French naturalist, M. Balansa, confirmed by the consuls who had visited him, he showed an appreciation of the advantages of civilised life, and a desire to maintain friendly relations with Europeans.

Thus wealthy, powerful, and feared, this man might have maintained his authority unbroken, but that by a continuous course of oppression and cruelty he at length

stirred up the spirit of resistance amongst his own people.
Vengeance, however atrocious, for acts of revolt is so fully
an admitted right of men in authority in Marocco that it
did not seem to count for much in the indictment against
him that on one occasion he inflicted on several hundred
—some said a thousand—insurgent prisoners the horrible
punishment of the ' *leather glove.*'   A lump of quicklime
is placed in the victim's open palm, the hand is closed over
it, and bound fast with a piece of raw hide.   The other hand
is fastened with a chain behind the back, while the bound
fist is plunged into water,   When, on the ninth day, the
wretched man has the remaining hand set free, it is to
find himself a mutilated object for life, unless mortification
has set in, and death relieves him from further suffering.
But, in addition to such acts as these, the Kaïd of Haha
was accused of capricious deeds of ferocity that revolted
the consciences of his people.   Among other stories of the
kind, we were told that on some occasion when he was
having a wall made round his garden, he happened to see
a young man jump over the low unfinished fence.   Feeling in some way annoyed at this, he had the unfortunate
boy's right foot struck off, as a lesson not to repeat the experiment.

In such a country, where the danger of revolt is so
terrible, the discontent among the people of Haha might
long have slumbered, but for the occasion given by the war
with the neighbouring province of Mtouga.   The spirit of
resistance spread rapidly, and it soon become apparent
that the position of the old Kaïd was becoming untenable.
At last he resolved upon flight, after previously securing
the aid and protection of his neighbour, the Governor of
Shedma.   Departing at night, with a train of women and
slaves, and with twenty-two mules laden with treasure, he
reached by daylight the borders of Shedma, just in time to
forestall pursuit from his outraged subjects.   He escaped
unharmed—although a bullet intended for him by the
pursuers struck in the hand his protector, the Governor of

Shedma—and continued his journey to the city of Marocco. On reaching the capital, he at once placed himself under he protection of the Viceroy,[1] and judiciously sacrificed half of his wealth as an offering to the imperial treasury. He was received with favour ; a handsome house was assigned to him as a residence ; and for anything we know to the contrary, he may be still enjoying, what is seldom granted to a high functionary in Marocco, a tranquil old age.

When the flight of the Governor was noised abroad in Haha, the people of the country proceeded, according to custom, to pillage and destroy the castle of their oppressor. Among other things brought to light were two skeletons built into the wall of one of the inner chambers. The Kaïd had two nephews, who were, or might have become, dangerous rivals, and it was in this way that he had disposed of them.

Among the stores found in the *kasbah* were several large earthen jars of butter, and others of honey, and these furnished forth a feast for the unbidden guests. The Kaïd was a thoughtful man, and even in the hurry of his departure he had not forgotten his disobedient subjects. The feast was not well over when the effects began to be apparent, and a large number of those who partook of it died in agony. The Kaïd had mixed a large quantity of arsenic with the delicacies which he had been forced to leave behind him.

Meanwhile, as we had seen, the relations between the people of Mogador and their neighbours had become very unsatisfactory. Indignant at the outrage committed before our departure, the Governor of Mogador thought it necessary to show his strength and enforce respect. He accordingly despatched 200 men, under the command of our old friend Kaïd el Hadj, to demand satisfaction for what had happened, and security for better conduct in the future. The result was exactly what might have been

The present Sultan of Marocco.

expected from the pious, but decidedly unwarlike, character of the leader. The Haha people explained to El
Hadj that they were more than a match for him and his
men, and that his wisest course was to return by the
shortest road to Mogador. The mind of the commander
was always open to prudent counsel, and he professed
himself convinced; and to save him and his men from
trouble by the way they were escorted to the gates of
Mogador by ten men from Haha.

Policy, however, soon effected what valour had failed
to achieve. A virtual blockade was established, and all
communication with Haha suspended. This may have
been inconvenient to the Mogador people; but, at the
worst, they could always obtain supplies from Shedma by
the road which we had followed from Aïn el Hadjar. To
the Haha people it soon became intolerable. Mogador is
their chief market. There they sell their provisions, and
it is the Mogador merchants who purchase their oil and
hides and other exports. An envoy was sent to re-estab-
lish friendly relations, and to entreat the foreign Consuls
to mediate in their favour. In token of entire submission,
they proposed that 2,000 men from the province should
come to make a peace-offering. The Governor judiciously
thought that number excessive—there was no knowing
what these wild hill people might do, if they fancied
themselves masters of the town—and agreed to receive a
deputation of 200 representatives of the Haha tribes. On
the appointed day they came, driving before them several
bullocks; and, on arriving before the Sultan's palace, proceeded to hamstring and slaughter them as a propitiatory
offering to the sovereign authority, whereupon friendly
relations were at once re-established.

While these events were undeveloped, the thoughts of
the Mogador people, European as well as native, were
fully exercised, although they felt secure from positive
danger. A part of the day was occupied by us in taking
leave of our escort and attendants, and in distributing

among them presents and rewards, well earned by some, and which could not be refused even to the less deserving. A great change had come over the appearance of many of our men since their return. Ambak and Hamed, who had made the journey on foot, with none but the scantiest and the poorest clothing, now appeared fresh from the bath and dressed in their best: Hamed looking especially dignified in a snow-white turban, and formidable-looking dagger stuck in his girdle. All seemed pleased and satisfied with the very moderate sums awarded them—to which in the case of the more deserving were added knives and other articles of English cutlery. The soldiers who had travelled all the way from Mogador received eight dollars (40 francs) each, while to the Marocco men were given five dollars each.

When the rest were disposed of, there remained the two officers. Towards Kaïd el Hadj, who had always maintained a friendly and respectful bearing, we had none but kindly feelings. When, in addition to the present of twelve dollars, Hooker handed him a silver watch, a large sheath-knife, and a few smaller presents, the old man was quite overcome; his eyes were filled with tears; and he took leave of us with many pious wishes for our future welfare. As regards El Hasbi the case was very different. He had from the very day of our departure from Marocco done his utmost to defeat our wishes and plans, and had not even the grace to veil his opposition beneath a civil exterior. Hooker decided on writing a letter to El Graoui, reporting our opinion of the conduct of his subordinate, and sending this to the great man with a sum of ten dollars, to be given to El Hasbi, or withheld, as he might decide.

The Haha troubles affected our pockets in an unexpected way. In ordinary times there is a constant demand for baggage animals at Mogador, and we were led to expect that we should sell at a profit the mules that we had bought at Marocco. The usual price at Mogador is from 8*l.* to 10*l.* for a serviceable beast. But in the present state of the neighbouring country the roads were considered unsafe, and traffic with the interior was almost completely stopped. We were therefore considered rather fortunate in selling our team of mules at about 5*l.* apiece.

In the course of the day we had a fresh illustration of native manners that somewhat amused us, for we were no longer in the frame of mind in which a slight abuse of authority could shock our European ideas. The Mogador brass-workers have a high reputation throughout Marocco, and during our first stay in the town we had ordered a variety of articles from the man who was considered the most skilful in his craft. The time fixed for delivering the goods was fully a week before the actual date of our return. When Mr. Carstensen heard that the order had not been executed at the appointed time, he sent a message informing the Governor of the fact ; and the latter forthwith had the man thrown into prison, and appointed a soldier to keep guard, and see that he did no other work than that promised to us. After two or three days the prisoner got some friend to intercede with Mr. Carstensen for his release, and at the instance of the latter the Governor relented. To-day the articles were all duly brought to the Consulate, and the maker seemed well content with the very moderate payment agreed upon.

We should have gladly made our return voyage in the comfortable steamer *Verité,* that had carried us from Tangier, but the date of her arrival on her return from the Canary Islands to Marseilles was uncertain. Meanwhile the *Lady Havelock* steamer, plying between the Marocco coast and London, had reached Mogador, and we resolved to take our passage by that conveyance to Tangier.

Farewell visits and the packing of our collections occupied the greater part of the day on June 6. At the wharf the port labourers were busy in shipping large bales of esparto grass, which chiefly comes from the adjoining province of Haha. This is now largely consumed by paper-makers in France and England. It is said that the greater part of what reaches England from Marocco is used in the paper-mills that supply the ' Times ' newspaper.

A caravan had lately reached Mogador from the Soudan, and we saw several bales of ostrich feathers lying on the wharf. They were imperfectly covered with coarse sacking ; and the outer layer of feathers, soiled and broken, seemed to be quite worthless, although the total value of such a bale must be very considerable. Some trade routes in Central Asia and elsewhere in the world involve terrible hardships to those engaged in the transport; but it seems that there is none nearly so formidable as that from Timbuktou to Marocco. In several directions the way across the Great Desert is facilitated by the occurrence of oases at moderate intervals ; as in the way from Tafilelt to Touat, and in the line followed by the Arab traders from Tripoli to Murzouk. But throughout the greater part of the way from Timbuktou to Akka there appear to be no true oases. Wells are few and far between, and the supply of water often miserably scanty ; and even when a caravan escapes all the dangers of the long way, and the bones of men and camels are not left to bleach upon the burning sands, the sufferings of the travellers must reach the verge of human endurance.[1] At the present day the regular caravans no longer attempt to reach Marocco or Fez by way of Tafilelt, the routes over the Atlas being too insecure. The bad reputation of the Oued Noun people is equally effectual in closing the coast route from the south to Mogador ; and the course adopted is by the oasis of Akka, lying south of the Anti-Atlas range, and about 100 miles east of Oued Noun. From Akka to Agadir the way

[1] The only European who is believed to have accomplished the journey is Caillé. There seems to be no reason to doubt that in some way he reached Fez from the south by land; but it is a question whether his account of the direct route from Timbuktou to Tafilelt is derived from native informants, or whether, in default of notes, a defective memory led him into errors and inconsistencies that throw a shade of doubt over his narrative.

rounds the western extremity of Anti-Atlas, through a hilly and populous country, which appears to be safe enough for any but Christian travellers.

Among other articles exported from Mogador is the brown gum arabic of commerce, which comes chiefly from Demnet, and elsewhere on the skirts of the Atlas east of Marocco. If native testimony is to be credited, the *Acacia* producing this gum is not different from that which we saw growing abundantly in Haha and elsewhere through the hilly country,[1] though we did not hear of any gum being exported from the western provinces. It seems unlikely that this plant, named by Wildenow *Acacia gummifera*, should be the only one of a group of allied species extending across Northern Africa that produces no gum. We were interested in finding that the parcel shown to us, sent for export from Mogador, was packed in the dry stems and leaves of a *Ceratophyllum*, a genus of aquatic plants not hitherto seen in Marocco.

During the afternoon the two natives who were despatched two days before returned safely from Haha, driving before them a donkey laden with the Agadir *Euphorbia*. Hooker's suspicion was at once verified. The plant of the coast region is quite different from the gum-producing species of the inland region, but in appearance it comes near the East Indian species, *E. officinarum* of Linnæus. Under the name *E. Beaumieriana*, the coast plant, which has yet not been collected in flower or fruit, along with *E. resinifera* from Demnet, and another new species brought by a native collector from the southern borders of Sous, has been carefully described by our friend M. Cosson.

In the course of the evening we went by invitation to a Jewish wedding, which was celebrated in the house of one of the chief Israelite families in the town. The proceedings were quite in accordance with the descriptions given by other travellers in Marocco. The bride, who sat cross-legged, arrayed in gorgeous attire, had regular fea-

[1] See Appendix D.

tures and large dark eyes, but seemed dazed and stupified
by the crowd, the noise, the glare of many lights, and the
heat of the close rooms, from which we were not sorry
soon to escape.

On June 7 the time for our departure arrived, and
towards sunset we went on board the *Lady Havelock*,
accompanied by our kind host, Mr. Carstensen, and by the
late Mr. Grace, the representative of one of the chief English
mercantile houses engaged in the Marocco trade, to whom
we were indebted for numerous marks of attention during
our stay in South Marocco.

We found in Captain Bone, who commanded the *Lady
Havelock*, an old acquaintance ; for it was in this steamer
that, on our first arrival, we crossed the Straits from
Gibraltar to Tangier. He had warned us not to expect
much good from Marocco or the Moors, and was always
well pleased when in the course of conversation he was
able to extract any facts to confirm his unfavourable pre-
possessions.

The north wind had been blowing freshly all day, but,
as usual, it fell at nightfall. The moment for heaving the
anchor had arrived ; we took leave of our Mogador friends,
and soon found ourselves once more gently rolling on the
broad Atlantic waves.

On the morning of June 8 we were before Saffi. Mr.
Hunot, the British Vice-Consul, soon came on board, and
we gladly accepted his courteous invitation to spend the
day ashore. Through his brother, who had so kindly
assisted us during our diplomatic struggle with the authorities in the city of Marocco, Mr. Hunot already knew of
our journey in the interior, and kindly interested himself
in forwarding our wish to make the best use of our time
at Saffi. This place is considered to be much hotter in
summer than Mogador ; yet, during an excursion of several
hours, we found the heat much less oppressive than it
commonl is at the same season on the shores of the
Mediterranean.

The form of the land here at once dictated the best course to be taken by a party of naturalists. On the north side a range of lofty, almost vertical cliffs rises from the sea beach, and leaves no space for any but a few marine plants. On the landward side of the town the hills show gentle slopes, in great part under tillage, and bare of trees. Along the shore southward the coastline is formed by reefs of friable tertiary rock, rising from thirty to fifty feet above the water's edge, and forming a shelf of level land, in great part overgrown with shrubs and small bushes. On the landward side the hills rise at first with a gentle slope, and then more steeply, until, about four miles from the town, they show a steep escarpment of limestone rock, locally known as the Jews' Cliff, and this was fixed upon as the limit of our excursion.

Mr. Hunot, who had accompanied us, had kindly provided horses for our use; but we found so many objects to interest us that the greater part of the way was made on foot.

We had not gone far along the sea rocks when we found them plentifully covered with a species of *Zygophyllum* altogether new to us. Though evidently allied to the *Z. album*, so common in Egypt and elsewhere in North Africa, this differed at first sight in the much greater size of the thick succulent branches and leaves. It turned out to be the *Z. Fontanesii* of Webb, a plant hitherto known only in the Canary Islands. Another characteristic plant of those islands was *Helianthemum canariense*; but this had already been found by us near Aïn el Hadjar.

When we reached the foot of the Jews' Cliff, which rises some 300 feet above the shelf of land at its seaward base, we resolved to divide our forces. From fragments already picked up, and from Mr. Hunot's information, we were led to think that some of the exposed beds of limestone must abound in fossils. Hooker resolved to make a search; and, with Mr. Hunot, ascended the face of the cliff, which is easy enough of access, and ultimately reached the top. He was rewarded by getting good specimens of several fossils, the most abundant being Echinoderms of a new type, since described by Mr. Etheridge in the ' Quarterly Journal of the Geological Society.' [1]

Ball meanwhile was engaged in botanising on the sea rocks and among the bushes at the foot of the cliff. In the former habitat he found two species heretofore known only at Mogador, *Andryala mogadorensis* and *Frankenia velutina*. Among the bushes there were no plants of special interest, but he nevertheless had an unexpected encounter. While searching about among the bushes a rustling in the dry grass caught his ear; he looked down, and there, within a yard of his feet, was *el efah*, the dreaded ' two minutes' snake,' nearly as thick and about as long as a man's arm. As the enemy was retreating, gliding gently among the bushes, there was no occasion to move, and he watched it for a few seconds till it disappeared. The glistening scales were of many colours, forming a sort of mosaic on a ground of pale brown, very much as represented in the plate to Jackson's ' Account of Marocco.' Even supposing that the virulence of the poison in the bite of this snake has not been exaggerated by popular report, it can scarcely be thought formidable to strangers unless they happen to be botanists. It keeps habitually to the cover afforded by the numerous small bushes of the coast region, and its form is so ill fitted for active motion that it can only strike a near object. The only danger arises from the chance of inadvertently hurting it while moving about in the places which it frequents.

About 4 P.M. we returned to dine at the British Consulate. At Mr. Hunot's table we met Mr. Jordan, the son of a British merchant engaged in the Marocco trade. He had been brought up in the country, spoke Moorish-Arabic familiarly, was used to dress as a Moor, and had established

---

[1] See the description of *Rotuloidea*, in Appendix F.

intimate relations with many of the natives. Emboldened by custom, he had on one occasion joined a party of Moorish merchants bound for Tarudant in Sous, and safely reached that place. Something, however, either in his appearance, or accent, or gesture excited suspicion; it was noised abroad that a Christian was in the town, and an excited crowd soon gathered round the house in which he and his companions were lodged. As the demeanour of the people became more and more threatening, the travellers barricaded the entrance, and prepared to defend themselves by force. After some hours, as evening was coming on, the assailants became more determined, and proceeded to pile up faggots round the building with the obvious intention of burning the house with its inmates. Just as matters were looking very serious, the Governor of the town made his appearance with a party of soldiers; the doors were opened, and the Governor said to Mr. Jordan: ' You have a horse, and you have from this till to-morrow morning to put a wide space between you and Tarudant; you had better lose no time.' Protected by the soldiers, the Englishman rode out of the city, and made his way towards Agadir by night, thence returning safely to Mogador.

Tarudant was once a large and flourishing city, and its gardens were famous throughout Marocco; but, like the rest of the country, it has fallen off from its former condition, and is now a poor and decaying place.

It is clear that in that part of the empire increasing religious fanaticism has accompanied declining prosperity. In the sixteenth century Tarudant was resorted to by English and French merchants, and it was the seat of active trade, and of manufactures in copper which was extracted from mines in the neighbouring chain of the Great Atlas. The population was apparently then altogether of the native Berber stock. In the course of the continued efforts made by successive Sultans to establish their authority in the Sous province, the Moorish element became more and more predominant in the towns, and

to this we may reasonably attribute their subsequent decline.

When Leo Africanus travelled in Sous, early in the
sixteenth century, Tarudant was only one of many large
and flourishing towns, and was much surpassed in importance by Tagavost, a place whose very name has disappeared from memory, and whose exact site is unknown
to modern geographers.

Mr. Hunot, who is well acquainted with the city of
Marocco, estimated the population at about 40,000, but
admitted that there were no materials for an accurate guess
on the subject. Fully one-fourth of the inhabitants had
been carried off by the last visitation of cholera, from
which the coast towns, with the sole exception of Mogador,
had also suffered severely.

The main check to population in the greater part of
the empire arises, however, from the recurrence of famines.
These sometimes are caused by locusts, but are then of a
partial and local character; but those consequent on the
occasional failure of the winter and spring rains are not very
unfrequent, and are of terrible severity. Among the means
resorted to at such times for supporting life, we learned
that the roots of a small plant of the *Arum* tribe are much
used. This, known to botanists as *Arisarum vulgare*, is
very common throughout North Africa, as well as in many
parts of the south of Europe. It flowers in this country
in winter, and the leaves wither and disappear in the
spring. The root, which is not so large as an ordinary
walnut, contains, as is usual in the Aroid tribe, an acrid
juice, which makes it quite uneatable in the natural state.
This, however, is easily removed by frequent washing
of the pounded roots, and the residue is innoxious and
nutritive. The same process has been applied with success
to the common European plant, *Arum maculatum*, as
well as to many exotic species of the same tribe.

Among the many difficulties that beset commercial
intercourse with Marocco, the frequent interruption of

internal traffic arising from frequent petty warfare between neighbouring tribes is not to be forgotten. A merchant may purchase a quantity of produce at what appears a remunerative price; but if he be unable to have it conveyed within a convenient time to the port whence it is to be shipped, his bargain may turn out a very bad one. At Mogador we had left things in a condition foreboding a complete suspension of communication with the interior; we now heard that owing to some local troubles the coast road from Saffi to Mogador was temporarily closed.

At nightfall we returned to our steamer, but found that we were to remain for the night in the roadstead of Saffi. On the next morning our obliging host, Mr. Hunot, again came on board, and we enjoyed his agreeable conversation until the time came for starting on the short run to Mazagan. We reached that place in the afternoon of June 9, and landed with Captain Bone at a wharf beside the Castle built by the Portuguese. It was proposed that we should go through the town, and visit the great cistern which was constructed during the prolonged Portuguese occupation of this place, and which enabled them to resist successfully the frequent sieges undertaken by the Moors. We preferred, however, to make use of the short time at our disposal in examining the vegetation near the shore on the north side of the town.

The net result of our short excursion was not very large or brilliant; but, in the case of a country so little known as Marocco, the interest of his collections to a naturalist does not mainly depend on the rarity or novelty of the objects he may happen to meet. Each plant or animal carried away contributes an item of information respecting the distribution of the organised world, the value of which it is impossible at the time to estimate. Travellers who happen to visit little-known countries would do well to remember that, with the most trifling expenditure of trouble, they may make useful contributions to natural science by preserving specimens of even the most insignificant-looking objects, provided always that these are afterwards placed in the hands of competent naturalists.[1]

We returned on board about sunset, but did not leave the roads of Mazagan till about 10 P.M. When we came on deck next morning, June 11, we were nearing the coast opposite Casa Blanca, and cast anchor soon after 7 A.M. We here found the *Sydney Hall*, belonging to the same owners as the *Lady Havelock*. She had left London on June 2, reached Casa Blanca on the 10th, and soon after our arrival started again on her outward voyage to Mogador and the Canary Islands. We had the pleasure of again seeing Mr. Dupuis, the active British Vice-Consul at Casa Blanca; but as our stay was to be short, and we had already made an excursion ashore, we did not now attempt to land.

During our return voyage our minds were once more exercised by the peculiar climatal conditions of this portion of the African coast. It did not appear that the cool temperature which had prevailed since our return to the neighbourhood of Mogador on the 1st inst. was considered in any way remarkable or unusual, although travellers who have visited the city of Marocco at this season speak of a temperature of 90 F. in the shade as not uncommon; and at Fez, though in the immediate vicinity of high mountains, still higher temperatures have been recorded. The direction of the wind on the coast in summer, which to the south of Cape Cautin is constantly between the north and north-east, is less uniform to the north of that limit; but the prevailing sign is NE., and this no doubt is the most important factor in determining the climate.

There is, however, another element that cannot be

---

[1] We have lately received a parcel not much larger than an ordinary pocket-book, containing specimens, or fragments, of about twenty plants, picked up by the commander of a Spanish ship of war, who landed on the African coast south of Oued Noun. Most of them are of great scientific interest and value.

overlooked. When we examine the chart exhibiting the oceanic currents in the North Atlantic, compiled at our Meteorological Office, and fix our attention on the portion lying between the 30th and 40th degrees of latitude, and extending from the coasts of Portugal and Marocco to 20° of west longitude, we find that the currents throughout this large area constantly move in a direction between SE. and SSE., with an average velocity which increases from about five miles per day in the longitude of Madeira, to at least ten miles as we approach within 100 miles of the shores of Europe and Africa. This velocity again diminishes with a nearer approach to land; and, from a few observations, it would seem that along the Marocco coast the current is deflected in a SW. direction, parallel to that of the coast line.

It is clear that in this continual flow of cool water from the north-west we have a cause which cannot fail to produce its effect on the climate of the adjoining coasts. It would be a matter of interest to ascertain how far the direction and velocity of the ocean currents are modified by the prevailing winds, which here set in nearly opposite directions in winter and in summer; but an answer to such an inquiry will require much time, and the accumulation of a large number of careful observations.

We completed our cargo by taking on board at Casa Blanca considerable quantities of maize, beans, oil, goat-skins and wool; and our captain resolved not to touch at Rabat, but to run direct to Tangier. Eighteen hours under steam carried us past Cape Spartel, and in the afternoon of June 12 we lay off Tangier.

Hooker's numerous and pressing engagements in England made him resolve to forego the pleasure of revisiting the neighbourhood of Tangier, and comparing the summer vegetation with that which we had admired two months before; he therefore determined to reach Gibraltar as soon as possible, with the hope of there catching the Peninsular and Oriental Mail Steamer for England. Ball

could not deny himself the opportunity for a full day's
botanising on ground so attractive, and therefore removed
his baggage ashore; while Hooker returned on board the
Lady Havelock, which was to cross the Strait during the
night.

On arriving at the Victoria Hotel, we learned that Sir
J. D. Hay had taken up his residence at his charming
villa on the Djebel Kebir, but we found awaiting us a
kind note enclosing a welcome packet of letters from
England. After a hasty dinner at the hotel, the time for
parting came, and Hooker got out through the sea gate
just before it was closed for the night. The mail steamer
had left Gibraltar for England on the same day that we
returned to Tangier; but on the following morning
Hooker found the steamship *Burmah*, bound from Bom-
bay to London, about to depart from Gibraltar, and after
a rather slow voyage he reached the Thames on the morning of June 21.

Ball enjoyed a capital day's plant-hunting at Tangier.
The morning was given to the sandy tract near the shore
and the course of the stream that passes by the east side
of the town. This now made a much more brilliant show
than it had done in the month of April. Many fine
Umbelliferæ and *Labiatæ*, then barely in leaf, were now in
full flower and fruit. Of these the queen was *Salvia
bicolor*, a magnificent species, usually four or five feet, but
sometimes eight or even ten feet high, much branched,
with leaves of varied form from twelve to eighteen inches
long, and great interrupted spikes of large blue and white
flowers.

The slopes of the Djebel Kebir, which had been so
brilliant in the spring, had now lost their splendour. The
gum cistus, the golden *Genista* and *Cytisus*, the heaths,
and many other ornamental species had long since shed
their petals, and had been succeeded by new comers, most
of them with comparatively inconspicuous flowers. For
the botanist, however, the fruit is often more important

than the flower, and the afternoon was not long enough to collect all the interesting species that presented themselves.

On June 14 Ball crossed the Strait from Tangier to Gibraltar in the ordinary small steamer. While awaiting conveyance to England he was detained three days, which were made short and agreeable by the hospitality of Sir W. Fenwick Williams, then governor of the fortress, and returned to England by a steamer bound from Calcutta to London, *viâ* the Suez Canal.

Our large collections reached England by the *Lady Havelock*, which arrived only about the end of June, and these, as well as the cases containing sundry purchases made in Marocco, were all in good condition.[1]

---

[1] As a general rule packages sent by English ships are rarely tampered with, unless they happen to contain wine or spirits, in detecting which the British seaman shows a marvellous readiness. When leaving England Hooker had carried out a nest of wooden cases intended for sending home living plants. In the innermost of these he had with his own hands placed two bottles of brandy as a provision for the journey. The lid of the inner box was screwed down, and this placed within the next, which was also screwed down, and this again within another. When the cases, seemingly untouched, were opened at Mogador, the brandy had disappeared.

CHAPTER XIV.

Resources of Marocco—Moorish Government a hopeless failure—
Future prospects of Marocco—Objections to European interference
—Answers to such objections.

SCIENTIFIC travellers, whose attention was mainly engaged
in their own special pursuits, and whose opportunities
for gaining information were restricted by ignorance
of the native languages, have no claim to speak with
authority of the condition and prospects of a country so
extensive as the Marocco Empire. But it would be strange
if we had failed to derive some conclusions from the results
of our personal observation and the information gained on
the spot.

Of the material resources of Marocco it is difficult to
say too much. Even under existing conditions, a great
portion of the territory is extremely fertile, and supplies
for export a large amount of agricultural produce. The
two natural disadvantages with which it has to contend
are, occasional deficient rainfall and the ravages of locusts.
For the first, the remedy is to be sought in irrigation.
The unfailing streams from the Atlas already serve to a
limited extent ; but the area of productive land might by
intelligent management be very largely increased. We
have seen an estimate of the quantity of water discharged
by the five principal streams that fall into the Atlantic
north of the Atlas, which fixes the amount at 9,000 cubic
feet per second ; and if to these were added the Moulouya,
which falls into the Mediterranean, and the Siss, the
Draha, the Asakka, and the Sous, which drain the southern
slopes of the main chain, we should probably double the

above estimate, and find an aggregate amount sufficient to
irrigate three millions of acres. These figures must of
course be considered as mere guesses; but there can be no
doubt that they indicate a very large reserve of unused
natural resources. With an almost unequalled climate,
there is scarcely any one of the productions of the warmer
temperate and subtropical zones that may not here be
obtained. Besides grain, the country now supplies large
quantities of olive oil, dates, oranges, and almonds, with
a little cotton. The latter may be largely increased; and
there seems to be no reason why coffee, tea, sugar, indigo,
and other valuable exotic produce, should not be raised in
the southern provinces.

There can be no doubt of the existence of mineral
wealth in the Great Atlas. We have the direct testimony
of Leo Africanus to the working of mines of copper and
iron in the districts visited by him; and specimens brought
by Shelluh mountaineers show that ores of lead, silver,
nickel, and cobalt are likewise to be found. The forests
of the Atlas would, if saved from wanton destruction, be a
further important source of national wealth.

Rich in all the material elements of prosperity, this
great territory, whose area may be roughly estimated at
190,000 square miles, is cursed by a Government which
has in the past wrought nothing but ruin and degradation,
and whose continued existence forbids the faintest hope of
future improvement.

Nothing seems to be more clear than the decadence of
the race who now represent the Arab conquerors of Mauritania. In their better days they united to martial vigour
and skill some aptitude for progress in arts and learning.
Works of public utility were not unknown; and, at a time
when nearly all Europe was plunged in intellectual darkness, Fez was one of the chief centres of Arabic culture.
The history of the last four centuries in Marocco has been
one of continuous and uninterrupted decline. Unable to
establish their authority over the larger portion of the

region which they claim to govern, the Sultans have left
to anarchy the mountain region into which the best part
of the population was compelled to retire when driven
from the fertile lower country. Over the provinces wherein
they are able to enforce it, the rule of the Moorish Sultans
is little else than an organised system of extortion, in
which unchecked license is given to the agents of the
central authority, on the sole condition of making this the
final depository of whatever wealth the country can produce. The springs of industry and enterprise are broken ;
no man can dream of improving his own condition or that
of his family, unless by elaborate fraud and concealment
he can hoard up wealth, which he dare not employ in any
way useful to the community.[1]

When we inquire what prospect there may be of any
escape from the miserable condition to which Marocco is
now reduced, no hopeful answer can be found. The most
sanguine believer in the future of the Mohammedan races
can suggest nothing better than the chance of the appearance of a Sultan, intelligent and energetic, and powerful
enough to revive the traditions of the better days when
rulers took some thought for the welfare of their subjects,
and who might initiate an era of security and progress.
But, to say nothing of the improbability of the appearance
of such a man in a family that by frequent intermixture
with the black race has become more Negro than Moorish,
it seems a pure illusion to imagine that even an extraordinary man seated on the throne of Marocco, and surrounded by such agents as he would have at hand, could
accomplish salutary reforms, and, more than that, to suppose that these could have any permanence. It is conceivable that if the Moor and Arab did not stand in the
way, and the Berber stock were restored to their original
inheritance, a great ruler might overcome their fatal tend-

---

[1] The stories and fables given in Appendix G afford a striking
commentary on the working of the existing system of so-called government in Marocco.

ency to tribal decomposition, weld them into a nation, and set them on the path of progress. History affords examples of some such transformation among vigorous barbarians or semi-savages. But with an effete race, corrupted by luxury, who have lost the spirit, but preserved many of the traditions, of a decayed civilisation, no such miracle is to be worked. Men of great powers, such as one cannot expect to see on the throne of Marocco, have ere now failed in the attempt, or the little they have effected has died with them.

No rational believer in progress can cling to the belief that this is the spontaneous tendency of all branches of the human race, the ultimate condition to which, with whatever delay, all must conform. Far from this, all history shows that the task of leading mankind on the onward road has always been the privilege of a few races only. The larger part of the earth is even now inhabited by people either in a stationary or a retrograde condition, and of the latter state Marocco affords one of the most striking examples.

The one reasonable prospect of improvement in the condition of Marocco is to be sought in its passing under the control of a civilised State, strong enough to overcome speedily the inevitable resistance of the Moorish ruling class, and advanced enough to consult the welfare of the people it undertakes to govern. If we ask what European State is by character and circumstances best fitted for such an undertaking the answer must be—France. Having already achieved with tolerable success a similar task in the adjoining region of Northern Africa, the French have every motive to add to their possessions a territory offering far greater natural advantages; and it is probable that they would have already effected the conquest, but for the inevitable jealousy of other European Powers. The French are not successful colonists ; nor have the economic results of their annexation of Algeria been as brilliant as might have been expected. But in Marocco colonisation

is not to be sought or desired.   Under a government
affording security for industry the Berber would settlein
the unoccupied lands of the lower country, and carry out
under intelligent control the works which would fit them
for a large increase of population.

Many readers who hold to the traditional political
ideas of the past will shrink from the conclusion here
expressed.   Not concerning themselves with the results of
such a change on the future condition of Marocco, they
will urge that such a great territorial extension of the
French possessions in Africa would increase to a formidable
extent the power of our ancient rival throughout the
Mediterranean region; and they will with justice argue
that it is not for the general interest of the civilised
world that any single Power should obtain a preponder-
ating influence over the rest.

Experience seems to supply an answer to these objections.   If extensive foreign possessions be in some respects
a source of strength to a country, they not less certainly
are a cause of weakness in others; and in the case of a
Power not holding maritime supremacy, the possession of
valuable dependencies easily assailable by sea acts as a
weighty check on the aggressive tendencies of the people.

A cordial acquiescence in the extension of French
territory westward, might reasonably obtain in return a
diminution of the jealousy with which our neighbours
view the increase of English influence on the east side of
the African continent.   A cynic may remark that the
policy here suggested would resemble an agreement between freebooters for the division of spoil; but, in truth,
we believe it to be a mistake to suppose that there would
be any spoil to distribute.   It is more than doubtful
whether any future extension of the African possessions
either of France or England would more than pay the
necessary expenses of occupation and administration.   The
gain to both countries would be of a different order—the
outlet provided for the healthful play of energies cramped

within the limits of an old society, and the sense, invigorating to the whole nation, of accomplishing a useful part in the world's development.

Objections to every attempt on the part of a modern civilised State to undertake the government of inferior races have been urged on various grounds by writers of the highest reputation.

The barriers established by differences in mental condition, in traditions, and inherited ideas, between peoples in a different stage of development, are easily shown to create formidable difficulties in the way of mutual understanding and appreciation, which must precede all useful efforts to carry the less advanced races along the path of progress. The history of British India where, at least during the present century, the experiment has been tried on the largest scale, and with the most genuine regard for the welfare of the governed populations, supplies many an example of the errors inevitable in so difficult an enterprise. Measures devised with the best intentions have sometimes failed altogether in achieving the expected effect, or, when this has been attained, have created discontent, because not corresponding with the ideas of the native population.

How much better it would be, say objectors of this class, to let these backward races work out for themselves the problems of material and mental development, in conformity with the conditions which nature and history have imposed, than to attempt, in the face of your own admitted ignorance, to play the part of Providence towards them.

If the discussion were to turn upon the destiny of a country wherein the elementary conditions of social order had been secured, wherein progress of some kind, at however slow a pace, was not an impossibility, it might be possible to admit the force of these arguments. But it is forgotten that in point of fact most barbarous countries have failed to reach this indispensable preliminary stage.

However diverse the conditions and the ideas of the human race, the primal requisites for social order are everywhere the same. Security for person and property, the protection of the weak against the strong, tribunals before which justice can be obtained without fear or favour—where these do not exist, the Power, whatever it may be, that confers them on a people is a beneficent one, and for the sake of these any errors that it may commit in its government will be condoned by posterity.

These remarks apply with especial force to such a country as Marocco, where, under the yoke of invaders, the greater part of the population has been for centuries constantly declining in material and mental condition.

When all has been said, it must be felt that theoretical considerations are little likely to prevail against that which history declares to be the uniform condition of human progress. As a general rule the most vigorous nations are those in which the increase of population is most rapid, and extension into new territories is their inevitable destiny. Statesmen and rulers may to some extent guide and control, but they are powerless to prevent the operation of natural laws. The choice, in regard to the inferior races, seems to be whether they shall fall under the rule of the stronger, and be gradually modified by the influence of new ideas and institutions, or whether they shall disappear altogether and give place to the new comer. Where, as has too often happened, the latter process is effected by injustice and violence, the evil to the world arises not so much from the loss of a race unfitted to bear the strain of competition, as from the moral deterioration that ensues to the invaders.

Amongst the opponents of the extension of European rule over the adjacent continents must be reckoned those who base their objections mainly on economic grounds. If the question of the French occupation of Marocco should arise in a practical shape, it is little likely that French statesmen will be withheld by considerations which, even in England, have not obtained wide acceptance; and it would be out of place to discuss them here. It is, indeed, impossible to deny that there is a share of truth in the views of those who hold that colonies and foreign possessions do not, as a general rule, directly add to the prosperity of a country. If the aim of any nation were merely to attain a high level of material well-being, and it could either restrain its citizens from intercourse with less civilised people, or be content to forego the duty of protecting them, it might be possible to avoid entering on the path which inevitably leads to extension of territory. Fortunately for the human race, such ideas never have prevailed among those nations which have played any important part in history. If the instinct of adventure, that has brought the more advanced races into contact with the barbarian and the savage, has always been alloyed by association with the baser passions of some, it has also been ennobled by the higher aims of others. To lay, in new regions, the foundations of civil society; to establish, more or less imperfectly, the reign of order and justice, and to secure the protection of the weak against the strong—these have been the tasks hitherto achieved by the ruling races of the world; and as knowledge has increased, as the difficulties of social progress have become better understood, and a stricter code of justice in dealing with the inferior races is gradually becoming established, it is allowable to hope that the inevitable changes may be accomplished with less of human suffering and with better success.

Rome might have been a happier State if its citizens had confined their ambition to make it a commercial emporium for the neighbouring tribes of middle Italy, and, content with self-defence, had refrained from all distant enterprise; Carthage need not have tempted her fate, if she had been satisfied with her own corner of Africa; but then the world would have had no history, and the series of changes from which modern civilisation has been developed would have been for ages, it may be for ever, delayed.

Of the entire African continent it may be truly asserted that, with the exception of the small portions held by England and France, its condition, for at least thirty centuries, has been either stationary or positively retrograde. The main cause that has maintained unbroken the long night of barbarism throughout so vast a region must be sought in the physical obstacles that prevented the ruling races of the world from extending their power southwards from the shores of the Mediterranean. For a time it appeared that the Saracen conquerors of North Africa were destined to spread the light of a relatively high civilisation over a great part of the continent. But that race is effete ; it is gradually losing ground ; and it remains for the nations that claim to lead the van in the onward march of the human race to undertake the work that awaits them.

APPENDICES.

APPENDIX A.

Observations for determining Altitudes of Stations in Marocco.

BY JOHN BALL.

THE instruments provided for the measuring of heights during our journey in Marocco were, in the first place, two mercurial barometers belonging to Sir J. Hooker, which were unfortunately left behind at the last moment by his attendant who had them in charge. Mr. Ball carried an aneroid barometer, by Secrétan of Paris, which, during many mountain journeys before and since, has performed very satisfactorily; and Mr. Maw had a small pocket aneroid of ordinary construction, not deserving of much confidence.

At Sir J. Hooker's request, Mr. Carstensen, then British Vice-consul, and M. Beaumier, French Consul at Mogador, both recorded observations of the barometer and thermometer twice daily (at 10 A.M. and 4 P.M.) during the period of our stay in and near to the range of the Great Atlas. Mr. Carstensen's instrument was a mercurial barometer, apparently a moderately good instrument; but, inasmuch as it showed itself more sluggish than M. Beaumier's instrument, and the amplitude of its variations was less considerable, its records do not appear to deserve equal confidence. On rendering the measurements into millimetres, and correcting both instruments so as to bring the indications to 0° C. at the sea level, the observations with the mercurial barometer fall short of those of the other instrument by a mean difference of 5·5 mm., the chief cause of the discrepancy being apparently due to the scale of the former being unduly low. Comparing corrected observations for ten days of very settled weather, during which the utmost range of either instrument did not exceed $2\frac{1}{2}$ millimetres, we have the mean pressure

By Carstensen's instrument = 755·30 mm.
    ,, Beaumier's   ,,   = 760·80 mm.

The error of the last-mentioned instrument does not probably exceed 1 mm. in excess of the true pressure, and, if the observations of the mercurial barometer were used, it would be expedient to apply to them a correction of + 4·5 mm. But, in addition to the circumstances already mentioned, it must be noted that Mr. Carstensen's observations extend over but eighteen days; while M. Beaumier's record covers twenty-six days, from May 11 to June 5 inclusive. For these reasons it has appeared best to make use exclusively of the record supplied by M. Beaumier. His instrument was an aneroid barometer of the construction adopted by its maker (Leja), called in Paris *baromètre holostérique*. The readings were recorded daily at 10 A.M. and 4 P.M., and are carried to intervals of the quarter of a millimetre.

The first questions that arise in applying observations to the determination of altitudes relate to the corrections applicable to each instrument. The corrected readings of Secrétan's aneroid at Tangiers, and during the voyage between that place and Mogador, varied from 760 mm. to 761·5 mm., and may safely be assumed to be nearly correct; but, on arriving at Mogador, they fell considerably. Inasmuch, however, as on comparison with the mercurial barometer at the British Consulate the difference was inconsiderable, the fall was attributed to the condition of the weather at that time. It was only on our return to Mogador from the interior, when a direct comparison between Secrétan's and Leja's instruments disclosed a difference of 7·3 mm. between the readings, and a further comparison between the recorded observations of the mercurial barometer and M. Beaumier's instrument showed a difference of about 5·5 mm. between the scales of those instruments, that it became clear that Secrétan's aneroid had suffered some change at or about the time of landing at Mogador. A careful comparison of all the observations leaves no ground for supposing that this arose from any gradual process; and it seems almost certain that' by one of those accidents to which the best aneroids are exposed, a casual blow, received about the time of landing at Mogador, caused the fall of 7 or 8 millimetres which was then observed. It is quite possible that more complete accuracy would have been attained by applying a correction of — 1 mm. to M. Beaumier's observations ; but it was thought more convenient to treat the discrepancy between the instruments as altogether due to error in Secrétan's instrument, and to apply to its readings in South Marocco the correction + 7·3 mm. So far as regards the altitudes determined by comparison with the Mogador observations, the difference between the method adopted and that above suggested is quite insensible ; but with respect to the altitudes given in the following table as determined between April 29 and May 10, wherein the barometric pressure at Mogador is assumed at 760 mm., it is clear that, if the error of Secrétan's instrument has been overcorrected to the extent of 1 millimetre, the altitudes given in the table should be increased by some 12 or 13 metres.

The next corrections requiring consideration are those arising from the temperature of the instrument at the time of observation, and in reference to this point the best makers of aneroid barometers are much open to criticism. They assert, and with approximate accuracy, that in the best instruments compensation for the effect of temperature on the instrument is provided; but they forget that in order to compare the indications of the aneroid with those of the mercurial barometer, or to apply to them any of the formulæ used for calculating altitudes, it is necessary to know at what temperature the column of mercury stands, the length of which is assumed to be shown by the scale of the aneroid. In point of fact, the scale of the latter instrument, when carefully laid down, is determined by direct comparison with the mercurial barometer under varying pressures, and the proper course would be to inscribe on the case of the aneroid a record of the temperature at which that comparison was made. From inquiries made of some of the best makers it seems probable that the best approximate correction is obtained by assuming the reading of the aneroid to correspond with that of the mercurial barometer at the temperature of 15° C., and this has been applied in the annexed table.

The height of M. Beaumier's instrument above the sea level

being about 10 metres, a small correction of + 0·9 mm. has been made in order to obtain the corrected reading adopted in the following table for the ' Mogador barometer.'

As most of our observations at stations in South Marocco were necessarily made either early in the morning or late in the evening, while those at Mogador were registered at 10 A.M. and 4 P.M., the pressure at the latter place corresponding to the hour of each of our observations has been found by intercalation. There is of course an obvious possibility of error here; but, except on a few occasions, when the changes of pressure were considerable and rapid, the amount is probably trifling.

It is familiar to all who have given attention to this subject, that one of the chief causes of error in the results obtained from barometric observations for altitude arises from the impossibility, in the present state of knowledge, of obtaining with tolerable accuracy the temperature of the stratum of air lying between the lower and the higher stations. This is especially true in climates such as that of South Marocco, where the sky is commonly clear, and the air relatively dry. The cooling of the surface at night, and the heating in the sunshine by day, have an effect on the layer of air in contact with that surface, and still more on the traveller's thermometer, which at the best is imperfectly protected from radiation, out of all proportion to the actual cooling or heating effect on the air not in immediate proximity to the soil. As far as circumstances permitted, it was sought to take observations about an hour after sunrise and very soon after sunset, so as to diminish to the utmost this source of error.

It remains true, in the writer's opinion, that when all these sources of error in the determination of heights by means of the barometer have been put together, there remains one surpassing all the others in amount which altogether escapes our means of correction. The formulæ employed for the reduction of observations to numerical results are, and must be, based on the assumption that a condition of equilibrium between the forces acting on the instruments at each station has been attained; whereas the utmost that can be asserted is that there is a continual tendency towards such equilibrium, requiring a variable time to effect it. But before equilibrium can be attained new changes occur, and the process of adjustment recommences. Even as regards stations near enough to be within sight of each other, repeated observations, however carefully corrected, give sensibly different numerical results, and when the stations are widely separated the discrepancies become serious in amount. The best course for a traveller in a mountain country is to endeavour to ascertain as nearly as possible the altitude of some fixed station by taking the mean of several observations compared with his distant station, and then to determine the altitude of the higher points reached near to such fixed station by comparison with an assumed reading of the barometer at the latter as derived from intercalation.

The altitudes of the stations at Hasni and Iminteli, given in the following table, derived from several comparisons with the Mogador readings, are probably nearly correct. That of Arround, as derived from comparison with Mogador at a time when the oscillations of pressure were relatively great and rapid, does not deserve much confidence; and the mean of two comparisons with Hasni has been preferred, the more readily as this nearly agrees with the result obtained from a boiling-water observation.

For the reduction of our observations the formula proposed by Count St. Robert, and first published in the *Philosophical Magazine* for 1864, has been preferred, and, for convenience, the tables based on that formula, published by the same author in the Memoirs of the Academy of Turin for 1867, have been used.

It is true that in the construction of the latter tables a value has been assumed for the constant expressing the rate of diminution of density in the atmosphere corresponding to uniform increase of altitude that is not constantly correct; but it would appear that the error resulting from this is but trifling. In regard to the greatest elevation attained by us in the Atlas, the difference in the measurement obtained by using the tables from that ascertained by accurate computation from the formula does not exceed 5 metres.

It may here be remarked that the altitudes inserted in some letters from Sir J. Hooker to the late Sir Roderick Murchison, which were published in the Proceedings of the Royal Geographical Society for 1871, and also most of those given by Mr. Maw in a paper presented to the Geological Society in January, 1872, were roughly calculated at the time when the party were

TABLE OF ALTITUDES, DEDUCED FROM BAROMETRIC OBSERVATIONS IN MAROCCO.

Date, 1871	Hour	Place of Observation	Observed Barometer	Corrected Barometer	Corrected Mogador Barometer	Thermometer in Air, Fahr.	Altitude in Metres	Altitude in English feet	Observations
April 11	10 P.M.	Nahum's house, Tetuan, second floor	754·4	752·6	—	61	84·8	278	Assumed pressure at sea level 760 mm.
,, 12	6 A.M.	Do.	756·2	*753·8	—	55	70·0	230	* As the river near its mouth can be only about 3 m. above the sea, a correction of −0·6 mm. is inferred for April 12.
,,	7 A.M.	Ford, Tetuan River	762	*759·6	—	60	4·78	16	
,,	10 A.M.	Upper limit of Chamaerops	780	*727·6	—	61	374·0	1,227	
,,	3 P.M.	Ridge of Beni Hosmar	685·2	*682·8	—	60	926·0	3,038	
,,	10 A.M.	Douar Arifi	748·5	†754·5	—	70	64·5	212	† See note as to corrections of Secrétan's instrument after arrival at Mogador. Assumed special correction for April 29 + ·05 mm.
,, 30	9.15 A.M.	Souk el Tleta	724	729·6	—	76	360·3	1,182	Assumed pressure at sea level 760 mm.
May 1	5 A.M.	Camp, Ain Oumast	724	729·6	—	54	345·5	1,134	Do.
,,	1 P.M.	Well under Hank el Gemmel	720	725·6	—	77	410·1	1,345	Do.
,, 2	2 P.M.	Summit of Hank el Gemmel	712·6	718·2	—	78	502·4	1,648	Do.
,, 2	6.30 A.M.	Camp, Sheshaoua	724	729·6	—	57	347·8	1,141	Do.
,,	1.45 P.M.	Ain Beida	720	725·6	—	80	412·4	1,353	Do.
,, 4	2 P.M.	Marocco: Palace of Ben Dreïs, 40 ft. above Piazza	712·5	718·1	—	78	508·6	1,652	Do.
,, 5	4 P.M.	Do.	713	718·6	—	79	499·3	1,638	Do.
,, 6	7 P.M.	Do.	710·5	716·1	—	73	523·7	1,718	Do.

Date		Time	Place							Remarks
May	7	7 A.M.	Do.	710	715·6	—	72	529·0	1,736	{Assumed pressure at sea level 760 mm.
"	8	5 A.M.	Do.	712	717·6	—	72	504·0	1,654	Do.
			Do.	—	—	—	—	511·9	1,679	Mean of five observations.
"	9	11 P.M.	Camp at Mesfioua	694·0	699 6	—	78	733·2	2,406	{Assumed pressure at sea level 760 mm.
"	9	6 A.M.	Olive Grove below Kaïd's house, Tasseremont	698·5	699·1	—	58	729	2,392	Do.
"		Noon	Camp by Ourika river, below village of Achliz	667·5	673·2	—	69	1,077·1	3,534	Do.
"	11	10 P.M.	Do.	681·0	686·7	—	59	874·2	—	Do.
"	11	Noon	Do.	681·5	687·2	—	71	887·0	—	Do.
				—	—	—	—	880·6	2,889	Mean of two observations.
"	10	4 P.M.	Camp, Ourika valley (Assghin)	669·5	675·2	—	72	1,044·4	3,427	{Assumed pressure at sea level 760 mm.
"	11	6 P.M.	Summit of pass to Reraya	664·0	670·05	759·7	62	§1,094·3	3,590	§Comparison with Mogador observations reduced to sea level.
"	12	6 A.M.	Camp Tassilunt, Reraya	674·0	679·35	760·4	56	§963·1	3,160	
"	13	7 A.M.	Camp Hasni, in Aït Mesan valley	652·0	656·5	761·45	57	§1,263·3	—	
"	14	10 P.M.	Do.	651	656·04	760·9	59	§1,274·0	—	Do.
"	15	9 A.M.	Do.	617	653·77	758·8	56	§1,297·4	—	Do.
"	17	10 P.M.	Do.	651	654·83	762·2	60	§1,292·6	—	Do.
				—	—	—	—	1,281·8	4,205	Mean of four observations.
"	13	Noon	Adjersiman, village in Aït Mesan valley	622	626	762·1	59	§1,687	5,535	§Comparison with Mogador.
"	15	2 P.M.	Village of Arround	602	606·16	762·1	52	§1,950	—	Do.
"	15	6 P.M.	House, Arround	597	605·5	756·6	49	§1,947·5	—	Do.
"	16	6 A.M.	Do.	598	606·5	756·6	46	§1,926·4	—	Do.
				—	—	—	—	1,941·3	6,370	{Mean of three observations.
"	13	2 P.M.	Arround, by comparison with Hasni	—	—	—	—	1,968	6,463	{Differences by St. Robert's method 686·2 m. and 690·2 m. respectively.
"	15	6 P.M.	Do.	—	—	—	—	1,972		

TABLE OF ALTITUDES—*continued.*

Date, 1871	Hour	Place of Observation	Observed Barometer	Corrected Barometer	Corrected Mogador Barometer	Thermometer in Air, Fahr.	Altitude in Metres	Altitude in English feet	Observations
May 16	8 P.M.	Arround, by boiling-water observation, at 202·2 Fahr.	—	—	757·1	40	§1,976·1	6,483	§ Comparison with Mogador.
		Same, height adopted on comparison of all observations	—	—	—	—	1,970	6,463	§ Comparison with Mogador.
,, 15	1 P.M.	Highest olives in Aït Mesan	632·5	640·3	757·6	59	§1,488	4,882	§ Comparison with Mogador.
,, 16	9 A.M.	Saint's tomb below Tagherot pass	568	—	—	39	†2,393·2	7,852	† By comparison with Arround.
,,	2.30 P.M.	200 ft. below summit of Tagherot pass	498·5	—	—	25	†3,439·4	11,284	Do.
,,	,,	Estimated altitude of Tagherot pass	—	—	—	—	†3,500·4	11,484	Do.
,, 18	8 P.M.	Camp, Sektana	646	684·1	764·3	58	§1,378·2	4,525	§ Comparison with Mogador.
,, 19	10 A.M.	Do.	647	649·3	764·1	65	§1,379·1	—	Do.
,,	10 P.M.	Camp, Amsmiz	672	675·3	762·7	58	1,018·8	—	Do.
,, 20	9 A.M.	Do.	672	674·9	763·1	72	1,049·0	—	Do.
,, 21	9 A.M.	Do.	672·5	676·55	761·85	70	1,024·2	—	Do.
							1,030·7	3,882	Mean of three observations.
,,	2 P.M.	Iminteli, Amsmiz valley	647	651·4	761·6	64	§1,845·2	—	§ Comparison with Mogador.
,, 22	6 A.M.	Do.	646·5	651·0	761·5	60	§1,344·0	—	Do.
,, 23	8 A.M.	Do.	646	651·6	760·15	65	§1,348·1	—	Do.
							1,345·8	4,415	Mean of three observations.
,,	—	Do.	—	—	—	—	**1,348·8	4,425	{** Altitude by comparison with Amsmiz on May 21.

Date	Time	Station					Altitude adopted.	Mean of four results.	Remarks
May 28	—	Iminteli	—	—	—	—	††1,846·5	4,418	{ †† Altitude adopted. Mean of four results.
„ 22	9 A.M.	Halt at base of Djebel Tezah	629	—	—	64	§§1,708	5,604	§§ Comparison with Iminteli.
„ „	2.30 P.M.	Summit of Djebel Tezah	512·5	518·0	760·85	60	§§3,359·7	11,023	§§ Comparison with Mogador.
„ „		Do. do. do.	—	—	—	—	§§3,340·5	10,961	§§ Comparison with Iminteli.
		Do. do. do.	—	—	—	—	3,350·1	10,972	Altitude adopted.
„ „	4 P.M.	Upper limit of Quercus Ballota	565	—	—	59	§§2,490	8,170	§§ Comparison with Iminteli.
„ 24	5.30 A.M.	Kasbah at Amsmiz	666·5	672·3	759·9	61	§1,063·5	3,489	§ Comparison with Mogador.
„ 25	8 A.M.	Kasbah at Mzouda	694	699·5	760·1	65	§721·3	2,367	Do.
„ 26	2 P.M.	Halt near Kasbah, Keira	688	692·7	760·1	70	§814·3	2,671	Do.
„ 27	8 A.M.	Camp Seksaoua	682·5	686·9	761·4	64	§879·5	—	Do.
	6 A.M.	Do. do.	683	687·2	761·6	60	§868·5	—	Do.
		Do. do.	—	—	—	—	874	2,867	Mean of two observations.
„ 29	6 A.M.	Camp below Milhaïn	670	674	760·8	58	§1,085·3	3,397	§ Comparison with Mogador.
„ „	5.30 P.M.	Watershed between Aïn Tursil and Mtouga	656·0	661·15	760·7	52	§1,190·2	3,905	Do.
„ 30	7 A.M.	Camp by Kasbah, Mtouga	677·0	681·4	761·4	58	§§940·3	3,085	Do.
„ „	8 P.M.	Camp, Mskala	716·0	719·8	761·9	60	§§466·3	—	Do.
„ 31	8 A.M.	Do. do.	715·5	718·5	762·7	65	§§486·3	—	Do.
		Do. do.	—	—	—	—	476·3	1,562	Mean of two observations.
„ „	5 P.M.	Room 6 m. above Court of Kasbah of Shedma	720	722·8	762·9	68	§437·4	—	§ Comparison with Mogador.
June 1	8 A.M.	Do. do. do.	720·5	722·0	764·35	67	§446·1	—	Do.
		Do. do. do.	—	—	—	—	441·8	1,449	Mean of two observations.
	—	Court of the Kasbah	—	—	—	—	436	1,430	§ Comparison with Mogador.
„ 2	6 A.M.	Camp, Aïn el Hadjar	744·5	746·4	763·7	58	§154·8	—	Mean of two observations.
„ 3	5.30 A.M.	Do. do.	744·0	746·6	763·0	56	§152·1	504	Comparison with Mogador.

travelling in Marocco, and before the necessity for a considerable correction to the readings of Secrétan's aneroid had become apparent. The difference arising from this and other corrections applicable to the highest points reached by us is considerable, and requires a deduction of about 500 feet from the estimated height of the Tagherot pass, and about the same from the calculated altitude of Djebel Tezah. The corresponding error in the calculated altitudes for the low country stations, e.g. those between Mogador and Marocco, averages about 200 feet.

J. B.

APPENDIX B.

Itineraries of Routes from the City of Marocco through the Great Atlas.

THE information respecting the routes here given was supplied by a Jew named Salomon ben Daoud, an inhabitant of the city of Marocco engaged in trading operations with the natives of the portions of the Great Atlas wherein the authority of the Sultan is recognised. In the absence of more accurate reports, it appears desirable to publish this slight contribution to the topography of a country altogether unknown to Europeans, excepting so far as we were able to visit a few of the places enumerated. To assist those who may hereafter seek to follow any of these routes, the names of places inserted in the French map of Marocco by Captain Beaudouin, or in the map annexed to this volume, are distinguished by an asterisk. The distances are reckoned by hours, one of which may be counted as equivalent to four miles in the plain, and to a somewhat lesser distance in the mountain. A day's journey usually varies from eight to ten hours. The spelling of the names is made to agree with that adopted throughout this work, the vowels having the same sounds as in most European languages, and not those peculiar to our country.

ROUTE 1.

Marocco to Demenet, and Excursion from Demenet to places in the neighbourhood.

Marocco to Aïn el Berda	. . . . .	3 hours.
„	*Sidi Rahal	5 „
„	*Oued Tessout (ford the river) .	4 „
„	*Tidli (a mountain)	2 „
„	Draha	1 hour.
„	*Demenet	3 hours.
		18 „

This road is undulating, with hills and valleys, or hollows.

From Demenet cross over the river Emhasser, and proceeding for one hour on the mountain you will reach a place called Iminifri, on a high mountain, which contains an opening or pass only just large enough for one person to creep through on his hands and knees, the length of the pass being about 100 yards; and when through it you will find open ground on the top of the high mountain. There will be seen remains (ruins) of old Christian buildings, in which live many birds. From the upper part of this mountain overhanging parts (or cliffs) branch out downwards against the mountain, towards the River Tor, of 500 yards long; but these overhanging parts do not reach the water of the river.

[The places here spoken of apparently lie N.W. of Demenet —the El Acchabi of the French map. The river Tor is probably the Oued Lakdeur of the same map.]

ROUTE 2.

Demenet to the Sources of the Oued Tessout.

Demenet to Aït Cid Hassan (between mountains)	.	4 hours.
„	*Aït Emdoual	1 hour.

In Aït Emdoual is a river one day's journey long. There are inhabitants along the river. From this to Aït Affan one day's journey over barren desert ground uninhabited.

From *Aït Affan to Ansai (contains some inhabitants)	.	? hours.
„	„	Aït Kassi	2 „
„	„	Tel Khedit	2 „

Tel Khedit is a mountain, and contains the source of the river Tessout; and on this mountain the snow remains both in summer and winter.

[This route agrees in many respects with the indications of the French map. The name Aït Chibatchen, there laid down south of Demenet, is probably the Aït Cid Hassan of the itinerary. It appears, however, that the importance of the mountain chain on the north side of the upper valley of the Tessout must be much exaggerated by the hill-shading on the map. The main chain of the Atlas is undoubtedly that on the southern side of that valley. The head of the valley is, on the French map, united to the province of N'tifa. *A priori* probability and the wording of the itinerary suggest that it all belongs to Demenet.]

ROUTE 3.

Marocco to N'tifa.

Marocco to Zourt ben Sessy .	.	.	.	.	2 hours.
„ Ras el Aïn	.	.	.	.	2 „

[Here is a mountain called Bou Surkar, or stony.]

„ *Tamlelt	.	.	.	.	3 hours.
„ *N'tifa	.	.	.	.	1 day.

[None of these places seem to be laid down on the French map, unless Tendalet be the same place as Tamlelt. N'tifa is properly the name of the province. The particular place so named by our informant, is probably the residence of the Kaïd, or Governor. It seems likely that this is very near the place marked Bezzou on the French map.]

ROUTE 4.

Demenet to N'tifa.

Demenet to Aït Mazan (valleys and hills)	.	.	3 hours.
„ *N'tifa	.	.	.	3 „
				6 „

On this road is found the Gum Euphorbium plant, or tree, and the trees producing the brown gum arabic. From N'tifa forward is the country of the tribe Aït Attab where there is little or no government among the people.

ROUTE 5.

Tour in the Mountains East and South-east of Marocco.

Marocco to Zourt ben Sessy	. . . . .	2 hours.
„	Ras el Aïn	8 „
„	*Tagana	2 „
„	Aït Zehad (on the mountain in Mesfioua)	4 „
„	Iminterrat	2 „

[Here are found some ruins of Christian buildings of old times.]

Marocco to Tasselt .	. . . . .	2 hours.
„	Tel Eizrat	2 „
„	Tighidoun Idioum	4 „
„	Aït Izzel (high mountain) . . .	4 „
„	Assefrag (Lasfaour) . . .	6 „
„	Imin Gagar	6 „
„	Imin Zadin	2 „
„	Tasghinout (Tasseremout ?) . .	3 „
„	Aït Absalem	3 „
„	Tidiren	2 „
„	Ohamma	? „
„	Aïn Hehia	2 „
„	Gries (Gers ?)	3 „
„	Ohida	3 „
„	Tigardoun	2 „
„	Tigola (Tougla ?)	2 „
„	Tabia (Aït Tieb ?) . . .	2 „
„	Tamzart	4 „

These being the mountains of the Mesfioua country.

ROUTE 6.

Tour in the Mountains South of Marocco.

Marocco to Amreen (plain, orchards)	. . .	4 hours.
„	Resmat	2 „
„	*Ourika	? „

There is the river Ourika passing between mountains. Crossing, and going up to the left, the first village is Achliz. The chief, or sheik, who governs Ourika lives here, he being under the orders of his superior, Ibrahim el Graoui, who lives in Marocco. From Achliz you go to Azrou Miloul, and from this to Tourit. Here are salt wells or springs. From Tourit you go to Agadir,

which is on the top of the mountain, from this to Timluzen,
and from this to the Zaouia. The before last stages from
Ourika are all half an hour's distance one from the other.
From Tourit to Sissag on high mountains. [Apparently Sissag
is the name of the Zaouia.] These are the villages on the
left side of the river until Sissag. The villages on the right
hand side of the river are as follows :—The first is Alzli; from
this you go to Tafzhia, and from this to Anrar, and from this to
Amsin; from this to Assgher, and from this to Arzballo; from
this to Egremon, and from this to Ashni, and from this to
Esurgraf; this mountain is covered with snow. These are the
Ourika mountains near to Marocco town, besides the higher
mountains which are above these we have mentioned.

[All the places here mentioned are in the Ourika valley.
The left and right sides are those on the left and right of the
stream to a person ascending the valley, contrary to the usage
in European countries where those terms are supposed to refer
to one following the course of the stream.]

ROUTE 7.

Description of the Roads of Ghighaya.

Marocco to Tahanout	. . . . .	6 hours.
„ Tasslamat	. . . . .	½ hour.
„ *Souk el Ad (of Moulai Ibrahim)	.	2 hours.

From this to El Anraz; here is a village called Amareen,
three hours distant—it leads to Immaregen ; and from this one
hour's journey will bring the traveller to a place called Agadir
Tagadurt el Bour, and from this last is the commencement of
the road or highway to the province of Sous.

[There can be no doubt that the district here rendered
Ghighaya from the Hebrew, is the same as we wrote down as
Reraya, the r in the latter name having a guttural sound without an equivalent in any European language known to us. The
Souk el Ad, or Sunday Market, is of course somewhere near to
the Sanctuary of Moülai Ibrahim. Although we fail to identify
any of the villages named above, it seems probable that the
place spoken of as Agadir Tagadurt el Bour, is the same as
Arround, where we passed two nights.]

ROUTE 8.

Description of the Road between Ghighaya and Ourika.

Marocco to Tahanout.	.	.	.	.	.	6 hours.
„ Tedroura .	.	.	.	.	.	1 hour.
„ Ourika	.	.	.	.	.	1½ „

This being a road to a mountain containing snow, in the country of Ghighaya. From Tranghert, six hours' journey to a mountain called Ousertik, within the jurisdiction of the Governor, Kaïd Ibrahim el Graoui.

[The writer here gives an alternative route from Marocco to Ourika, slightly longer than the direct way given in Route 6, and then refers, obviously not from personal knowledge, to a mountain path connecting Ourika with one or other of the two valleys included in the district of Reraya or Ghighaya. Tranghert is probably a village in the western branch of the Ourika valley.]

SALOMON BEN DAOUD.

APPENDIX C.

Notes on the Geography of South Marocco.

By JOHN BALL.

SOME remarks upon the geography of South Marocco seem to be called for from a writer who has ventured to put forth a new map, largely differing from those hitherto published; but the subject is encompassed with so much difficulty, and the amount of accurate information available is so limited, that a prudent writer must be content to regard most of his own conclusions as merely provisional, and liable to be modified or set aside by the results of further exploration, whenever this shall become practicable. In the mean time, some good may be effected by clearing the ground of some received errors that are absolutely disproved by facts now ascertained.

Little need here be said of the slight contribution to the knowledge of South Marocco that can be gleaned from the writers of antiquity. The earliest document bearing on the

subject was doubtless the record of the voyage of Hanno, set
up in the temple of Saturn at Carthage.   This is known to us
only by the version, rendered by an unknown hand into Greek,
which, with all the accumulated errors of the translator and the
subsequent transcribers, has reached us under the title of the
Periplus of Hanno.   From this record the particulars to be
gleaned regarding this part of Africa are scanty and of an
uncertain character.   Commentators have, with much probability, identified the Solois promontory of Hanno (Λιβυκὸν
ἀκρωτήριον λάσιον δένδρεσι) with Cape Cantin.   But what are
we to make of the next statement that, having passed the cape,
they sailed for half a day east, or south-east (πρὸς ἥλιον ἀνί-
σχοντα), before reaching the great marshy lake, 'where elephants
and other wild beasts abounded'?   True it is that south of Cape
Cantin there are two slight indentations, mere coves, where the
land for a short distance trends to the south-east; but the
general direction for a mariner along this part of the coast
is SSW., as far as Mogador.   Agreeing with the commentators
that the 'great marshy lake' was probably near the mouth of
the Oued Tensift, we are led to believe that Hanno disembarked
settlers at no less than five stations on the coast of what is now
the province of Haha.   If we may rely on the correctness of
the Greek text we must infer that these were settlements
established by the Carthaginians before the date of Hanno's
expedition.[1]   The next place reached by Hanno was 'the great
river Lixus, flowing from Libya, about which dwelt a nomadic
people,' who are called in the text Lixitæ (Λιξίται).   It is
further stated that the river is said to flow from great mountains
in the neighbourhood, around which dwell the Troglodytes, referred to in our text, p. 301.   The only assertion that can be confidently made about the Lixus of Hanno is, that it was quite a
different stream from that afterwards known to the Romans by
the same name, the latter being the modern Oued el Kous,
falling into the sea at El Araisch, and which Pliny makes fifty-
seven Roman miles from Tangier.   The learned commentator,
C. Müller, identifies the Lixus of the Periplus with the Draha;

---

[1] The phrase used is κατῴκίσαμεν πόλεις πρὸς τῇ θαλάττῃ καλουμένας
Καριχόν τε Γύττην καὶ Ἄκραν καὶ Μέλιτταν καὶ Ἄραμβυν.   When the
author speaks of Thymiaterium, founded by Hanno in this expedition,
he says, ἐκτίσαμεν πρώτην πόλιν.

but, unless we assume that great physical changes have occurred during the interval, this supposition is scarcely compatible with the existence of a numerous population near the mouth of the river. It may possibly have been the river Akassa (the native name of the river of Oued Noun); but it appears far more probable that it was the Sous, the only one of these rivers which is believed constantly to discharge a large volume of water into the sea. It may be, indeed, that there is an etymological connection between the names Sous and Lixus, as there undoubtedly is between some names still current and those used by the Romans.

After Hanno, the next voyager along this coast of whom we known anything was Polybius. The original record of his voyage has, unfortunately, not come down to posterity, but a few particulars have been preserved by Pliny.[1] We learn incidentally that the Romans called Cape Cantin promontorium Solis, a name evidently suggested by the earlier name Solois of the Carthaginians, afterwards rendered in Greek by Ptolemy ἡλίου ἄκρον. Whether Polybius succeeded in reaching the Senegal, or some other river within the tropics, may be uncertain; but he undoubtedly visited many places on the Atlantic coast of Marocco. We hear for the first time of the rivers *Subur* (modern Sebou), and *Salat* (the Bouregrag, which falls into the sea at Sallee). He touched at the port of *Rutulis*, said to have been eight Roman miles beyond the mouth of the river Anatis, which was 205 Roman miles from Lixus (El Araisch). The river is doubtless the modern Oum-er-bia, and the port was the same which the Portuguese named Mazagan. The next port touched by Polybius was named *Risadir*, which has been with much probability identified with Agadir.[2] As for the rivers named by Polybius on the coast south-west of the Atlas, their identification with any known to modern geographers is purely conjectural.

[1] See Pliny, V. 1, § 8. His account is vague and confused, and the distances not to be reconciled with those given by him elsewhere.

[2] Not content with the indication afforded by the identity of the two terminal syllables in each name, C. Müller conjectures that the ancient name of the promontory near Agadir was Râs adir, Râs being the common Arabic designation for a headland. He apparently supposes that the natives spoke Arabic in the time of Polybius. Even now none of the headlands on this coast have the designation Râs.

Of Roman writers Pliny is the only one from whom any positive information as to the geography of this part of Africa is to be gained; but even this is very limited.[1] He complains that the reports as to the region beyond the narrow limits within which Roman power was established in his day were most fallacious, and censures the Roman authorities for indolently giving circulation to mendacious stories, instead of investigating the truth for themselves. In his day Sala (modern Sallee) was the most southern of the Roman settlements in Marocco. He describes it as 'a town standing on a river of the same name, on the confines of the desert (*solitudinibus vicinum*), which was infested by herds of elephants, and still more by the tribe of the Autololes, through whose territory lay the way to the great mountain of Africa, the many-fabled Atlas.' It appears elsewhere that Pliny had access to the manuscripts left by Juba, which, unfortunately, have not come down to posterity. That accomplished prince appears to have held control over the whole territory of Marocco as far as the base of the Atlas. It is to these lost pages of Juba that we probably owe the only fragment of moderately correct information as to South Marocco which is to be found in Pliny's work.[2] The river Asana, whose mouth is said to be 150 Roman miles beyond Sala, is doubtless the Anatis of Polybius, and the Oum-er-bia of the Moors. The next river, which he calls Fut, is the Tensift. The distance assigned for the interval between the mouth of the Fut and the Atlas is excessive; but not largely so if Agadir be intended, that being the first place on the coast from which the high summits of the Atlas are habitually visible. The statement as to the existence of remains of vineyards and palm-groves about the ruins of ancient dwellings seems to lend probability to the belief that the Carthaginian

[1] I am indebted for information as to several passages in Pliny's writings to my friend, Mr. E. Bunbury, who will doubtless throw further light on the subject in an important work, 'An Historical View of Ancient Geography' which he is preparing for publication.

[2] 'Indigenæ tamen tradunt in ora ab Sala CL m. p. flumen Asanam, marino haustu sed portu spectabile: mox amnem quem vocant Fut: ab eo ad Dyrin (hoc enim Atlanti nomen esse eorum lingua convenit) CC m. p., interveniente flumine cui nomen est Vior. Ibi fama exstare circa vestigia habitati quondam soli vinearum palmetorumque reliquias.'

settlements on this coast may have had a prolonged existence. The fall of the parent State would have had but an indirect influence on their destiny. Verbal resemblances are so often misleading that little weight can be attached to them ; but it is natural to compare the word Dyris, said by Pliny to be the native name for the Atlas, with that now used by the natives— *Idrarn*—this being the plural form of *Adrar*, which means generically a mountain, both in the Shelluh and in several other Bereber dialects.

Besides what Pliny may have learned from King Juba as to the geography of the coast of South Marocco, he had access to contemporary testimony as to some part of the interior of the country. Suetonius Paulinus (the same who at a later date played a conspicuous part in Britain) being appointed governor of the provinces of N.W. Africa, then recently incorporated in the Roman Empire, resolved to penetrate southward beyond the Great Atlas, whether with a view to intimidate the native tribes, or for the mere satisfaction of carrying the Roman eagles into a new region. He appears to have left a written account of his expedition, which, like so much else of ancient geographical literature, has been lost. The particulars preserved by Pliny are unfortunately so vague as to be almost valueless.

In ten days from his starting point, wherever that may have been, we are told that he reached the highest point of his march. He reported the mountain to be covered with dense forests of trees of an unknown kind, and declares the summit of the range to be deeply covered with snow, even in summer.[1] From the summit of the Atlas Suetonius descended, and marched on through deserts of black sand, out of which rose here and there rocks that had the aspect of being burnt, to a river called Ger. Although it was the winter season the heat of these regions was found intolerable. The neighbouring forests abounded in elephants and other wild beasts, and with serpents of every kind, and were inhabited by a people called Canarians.

The controversies to which this passage has given rise are not likely to be definitively decided. The balance of opinion

---

[1] This must have been from native report, as the expedition was made in winter. If he had said that the snow never quite disappears, and sometimes falls heavily, even in summer, his statement would have been accurate enough.

leans to the belief that Suetonius ascended the valley of the
Moulouya, and traversed the Atlas by the pass now called Tizin
Tinrout, leading to Tafilelt. This was the pass traversed by
Gerhard Rohlfs in 1864, and to his narrative alone we can
refer for information respecting it and the country extending
southward towards the Great Desert. The existence in that
part of W. Africa, on the south side of the Great Atlas, so far
from the influence of the Atlantic climate, of vast forests
capable of maintaining elephants and sheltering a native population, would apparently be irreconcilable with existing physical
conditions, and is not readily admissible in the Roman period.
Whatever vigorous vegetation exists in the region traversed by
Rohlfs adjoins the banks of the stream; and, though sand may
encroach here and there, and sun-burnt rocks are seen there, as
elsewhere on the south side of the Atlas, the description is not
what would occur to any one following the course of the stream.
It seems, further, highly improbable that a prudent general, such
as Suetonius Paulinus, would have undertaken to lead an aggressive military force along the tortuous valley of the Moulouya,
some 250 miles in length, enclosed for the most part between
lofty mountains; and it is also to be noted that at the period of
his expedition the Romans held no station in the valley of the
Moulouya, if indeed they ever penetrated far into it.

The few particulars quoted above lead to the conclusion
that the Roman general in his southward march beyond the
Atlas did not follow the course of a stream, but was compelled
to cross a tract of desert before reaching the river of which he
speaks, which, therefore, probably flowed from E. to W. On
the whole, it seems to me that the brief record is more easily
reconciled with the supposition that Suetonius Paulinus made
Sala (Sallee), the farthest Roman station in Western Africa, his
base of operations; that he marched thence across the open
country towards SSW., and gained the summit of the Atlas
range at the pass between Imintanout and Tarudant.[1] Between
the course of the Sous and that of the Akassa, or river of Oued
Noun, there are extensive tracts of sandy desert, where, even in
winter, his troops may easily have suffered from heat and thirst;
and the river (called Ger) may have been the main branch, or
one of the tributaries of the Akassa flowing from the range of

---

[1] Mentioned in the text at p. 294.

Anti-Atlas. The former existence of great forests, frequented by elephants, on the flanks of that range, is far more probable than on the parched southern slopes of the interior, where, as Rohlfs tells us, the rocks and hills are now absolutely bare of tree and shrub vegetation. Finally, it is more natural to look for the ancient Canarians in the country near the Atlantic coast than in the interior.

The solitary argument of any weight in favour of the Moulouya and Tafilelt route seems to be derived from the fact that in descending southward from the pass at the head of the Moulouya valley the traveller follows the course of a stream which now bears the name Gers, or Ghir. But it must be remarked that this name exists elsewhere in Marocco, there being at least three streams so denominated, and further that it is now-a-days borne by the river of Tafilelt only during a short part of its course. Rohlfs, who is here our only authority, tells us that the stream first met in descending from the pass of Tizin Tinrout is called Siss.[1] After following this for seven or eight hours, it is joined by another stream which he called *Ued Gers*. The united stream bears the latter name for a distance of some six hours' ride, and then resumes the name of Siss, which it bears throughout its subsequent course till it is lost in the sands of the Sahara.

The long period that intervened between the decline of Roman power and the establishment of Mohammedan rule in Marocco, is a blank to the historian and the geographer. It can scarcely be doubted that Roman authority and Roman institutions spread themselves throughout a great part of the open country between the Atlas and the Atlantic, although there is but little direct evidence to that effect.

[1] This is evidently the river Ziz of Leo Africanus ; and in his time, as at the present day, travellers going from Fez to Segelmese (modern Tafilelt) followed the course of the Ziz, or Siss. He also speaks of a river Ghir, which may possibly have been the affluent of the Siss mentioned by Rohlfs ; but the particulars given are vague and scanty. It is interesting to remark that in Leo's day the valley of the Siss was inhabited by a hardy and energetic Bereber tribe named Zanaga, probably the same as the Azanegues whom Cà da Mosto found about Oued Noun. They have since migrated across the Sahara, and still calling themselves Zanega, and speaking a Bereber dialect, are dangerous neighbours to the negro tribes of the Senegal.

Little reliance can be placed on the statement of Leo Africanus that the people of Barbary were converted to Christianity 250 years before the birth of Mohammed, or about A.D. 320, for, in a country so split up into independent tribes, the new faith must have made way irregularly and at various periods; while it is most probable that it never struck root among the mountain tribes of the Great Atlas. But the positive assertion of the same writer, that when the Arabs arrived in Marocco they found the Christians masters of the country, probably holds good of all except the mountain tracts.

Whether any reliable information as to South Marocco is to be gleaned from the writings of the eminent Arabian geographers who lived between the tenth and fourteenth centuries, I am unable to say; but it seems sufficiently certain that the period of European exploration leading to practical results commenced in the fourteenth century. The Genoese, the Catalans, and the Venetians appear to have despatched several expeditions along the coast, most of them intended to reach the gold-producing regions of tropical Africa. The Portuguese, who were destined to outstrip all their rivals in maritime exploration, were the first to establish themselves on the western coast of Marocco; and, at one time or other, they held most, if not all, the Atlantic seaports. Much information doubtless lies concealed among the mediæval records of Italy, Spain, and, especially, of Portugal; but up to the present time nothing has been published to show that any European was able, from personal knowledge, to give an account of the interior of Marocco, before Marmol, who, having been taken prisoner by the Moors, passed several years at Fez and elsewhere in North Marocco, about the middle of the sixteenth century. The earliest known document showing a moderately correct knowledge of the coast is a map (number 5 in the series), contained in the celebrated *Portulano* of the Laurentian Library in Florence, bearing the date 1351.[1] In this map, which, from internal evidence, must be of Genoese origin, the general outline of the Marocco coast is correct, and

[1] A portion of this map, containing the coast of Africa from the Straits of Gibraltar to the latitude of the Canary Islands, was published (in facsimile) by Count Baldelli Boni of Florence in his edition of Marco Polo, and is reproduced in Mr. Major's valuable work, 'The Life of Prince Henry the Navigator.' London, 1868.

the positions of the few places laid down unmistakable. The now abandoned town of Fedala (Fidalah), Mefegam (Mazagan), and Mogodor here appear for the first time. Of early Portuguese maps there must be many not now known to geographers, and it was certainly from Portuguese authorities that Gerard Mercator partly derived the materials used in both editions of his Atlas. In the *Atlas Minor*, published by Hondius in 1608, a map of South Marocco is given in page 567, wherein for the first time an attempt is made to represent the positions of cities and mountains, and the courses of rivers in the interior of the country. The outline of the coast is here less correct than that given in the much more ancient Medicean map; but there is far more of detail, especially as to places which were evidently well known to the Portuguese. Thus, as mentioned in the text, we here for the first time find the island of Mogador with the name '*I. Domegador.*' The places laid down in the interior appear for the most part to be taken (but with numerous errors) from the work of Leo Africanus; but the chartographer has spoiled his map by making the river Sous flow from SE. to NW., instead of from NNE. to SSW. Mountains are scattered pretty uniformly over the map; but what is made to appear as the loftiest mass, and is marked '*Atlas M.*,' with a town named Tagovast at its foot, stands S. of Tarudant about the western extremity of the range of Anti-Atlas. The accompanying letterpress, page 566, is to a great extent derived from Leo Africanus, but with additions from other sources. It is curious to read that Tarudant, now a place which no Christian stranger dare approach, was then resorted to by French and English merchants.

The name of the remarkable man, who stands almost alone as a geographical authority for the interior of Marocco, has already been mentioned; but it is impossible to dismiss him so lightly. Leo Africanus, to give him the name by which he is known to posterity, was a Moor of Grenada, born in the latter part of the fifteenth century, who, with his kinsfolk, fled to Fez at or about the time of the siege of Granada in 1492. In those days Fez was the head-quarters of Arabic culture; Leo was an earnest and successful student, and, as a man of learning and intelligence, was taken into favour by Mouley Ahmet, the founder of the dynasty still reigning in Marocco. Either in company with the new ruler, or with his protection and authority, he travelled through almost every part of the empire, as well as nearly all the rest of Northern Africa, and evidently made copious notes. He wrote, in Arabic, various works on history and grammar which have not been preserved, and, in the same language, the original version of his description of Africa. It would appear that he carried this with him, in manuscript, when, in 1517, he was made captive by Christian corsairs, who took him to Rome. Leo X., hearing that a learned Moor had been brought a captive to Rome, sent for him, and treated him with kindness and liberality. A suggestion that he should undergo the rite of baptism seems to have encountered no obstinate prejudices, for he soon complied, receiving at the font the Pope's own names, Giovanni Leone, and perhaps becoming as earnest a Christian as the Pontiff himself. He afterwards lived many years in Rome, acquired the Italian tongue, and translated his work on Africa into that idiom. This remained for some time unpublished, until it fell into the hands of Ramusio, who included it in his famous work 'Delle Navigationi et Viaggi,' of which the first edition, in three folio volumes, was printed in Venice in 1550. It is not easy to account for the numerous variations between the original text and the versions which appeared in various languages during the century following the original publication; but in the absence of satisfactory explanation it seems safest to accept the text of Ramusio as alone authentic.

Like most modern readers, the members of our party, when they resolved to visit Marocco, knew nothing of the work of Leo Africanus beyond the fact that he is occasionally referred to by writers on North Africa. The time for preparation was far too short for extensive reading, and we took with us only the works of Jackson and Gerhard Rohlfs. It has, however, since that time been a matter of frequent regret that we had not the opportunity, while travelling in the country, of referring to the only writer who had actually seen the greater part of it with his own eyes, and as to whose general truthfulness there is no room for suspicion. It is impossible here to enter into the many interesting details that abound throughout the text; but it is worth while to point out the more important changes that are disclosed between the condition of South Marocco as it was more than three and a half centuries ago, and that of the present day.

So far as regards the manners, ideas, habits, and mode of living of the inhabitants, the changes are quite insignificant, save in so far as these are affected by a general decline in material prosperity. The central authority was at that period much weaker, and the separate tribes led a more independent existence. Amongst the Bereber people of the mountains, and even in many of the larger towns, such government as existed was ordinarily of the democratic type. Thus we read that in Tarudant four chiefs were elected to manage the affairs of the city, holding office for only six months at a time.

If it were possible to doubt the results of the establishment of a system of grinding despotism, administered by officials who enjoy practical impunity so long as they satisfy the pecuniary demands of their master, the pages of Leo Africanus bring ample evidence. It is, indeed, true that a slight improvement has ensued as regards internal tranquillity. There is now rather less of habitual turbulence; the mutual encounters between neighbouring tribes may be somewhat less frequent; and brigandage, which appears to have been not uncommon in the open country, is now comparatively rare. It may be doubted whether this advantage, such as it is, is not as much due to diminished population as to the successful administration of the Moorish Sultans.

On the other hand, there is overwhelming evidence of a general and progressive decline in prosperity. Throughout the southern provinces, and especially in Haha and Sous, Leo Africanus found numerous flourishing towns, most of them visited and described by him. In each one of these he found people living in comparative ease, inhabiting good houses with gardens, and possessing, according to the standard of the age, some literary education. From the towns, and even from the inner valleys of the Atlas, students flocked to Fez, then the head-quarters of Arabic knowledge and civilisation. All the principal places were then local centres of production, the artificers being principally Jews.

It is notable that excepting the city of Marocco, then full of a numerous and active population, none of the towns mentioned owed their foundation to the conquering race. Leo, not likely to detract from the achievements of his own people, expressly attributes the origin of most of them to the ' antichi Africani,' by which designation he commonly speaks of the primitive Bereber stock ; and, as regards the smaller towns lying in the low country north of the Atlas, he frequently speaks of the population being harassed by the Arabs, then, as at this day, leading a semi-nomad existence in the plains.

If we confront his description with the present state of the country we find comparative ruin and desolation. In all the southern provinces we now find but two inland cities of any importance, Marocco and Tarudant, and these dwindled to a mere tithe of their ancient wealth and population. Where the traveller in the sixteenth century found thriving towns at intervals of ten or twelve miles, there are now miserable villages whose wretched inhabitants maintain a bare existence, and are often unable to pay the imposts which leave no surplus behind. It does not appear that in the great province of Haha there is now a single place that can be called a town except the ruined seaport of Agadir, destined by nature to be the chief port of South Marocco, but closed to trade by the caprice of a Sultan. Throughout the interior we saw or heard of but two places that could by courtesy be called towns, Amsmiz and Moulai Ibrahim. Although no statistics are available, it seems a moderate estimate if we reckon that the present population of South Marocco cannot exceed one-third of what it was when Leo wrote.

Along with the decay of wealth and population, we naturally find that of everything that could raise the people in the scale of existence. In Leo's day iron and copper mines were worked in many places in the Atlas, and various handicrafts exercised, of which there is now no trace. Education, such as it was, was widely spread ; and in some parts of the Atlas where it was absent, the traveller noted the fact as a proof of the low condition of the population. He notes as a curious incident that when he visited the mountain district of Semele, where the people were ignorant of reading and writing, they forced him to remain nine days, hearing and deciding all pending cases of litigation ; in doing which, as he records, he had to act both as judge and notary, there being no one competent to write down the decisions of the court.

Several incidental statements in the work of Leo Africanus suggest an inquiry of considerable interest. There is nothing in the published annals of the Portuguese wars with the Moors to suggest a belief that the former at any time established their authority in the interior of South Marocco, or even undertook any inland expeditions. From Leo's narrative it appears, however, that, at the beginning of the sixteenth century, they had, at least occasionally, penetrated much farther into the interior than has commonly been supposed, and that the authority of the Portuguese king was in some places paramount. At Tumeglast, a place in the plain of Marocco, probably not far from the present village of Frouga, Leo lodged in the house with a Moor, named Sidi Yehie, who had come in the name of the king of Portugal to levy tribute, the same Moor having been made by the king chief (capitano) of the district of Azasi. Elsewhere he relates that the king of Marocco sent an expeditionary force against an independent chief in the district of Hanimmei, forty miles east of the city of Marocco (apparently in the present province of Demnet), and which force was accompanied by 300 Portuguese cavalry. The expedition was unsuccessful, the Sultan's troops were defeated, and, according to the narrative, not one of the Christian horsemen returned from the disaster. It seems highly improbable that the Portuguese should have taken part in such an affair if their troops had not at the time been stationed somewhere in the interior.

After Leo Africanus but little of a definite kind is to be learned from subsequent writers as to the geography of South Marocco. In 1791 the reigning Sultan applied to General O'Hara, then Governor of Gibraltar, for the assistance of an English physician to treat his favourite son, Mouley Absalom, who was at the time governing the province of Sous. Mr. Lempriere, an army-surgeon, undertook the office, and travelled by the west coast to Agadir, and thence to Tarudant. After successfully treating his patient, he was partly induced, and partly forced, to travel to the city of Marocco, whence, after considerable delay and difficulty, he succeeded in returning to Gibraltar. Mr. Lempriere probably travelled across the Atlas by the road from Tarudant to Imintanout, but his narrative supplies little information to the geographer. He speaks of the distance from Tarudant to the northern foot of the Atlas as an easy journey of three days, and describes the track as leading beneath and along tremendous precipices.

Frequent reference is made in the text to Jackson's ' Account of the Empire of Marocco,' of which the first edition appeared in 1809, and the third in 1814. This is undoubtedly the fullest and most correct modern work on Southern Marocco. Jackson spent sixteen years in the country, chiefly at Mogador and Agadir; he acquired the familiar use of the Moorish Arabic, and seems to have obtained merited influence among the natives. Either because he had but little taste for exploration, or because he found the difficulties too serious, Jackson has added little to our knowledge of the geography of the country. His map, though it contains some corrections, is on the whole inferior to that of Chénier, published a century earlier.

A definite contribution to the slight existing amount of positive knowledge was made by the late Admiral Washington, then a lieutenant in the navy, who accompanied the late Sir J. Drummond Hay on his mission to the city of Marocco in the winter of 1829–1830. His paper, published in the first volume of the ' Journal of the Royal Geographical Society,' is frequently referred to in our text; and in the accompanying map the positions of several points in the interior of the country were accurately laid down from astronomical observation.

A most important step towards extending our knowledge of the entire empire of Marocco was made in 1848, when the French War Department published the map compiled by Captain Beaudouin. Whatever errors it may contain—and these were unavoidably numerous—this must be regarded as a monument of intelligence and industry. Recognising the fact that the greater part of the territory is likely long to remain inaccessible to Europeans, the author applied himself to obtaining information from natives who were personally acquainted with various portions of the country. Hundreds of such informants, as we were assured, were separately examined by Captain Beaudouin; the information supplied by each was laid down on a skeleton map; and by the careful comparison of the separate materials the general map was compiled.

Without noticing minor errors, which are, of course, inevitable in such a work, the most serious objection to be made to

GEOGRAPHY OF SOUTH MAROCCO.

this map is that the orography is exhibited in a fashion *primâ facie* improbable, and which has been to a great extent negatived by subsequent evidence. The main range of the Southern Atlas is represented as a nearly straight wall, over 400 miles in length, with few and short diverging ridges, and, parallel to this on the south side, another equally straight and narrow ridge is made to stretch for nearly 300 miles. From near the eastern extremity of the main range two other straight ridges are shown, diverging abruptly at an acute angle, and enclosing a trench-like valley that extends north-west for fully 120 miles. If this were even approximately correct, we should be led to conclude that the structure of the Great Atlas is quite unlike that of any other known mountain region. The tendency of mountain ranges to follow a uniform general direction is always modified by the numerous secondary causes that have helped to fashion the earth's surface.

The first recent traveller who succeeded in penetrating some considerable portions of the Marocco territory was M. Gerhard Rohlfs. Assuming the garb and professing the faith of a Mussulman, he traversed many districts where no Christian dare present himself; but the care necessary to prevent his real character from becoming known imposed severe restrictions on M. Rohlfs. Produced under conditions where it was impossible to be seen taking notes or using any scientific instrument, it is not surprising that the narrative of his adventurous journey is extremely meagre; but even for the little that he is told about a region so little known the reader is thankful. The chief geographical results of these journeys were embodied in the map annexed to G. Rohlfs' first work[1] by the eminent geographer, M. Petermann. The scale of that map is small and admits of little detail; but, so far as regards the mountain country, I am disposed to think that the direct evidence, supplemented in some points by native report, requires us to depart more widely from the orographic features of Beaudouin's map than M. Petermann has thought it fit to do.

In the map accompanying this volume I have ventured, in addition to the changes for which I had direct authority, to introduce a few others, avowedly conjectural, which must await further exploration before they can be either adopted or con-

---

[1] Afrikanische Reisen, von Gerhard Rohlfs. Bremen, 1867.

demned. It is difficult to believe that in an age when the
barriers that have closed the other least known regions of the
earth are successively removed, Marocco, so close to Europe
and so attractive, can alone resist the progress of modern
exploration.[1]

APPENDIX D.

ON SOME OF THE ECONOMIC PLANTS OF MAROCCO.

By Joseph Dalton Hooker.

Gum Ammoniacum.

OUR endeavours to obtain accurate information regarding the
Marocco gum ammoniac plant were ceaseless and fruitless.
Jackson, who gives a rude figure of a portion of a leaf and a
scanty description ('Account of the Empire of Marocco,' 136,
t. 7), says that it is the produce of a plant like Fennel, but larger,
and called Fashook in Arabic, and that it grows in the plains
of the interior provinces, abounding in the north of the city of
Marocco, in a sandy light soil. Jackson further states that
neither bird nor beast is seen where this plant grows, the
vulture only excepted, and that it is attacked by a beetle
having a long horn proceeding from its nose, with which it
perforates the plant, and makes the incisions whence the gum
oozes out. Under his description of the vulture, he states
that, with the exception of the ostrich, this is the largest bird
in Marocco; that it is common in all places where the gum

-----

[1] The scope of these remarks being limited to the geography of
South Marocco, I have not noticed several recent publications, not
devoid of interest and value, but in which no important contribution
is made to our geographical knowledge. We have referred in the text
to papers by MM. Beaumier, Balansa, and Lambert, in the Bulletin of
the French Geographical Society. A more considerable work, entitled
'Morocco and the Moors,' by Arthur Leared, M.D., appeared in 1874.
It contains much information carefully collected by the author, along
with a lively account of his own experiences, but circumstances prevented him from entering on new ground.

ammoniac grows, as in the plains east of El Araiche,[1] where he has seen at least twenty of these birds in the air at once, darting down on the insects with astonishing rapidity (p. 118). Jackson's figure (t. 8) of the so-called beetle apparently represents a dipterous insect resembling a Bombylius, with a very long straight proboscis.

Lindley ('Flora Medica,' 46) doubtfully refers Jackson's Fashook to the eastern *Ferula orientalis* L.; and Flückiger and Hanbury ('Pharmacographie,' 289) say that, according to Lindley, the *Ferula tingitana* yields a milky gum resin, having some resemblance to Ammoniacum, which is an object of traffic with Egypt and Arabia, where it is employed like the ancient drug in fumigations. The authors go on to say that there can be but little doubt that the Maroccan Ammoniacum is identical with that of the ancients, and that it may well have been imported by way of Cyrene from regions lying farther westward.

Pliny and Dioscorides say that the Ammoniacum is the juice of a Narthex growing about Cyrene and Lybia, and that it is produced in the neighbourhood of the temple of Ammon.

Dr. Leared ('Morocco and the Moors,' 356) was informed that the Fashook grows at a place two days' journey from Mogador, on the road to the city of Marocco,[2] but states that the exudation from the roots of specimens which he obtained differed from the African Ammoniacum. We, on the other hand, were persistently assured that it grew nowhere along that route, nor nearer to it than El Araiche, north of Marocco city. And this is confirmed by information obtained by Mr. R. Drummond Hay to the effect that it is found near Marocco, and chiefly around Tedla. The Moors who gave us this information at once recognised the figure by Jackson, and called the plant Kilch (Kelth according to Leared). The roots presented to Kew by the kindness of Dr. Leared did not make any indications of growth.

---

[1] Not El Araisch, SSW. of Tangier on the Atlantic coast, but some place in the interior, and N. of the city of Marocco.

[2] This is no doubt *Elæoselinum humile* (Ball), which we found near or at the above defined locality. Ball formed a very decided opinion that Jackson's plant, whether the true Ammoniacum or not, was a species of *Elæoselinum*.

The Maroccan Ammoniacum plant must not be confounded with the Persian *Dorema Ammoniacum*, or 'Ushak,' which is also bled by insects.

The Fashook gum is used by the Moors and by some Orientals as a depilatory, and in skin diseases ; it is exported to the East from Mazagan, *viâ* Gibraltar and Alexandria.

Euphorbium, Furbiune or Dergmuse.

Euphorbia resinifera.—Berg. und Schmidt, Officinelle Gerwächse, v. iv. (1863) xxxiv. d.; Flückiger and Hanbury, Pharmacographia, 502; Ball, in Journ. Linn. Soc. Bot. xvi. 661; Euphorbium, Jackson's 'Account of the Empire of Marocco,' 134, t. 6 (left-hand figure only).

We have little to add to the description of the Euphorbium tree given by Jackson, and that in the 'Pharmacographia' cited above. As stated in the body of this work it is confined to the interior of the empire, and the only living specimens we met with were from a garden in Mesfiouia (see p. 163). Jackson confounded two plants under this name; one, the true species, growing in the Atlas, with 3–4-angled branches, the other a sea-coast plant, with 9–10-angled branches, which is carried to Marocco for tanning purposes, and of which he says, that during the three years of his residence at Agadir he never saw any gum upon it.

The true plant is figured and described by Jackson as an erect tree, with a stout short woody trunk, and very numerous upcurved long sparingly divided branches, the whole resembling a candelabrum. The angles of the branches are armed with short spines, and the flowers are produced from the tips of the young shoots. The thorns adhere to everything that touches them, and he supposes them to have been intended by nature 'to prevent cattle from eating this caustic plant, which they always avoid on account of its prickles.' The juice flows from incisions made with a knife, and hardens and drops off in September. The plants, he says, produce abundantly once only in four years, and the fourth year's produce is more than all Europe can consume. The people who collect the gum are obliged to tie a cloth over their mouths and nostrils, to prevent the small dusty particles from annoying them, as they produce incessant sneezing.

The history of the Euphorbium as given in the 'Pharmacographia' is, that it was known to both Dioscorides and Pliny as a native of the Atlas, and was named in honour of Euphorbus, physician to the learned King Juba II. of Mauritania, himself the author of treatises on Opium and Euphorbium.

The prevalence of cactoid *Euphorbiœ* in Marocco, of which there are three species in the southern districts, is a similar instance to that of the Argan, of tropical forms advancing far north in the extreme west of the old world; and as the Argan has its nearest ally in Madeira, so have the Maroccan Euphorbiums close congeners in the Canary Islands. All these belong to the section *Diacanthium* of Boissier, of which the other species are Abyssinian, Arabian, Indian, and South African.

Gum Euphorbium was extensively used by early practitioners as an emetic and purgative, and was exported in large quantities; now, however, the trade in it is rapidly declining, and we were informed that it is chiefly used in veterinary practice, and as an ingredient in a paint for the preservation of ships' bottoms.

Euphorbia resinifera is in cultivation at Kew, where specimens may be seen both in the Succulent-plant House and Economic-plant House.

The Arar, Thuja or Gum Sandrac Tree.

Callitris quadrivalvis.—Ventenat, Nov. Gen. Decad. 10; Richard, Conif. 46, t. 8, f. 1; Endlich, Synops. Conif. 41; Parlatore, in DC. Prod. xvi. pars 2, 452; Ball, in Journ. Linn. Soc. Bot. xvi., 670.

Thuja articulata.—Shaw's 'Travels in Barbary,' 462, with a plate; Vahl, Symb. ii. 96, t. 48; Desf. Flor. Atlant. ii. 353, t. 252.

Frenela Fontanesii.—Mirbel, in Mem. Mus. xiii. 74.

This tree is a native of the mountains of North Africa, from the Atlantic to Eastern Algeria; but we are not aware whether its eastern limit has ever been accurately determined. It has no congener, its nearest ally being a South African genus of Cypresses (*Widdringtonia*), of which several species are recorded from the Cape Colony, Natal, and Madagascar, and which differ in having alternate leaves and many ovules to each scale.

The great interest attached to this plant arises from the beauty and durability of the wood, which, there is every reason to believe, was known to the ancients from the earliest times, under the name of Thuja. It is thus hypothetically, but probably correctly, identified with the θύϊον [1] of the Odyssey (ii. 6), with the θύϊον and θυία of Theophrastus ('Hist. Pl.' v. 5), and the thyine wood of the Revelations (xviii. 12). It is undoubtedly the Citrus wood of the Romans, and the Alerce of the Spaniards; the latter name being derived from the Moors of Marocco, for it is not a native of Spain.

The first botanical notice of the Callitris is in Shaw's 'Travels in Barbary,' where it is figured and briefly described as *Thuja articulata* (462); and for its identification with the Alerce we are indebted to the late Mr. Drummond Hay when Consul of Tangier, who, further, sent a plank of the wood to the Royal Horticultural Society.[2] At about the same time, the attention of a most intelligent traveller, the late Capt. S. E. Cook (afterwards Widdrington), was attracted by the wood of the cathedral of Cordova (formerly a mosque built by the Moors in the ninth century) called Alerce, which differed from any Spanish wood, or any other wood now used in Spain. Coupling this name with the communication made by Mr. Drummond Hay to the Horticultural Society, Capt. Cook was enabled to identify the Cordova wood with the Callitris, which, as he assumes, was brought from Marocco, to roof a mosque intended to be second in sanctity only to that of Mecca.

Except in a garden at Tangier, we saw no specimen of the Callitris approaching a large size, or capable of yielding the beams which we were shown in the ceilings and roofs of buildings in that town and elsewhere, and which are considered to be indestructible. On the contrary, most of the native specimens we saw in Southern Marocco resembled small Cypresses, with very sparse foliage and branches, and were apparently

---

[1] It is mentioned under this name by Homer in his description of the Island of Calypso. See Daubeny *On the Trees and Shrubs of the Ancients*, p. 42.

[2] See Cook's *Sketches in Spain*, vol. i. p. 5 (1834); and Loudon's *Gardener's Magazine*, Ser. ii. vol. iii. p. 522.

shoots from the stumps of trees that had been cut or burnt down, though possibly their impoverished habit may have been due to the sterility of the soil. The largest were in the Ourika valley, and were about thirty feet high (see p. 177). In many cases the stem swelled out at the very base into a roundish mass half buried in soil, which is said to attain even four feet in diameter, though we saw none approaching that size.

It is the basal portion, whether the result of mutilation or natural growth, that affords the wood so prized by ancients and moderns, and which forms a most valuable article of export from Algiers to Paris, where small articles of furniture, &c., are made of it and sold at very high prices.

Under the name of Citrus wood, it is alluded to, according to Daubeny, by Martial and Lucan, and by Horace (' Carm.' lib. iv. Od. 1), who suggests its employment as the most precious commodity that could be selected for a temple in which a marble statue of Venus should be placed :—

> Albanos, prope te, lacus
> Ponet marmoream sub trabe citrea ;

Also Petronius Arbiter, descanting upon the luxury of the Romans, seems to represent it as worth more than its weight in gold, when he says—

> Ecce Afris eruta terris
> Ponitur, ac maculis imitatur vilius aurum
> Citrea mensa.

For a detailed description of what was known of this tree to the ancients, and of its value, we must refer to the description in Pliny (' Nat. Hist.' book xiii. chaps. 29, 30). This author describes it as the thyion and thyia of Homer and the Greeks, and adds that its wood was used with the unguents burnt for their pleasant odour by Circe ; as also that Theophrastus awarded a high rank to it, the timber being used for roofing temples and being indestructible ; as also that it is produced in the lower part of Cyrenaica, and that the finest kind grows in the vicinity of the temple of Jupiter Ammon.

Pliny himself gives Mount Atlas as the native country of the wood ; in the vicinity of which, he says, is Mauritania, a country in which abounds a tree which has given rise to the mania for fine tables, an extravagance with which women reproach the men when they complain of their vast outlay upon pearls. He attributes the knots from which the tables are made to a disease or excrescence of the roots, of which the most esteemed are entirely concealed under ground, these being much more rare than those which are produced above ground, and that are to be found on the branches also.

The principal merits of the tables were to have veins arranged in waving lines, or forming spirals like whirlpools. The former they called 'tiger' and the latter 'panther' tables; whilst others, which are highly esteemed, have markings resembling the eyes on a peacock's tail. In others, again, called 'apiatæ,' the wood appears as if covered with dense masses of grain. The most esteemed colour was that of wine mixed with honey.

In respect of their size, Pliny gives a little over 4 ft. as the average maximum, though one that belonged to Ptolemæus, King of Mauritania, was $4\frac{1}{2}$ ft. in diameter and $\frac{1}{4}$ of a foot in thickness. It was formed of two semi-diameters so skilfully united that the joining was concealed. Another, made of a single piece, was named after Nonius, a freedman of Tiberius Cæsar, and was 4 ft. less $\frac{3}{4}$ in. in diameter, and $5\frac{1}{4}$ inches in thickness. And with regard to the price, Cicero paid a million sesterces (9,000l.) for one; two belonging to King Juba were sold by auction, one for one million two hundred thousand sesterces, and the other for somewhat less. Some of Pliny's statements are probably fabulous; as that the barbarians bury the wood when green, first giving it a coating of wax, and that the workmen, when it comes into their hands, put it for seven days beneath a heap of corn, and then take it out for as many more, after which it is surprising to find how much it has lost in weight. More apocryphal still is his statement that it is dried by the action of sea-water, and thereby acquires a hardness and density that render it proof against corruption; also that, as if created for the behoof of wine, it receives no injury from it.[1]

In Marocco, where no ornament or article of luxury is known, it need hardly be said that the Alerce wood is employed only for building purposes and fire-wood; though the resin

---

[1] See Bostock's translation of Pliny, vol. iii. p. 194, &c.

called Sandarach, which was once a reputed medicine, is collected by the Moors and exported from Mogador to Europe, where it is used as a varnish.

Gum Arabic.

Acacia gummifera.—Willd. Sp. Pl. iv. 1056; DC. Prod. ii.
455; Hayne, Arzneigew. x. t. 8; Benth. in Trans. Linn. Soc.
xxx. 509; Ball, in Journ. Linn. Soc. Bot. xvi. 442.
Mimosa gummifera.—Brouss. in Poir. Dict. Suppl. i. 164.
Acacia coronillæfolia.—Desf. Cat. Hort. Par. ed. ii. 207.
Mimosa coronillæfolia.—Pers. Encheirid. n. 44.
Sassa gummifera.—Gmel Syst. ex DC. l. c.

Of this plant very little indeed is known, and we were unfortunately unable to find either flower or fruit of the only Acacia which we met with on our visit to Marocco, and which we were assured was the Gum Arabic plant (Alk Tlah) of that country. It is interesting as representing the northern limit of distribution of the immense genus Acacia in Africa. Our specimens, such as they are, coincide perfectly with the description of *Acacia gummifera* in Willdenow, and with the excellent figure in Hayne, which was taken from specimens collected by Broussonet near Mogador. We found the plant abundantly in the lower region of Southern and Western Marocco, occurring as a thorny bush, along with *Rhus pentaphylla* and other shrubs. That it was the plant producing the Marocco Gum Arabic the natives consistently testified, though this could not be inferred from the description in Jackson's 'Account of the Empire of Marocco' p. 136, who says of the gum that it 'is produced from a high thorny tree called Attalet, having leaves similar to the Arar, or gum Sandarac tree, and the Juniper.' Jackson goes on to say:—

'The best kind of Barbary gum is procured from the trees of Marocco, Ras-el-wed, in the province of Abda; the secondary qualities are the produce of Shedma, Duquella, and other provinces; the tree grows abundantly in the Atlas mountains, and is found also in Bled-el-jerrêde. The gum, when new, emits a faint smell, and, when stowed in the warehouse, it is heard to crack spontaneously for several weeks; and this cracking is the surest criterion of new gum, as it never does so when old;

there is, however, scarcely any difference in the quality. The
Attaleh is not so large a tree as the Arar, which produces the
Sandarac gum, nor does it reach the size of the Auwar tree, which
produces the gum Senegal.   It has a low crooked stem, and its
branches, from the narrowness of its leaves (long and scanty),
have a harsh, withered, and unhealthy appearance at the time
it yields the most gum—that is, during the hot and parching
months of July and August; but although not an ornamental
tree, it is a most useful plant, and will always be considered
valuable.   Its wood is hard, and takes a good polish; its seeds,
which are enclosed in a pericarpium, resemble those of the Lupin,
yield a reddish dye, and are used by the tanners in the preparation of leather.   These seeds attract goats, who are very fond of
eating them.   The more sickly the tree appears, the more gum
it yields; and the hotter the weather, the more prolific it is. A
wet winter and a cool or mild summer are unfavourable to the
production of gum.'

As observed in the body of this work, the gum does not
seem to be collected in the western portion of its range in South
Marocco, but in Demnet, whence it is brought to Mogador;
and it may very well be that it is only in the hotter and drier
regions of the interior that the gum is produced in sufficient
quantities to be worth collecting.

It is remarkable that no notice whatever of *Acacia gummifera* occurs in Flückiger and Hanbury's invaluable 'Pharmaco
graphia' (1874), where the Marocco gum is supposed to be the
produce of *Acacia arabica* Willd., a plant which extends from
Nubia to Natal, and eastward to Central India, but which
is not known as a native of Marocco.   In another passage
of the above work (p. 211), the 'Marocco, Mogador, or brown
Barbary gum,' is described as consisting 'of tears of moderate
size, often vermiform, and of a rather uniform light dusky
brown tint.   The tears, which are internally glassy, become
cracked on the surface and brittle if kept in a warm room;
they are perfectly soluble in water.'

It is possible that the *Acacia arabica*, which is found in
Senegal, may extend to the Sus Valley, and be the source of
some of the Marocco gum; and that more than one species
producing gum are confounded together by the Moors; this
is the natural inference from Jackson's account, itself anything

but explicit. On the other hand, I am informed in a letter lately received from Mr. R. Drummond Hay, H.B.M. Consul at Mogador, who has kindly had inquiries made for me, that the *Acacia arabica* (Alk Awarwhal) is not found in Sus, no tree of the kind existing either north or south of the Atlas, but that its gum is brought from Soodan by Arabs, and is of inferior quality to that of the *Acacia gummifera*. Mr. Hay further informs me that the *Acacia gummifera* grows chiefly in the provinces of Blad Hamar, Rahamma, and Sus.

As stated above, the specimens which we collected of *Acacia gummifera* precisely accord with the published description and drawing; but we have others under this name from Mr. Cosson's collector, Ibrahim, gathered near Mogador and at Ouanyna, which differ in having very short spines, $\frac{1}{6}$ to $\frac{1}{4}$ in. long, whilst those of our plant are from $\frac{2}{3}$ to $\frac{3}{4}$ in. long and much stouter.

Very small plants of *Acacia gummifera* are living at Kew, raised from seeds obligingly presented by Mr. Cosson. They grow exceedingly slowly, and several have been lost by damping off. They are not in a state fit for exhibition.

The Argan Tree.

Argania Sideroxylon.—Roem. and Sch. Syst. Veg. iv. 502; Alph. DC. Prod. viii. 187; Hook. in Kew Journ. Bot. vi. (1854) 97, t. iii. iv.; De Noé, in Rev. Hort. 1853, 125; Ball, in Journ. Linn. Soc. Bot. xvi. 563.

Sideroxylon spinosum.—Linn. Hort. Cliff. 69 (excl. syn. et loc.); Correa, in Ann. Mus. Hist. Nat. viii. 393.

Rhamnus siculus.—Linn. Syst. Nat. ed. 12, iii. 227, excl. syn., non Bocc.

R. pentaphyllus.—Linn. Syst. Nat. ed. Gmel. 398, fid. Dryandr. excl. syn. Bocc.

Elæodendron Argan.—Retz Obs. Bot. vi. 26; Willd. Sp. Pl. i. 1148, excl. syn. Jacq. and Bocc.; Schousboe, Iagttag. over væxtrig. in Marocc. 89.

Argan.—Dryandr. in Trans. Linn. Soc. ii. 225.

This tree is rightly regarded as the most interesting vegetable production of Marocco, being confined to that empire and to a very circumscribed area in it, belonging to an almost exclusively tropical natural family, yielding a most important article of diet to the inhabitants, and a wood that for hardness and durability rivals any hitherto described. The earliest account of the Argan tree known to us is a brief one by the celebrated African traveller Leo Africanus, who visited Marocco in 1510. Speaking of some of the customs of the Moors, Leo Africanus says : 'Unto their Argans (for so they call a kind of olive which they have) they put nuts; out of which two simples they express a very bitter oil, using it for a sauce to some of their meats, and pouring it into their lamps' ('Purchas,' ii. 772). And in another passage he describes the oil correctly, as 'of a fulsome and strong savour.' The further history of the Argan tree is given in a very full and careful account by the late Sir W. Hooker, in the 'London Journal of Botany' for 1854 (vol. vi. p. 97, Tab. iii. iv.), which, as the work is of limited circulation, we here introduce.

'Through the kindness and by the exertions of the Earl of Clarendon, Chief Secretary for Foreign Affairs, the Royal Gardens of Kew have been put in possession of living plants and fresh seeds of a tree or shrub very little known in Europe, little known even to botanists, but highly esteemed by the Moors, in those parts of Marocco where it is a native, for its useful qualities, viz. the "Argan." Its economical properties are best explained by the copy of a letter which his Lordship did me the favour to communicate along with the plants and seeds, from Henry Grace, Esq., British Acting Vice-Consul at Mogador, addressed to J. H. Drummond Hay, Esq., Her Britannic Majesty's Agent and Consul-General at Tangier; both of which gentlemen spared no pains in procuring the information and seeds and living specimens; an example we should be glad to see followed by our consuls in other countries abounding in new and useful plants.

'"Mogador, November 7, 1853.

'"Sir,—The Argan tree grows more or less throughout the states of Western Barbary, but principally in the province of Haha, and south of this town. The soil in which it is found is light, sandy, and very strong; it is usually seen upon the hills, which are barren of all else, and where irrigation is impossible.

'"I should imagine, from the appearance of some of the trees, that they are from one to two hundred years old; and a remarkably large one in this neighbourhood is probably at least

three hundred. This individual measures 26 ft. round the
trunk; at the height of three feet it branches off; the branches
(one of which measures 11 ft. in circumference near the trunk)
rest upon the ground, extending about 15 ft. from the trunk,
and again ascend. The highest branch of this tree is not more
than 16 ft. to 18 ft. from the ground, while the outer branches
spread so as to give a circumference of 220 ft.: this is the largest
I am aware of.

'"The mode of propagation, in this vicinity, is mostly by
seed. When sowing this, a little manure is placed with it, and
it is well watered until it shoots; from which period it requires
nothing further. In from three to five years after sowing it
bears fruit, which ripens between May and August (according
to the situation of the tree). The roots extend a great distance
underground, and shoots make their appearance at intervals,
which are allowed to remain, thus doing away with the necessity
of transplanting or sowing. When the fruit ripens, herds of
goats, sheep, and cows are driven thither; a man beats the tree
with a long pole, and the fruits fall and are devoured vora-
ciously by the cattle. In the evening they are led home, and,
when comfortably settled in their yards, they commence chew-
ing the cud and throw out the nuts, which are collected each
morning as soon as the animals have departed upon their daily
excursion. I have heard it remarked that the nut passes
through the stomach; but this is only a casualty, and not a
general rule. Large quantities of the fruit are likewise collected
by women and children: they are well dried, and the hull is
taken off, and stored for the camels and mules travelling in the
winter, being considered very nutritious.

'"The process of extracting the oil is very simple. The nuts
are cracked by the women and children (and not a few fingers
suffer at the same time, owing to the want of proper tools, for
the nuts are very hard, and a stone is the only implement used);
the kernels are then parched in a common earthen vessel,
ground in handmills of this country, and put into a pan; a little
cold water is sprinkled upon them, and they are well worked up
by the hand (much the same as kneading dough) until the oil
separates, when the refuse is well pressed in the hand, which
completes the process. The oil is left to stand, and the sedi-
ment removed. The cake (in which a great deal of oil remains,

owing to the want of a proper press) is generally given to the milch cows or goats.

' " I never heard of any part being used as manure, but I have no doubt it would form an excellent one.

' " Some of these Argans grow in clusters, others are in single trees.

' " I have, &c.,

(Signed)   ' "Henry Grace.

' " *To J. H. Drummond Hay, Esq., &c. &c.*" '

'Except a brief notice of the exportation into Europe of Argan oil by the Danish Councillor of State, Georges Höst, who travelled in the kingdoms of Marocco and Fez during the years 1766–1768, the only published account of the uses of the Argan is given in a very little known Danish work, published by P. K. A. Schousboe, entitled "Iagttagelser over Væxtriget i Marokko. Forste Stycke. Kiobnhavn, 1800, 4, 7 Tab.," of which a German edition appeared in 1801, in 8vo, by J. A. Markussen. It gives an account of some Marocco plants: and, after an introductory sketch of the physical geography of Marocco, it contains descriptions of the plants of the country in Latin and German, with occasional observations in German. The account of the Argan under Retz's name of *Elæodendron Argan* is long: first comes a technical description, followed by a history of its synonymy, and then the following notes (kindly translated for us by Mr. Bentham) :—

' " It is surprising that this tree should hitherto have been so little known; as it is found in a country near Europe, and visited by many travellers, who speak in their diaries and descriptions of oil of Argan and of Argan trees, these last as constituting a considerable proportion of the forests of the country. It is, however, not to be met with in the northern provinces, but only towards the south. All those persons, from whom I have sought more accurate information on the subject, are unanimous in stating that it only grows between the rivers Tansif and Sus—that is, between the 29° and 32° N. lat.—and there constitutes forests of considerable extent. It flowers in the middle of June, and the fruit remains on the tree the greater part of the year. The young fruit sets in the end of July or beginning of August, and grows slowly till the rainy season

commences, towards the end of September. It now enlarges
rapidly and attains its full size during that season, so as that by
the middle or end of March it is ripe enough to be gathered for
economical uses. Both the fruit and the wood are serviceable,
but especially the former; for from the kernel an oil is extracted
which is much employed for domestic purposes by the Moors,
and is an important production of the country, as it saves much
olive oil, which can thus be thrown into commerce, and made to
bring money into the country. It is calculated that in the whole
Argan region one thousand hundredweight of oil is annually consumed, thus setting free an equal quantity of olive oil for exportation to Europe. Our countryman, Höst, in his 'Efterretninger
om Marokos,' p. 285, says that· the Argan oil is exported to
Europe, where it is used in manufactures. Such may have
been the case in former times when it might be cheaper; but
now there would be no advantage in doing so, as it costs almost
as much as olive oil. At present, no Argan oil whatever is
exported.

' " As the practice in preparing this oil is somewhat different
from that of common olive oil, it may be useful to enter into
some details on the subject. I have myself been present during
the whole operation, and consequently speak from experience.

' " In the end of March the countryman goes into the wood,
where the fruits are shaken down from the trees and stripped of
their husks on the spot. The green fleshy pericarp, which is
good for nothing else, is greedily eaten by ruminating animals,
such as camels, goats, sheep, and cows, but especially by the
first two. Therefore, when the Arab goes into the woods to
collect Argan nuts, he gladly takes with him his herds of the
above animals, that they may eat their fill of the green husks
whilst he and his family are collecting and shelling the nuts.
The horse, the ass, and the mule, on the contrary, do not like
this food. When a sufficient quantity of nuts are collected they
are brought home, the hard wooden shell is cracked between
stones, and the inner white kernels are carefully extracted. These
are roasted or burnt like coffee on earthen, stone, or iron plates;
in order that they may not be too much done, they are constantly
stirred with a stick. When properly roasted they should be all
over of a brown colour, but not charred on the outside. The
smoke, which is disengaged during the process, has a very

agreeable odour. As soon as the kernels have cooled, they are ground in a handmill into a thick meal, not unlike that of pounded almonds, only that it is of a brown colour, and the meal is put into a vessel in which the oil is separated, which is done by sprinkling the mass now and then with hot water, and keeping it constantly stirred and kneaded with the hand. This process is carried on until the mass becomes so hard that it can no longer be kneaded : the harder and firmer are the residuary coarse parts, the more completely is the oil extracted. At the last, cold water is sprinkled upon it, in order, as they say, to expel the last particles of the oil. During the operation the oil runs out at the sides, and is from time to time poured into a clean vessel. The main point to be attended to in order to extract the greatest quantity and the best quality of oil, is that it should be well kneaded, and that the proper proportion of hot water for the extraction of the oil should be used; it is always safer to be sparing of it than to be too profuse. The residuary mass, often as hard as a stone, is of a black-brown colour, and has a disagreeable bitter flavour. The oil itself, when it has settled, is clear, of a light brown colour, and has a rancid smell and flavour. When it is used without other preparations in cooking, it has a stimulating and pungent taste which is long felt on the gums. The vapour which arises when anything is fried in it, affects the lungs and occasions coughing. The common people use it generally without preparation; but in better houses it is the custom, in order to take off that pungency, to mix it previously with water, or to put a bit of bread into it and let it simmer before the fire.

' " The wood, which is hard, tough, fine-grained, and of a yellow colour, is used in house carpentry, and for other purposes." '

'We have been at some pains to distribute the seeds of this plant, with which we have been liberally supplied, to various parts of the East Indies, and to such of our Colonies as appeared suited to the growth of this tree, in respect of climate, &c. It is impossible for seeds to be in better condition; and though the surrounding hard portion of the nut is as thick and solid as that of hickory, those which we ourselves sowed sprouted in less than a month from the time they were put in the ground. The young trees bore the rough treatment of the voyage in midwinter remarkably well; and it is easy to see that this is a plant of ready culture in favourable climates.

'The value of the husks of the fruit as food for cattle, and the uses of the wood, are mentioned in the above extracts. The nature of the oil seems only to have been considered in relation to olive oil. But vegetable oils are now so much in demand, especially by Messrs. Price & Co., for their great candle-works at Vauxhall, as well as at Birkenhead, near Liverpool, that I was anxious to know the opinion of Mr. G. F. Wilson, the scientific director of those vast establishments, on the nature of Argan oil. Some seeds were consequently communicated to that gentleman, and he lost no time in experimenting upon them, and assuring me that "they contain a large percentage of a very fine oil. We have tried it in several ways, in each case with a favourable result. Some is now being exposed to a severe test, to show how the air acts upon it: I have, however, little fear but that it will answer. Our city friends are inquiring for us the best means of getting a ton or two of the nuts for experiments on a large scale. The only unfavourable point I see is the small weight of kernel to that of hard shell :—

6 Nuts gave—kernel 30 grains
   „   „   hard shell 350 grains
   „   „   outer husk 193 grains.

The hard shell probably should be sent home with the seed when the kernels are required to yield a sweet oil; for unless prepared with great care, hardly to be expected in a wild country, the oil would not be nearly so sweet if sent home expressed, instead of in its kernel and shell. Perhaps if the kernel is pounded and rammed tightly into casks, we might obtain sweet oil without great waste in freight."

'In a botanical point of view this plant is scarcely of less interest than in an economical. It has had the hard fate, often the consequence of being with difficulty procured, to be much misunderstood, and, except by Schousboe, to be imperfectly described; and references are given in works to plants as being identical which have no relationship with it; or to descriptions which, if the same, exhibit little or no resemblance.

'The first botanist who appears to have noticed this plant is

Linnæus, who, in the *Hortus Cliffortianus*, in 1737, described it, from dried specimens, under the name of *Sideroxylon spinosum.* " From Clifford's Herbarium," observes Mr. Dryander, " now in the possession of Sir Joseph Banks, the Argan was taken up by Linné in his *Hortus Cliffortianus*; though most of the synonyms are wrong, and consequently the *locus natalis* (*utraque India*) which is deduced from them. The specimen in Linné's Herbarium, under the name of *Sideroxylon spinosum*, is without flowers, and it is impossible to tell you with any certainty what it is. Clifford's Herbarium is therefore the only authority by which this species can be ascertained." Linnæus's *Rhamnus siculus*, in the Appendix to the third volume of the twelfth edition of the *Systema Naturæ*, is, we are assured by Mr. Dryander, " the Argan, or Olive-tree of Marocco (see Höst's ' Efterretninger om Marokos,' p. 284), as appears from the specimen in Linné's Herbarium, which has a ticket affixed, with the name of Argan of Marocco, and which I have also compared with specimens in Sir Joseph Banks's Herbarium from Marocco." The description, too, of Linnæus is very correct. He errs only in considering the plant to be the same as the *Rhamnus Siculus pentaphyllos* of Boccone (*Rhus pentaphyllum*, Desf.), which has *folia quinata*, which latter he introduces into the specific character, but not into the description; and he erroneously followed Boccone in giving Sicily as the native country in addition to Africa, and in adopting the specific name *Siculus.*

' In the *Species Plantarum* of Linnæus, Malabar alone is mentioned as the native country of the *Sideroxylon spinosum*. Nevertheless, with the exception of Willdenow, who rejects it altogether as " planta valde dubia, forte nullibi obvia," most of the older authors adopt this name for the Argan of Marocco. Under it, it appears in the first edition of *Hortus Kewensis*, with the reference to *Species Plantarum* of Linnæus, and to Commelyn, *Hortus Amstelod.* tab. 83, where, however, nothing is said of its native country, further than may be surmised by the name adopted from Breynius's " Lycio similis frutex Indicus spinosus, Buxi folio " (which, as already observed, Willdenow considered to be his *Flacourtia sepiaria*, from India), and of which the flowers and fruit were unknown to the author. If this were the Argan, it was in cultivation in Holland as early as 1697. At a period not much later, viz. in 1711, according to the *Hortus Kewensis*, it was introduced into England : " Cult. 1711, by the Duchess of Beaufort, Br. Mus. H.S. 141, fol. 39." It is indicated as a stove-plant.

' Sir James Smith, article *Sideroxylon spinosum* in Rees's " Cyclopædia " (1819), throws no new light upon the subject ; he omits the reference to Commelyn. Retz, in " Obs. Bot." vol. vi. p. 26, refers the plant to *Elæodendron*, in which he is followed by Willdenow, and by Schousboe, which latter author has given by far the fullest and best account of the plant botanically and economically.

' M. Corréa de Serra, "Annales du Museum d'Histoire Naturelle," 1809, tom. viii. p. 393, tab. v. f. l., has published a very good analysis of the fruit, with very brief characters and no observations. At length Mr. Brown, " Botanicorum facile princeps," in his invaluable Prodromus, under his Observations on Sapoteæ, says, " *Sideroxylon spinosum*, L., fructu valde diversum proprium hujus ordinis genus efficit ; " and, acting upon his suggestion, Rœmer and Schultes, " Systema Vegetabilium," vol. iv. pp. xlvi. and 502, have formed of this plant a new genus, *Argania*, in which they have been followed by Endlicher and Alphonse De Candolle. In this latter work a very full generic character is given, which need not here be repeated.'

' It is singular that no further allusion to this tree should appear in Jackson's " Account of the Empire of Marocco " than the following : " Oil Arganic is also in abundance in Suse ; it is much used for frying fish and burning-lamps. When used for frying fish, a quart of it should be boiled with a large onion cut in quarters ; and when it boils, a piece of the inside of a loaf, about the size of an orange, should be put in ; after which it should be taken off the fire and let stand to cool, and when quite cold should be strained through a sieve ; without this precaution it is supposed to possess qualities which promote leprosy." —Dr. Barretta.¹

The limited distribution of the Argan is one of its most noticeable features, for as a genus it is not far removed from *Sideroxylon*, a very widely spread tropical and subtropical genus of both hemispheres, and which reaches its northern limit in Madeira (in the same latitude as that attained by the Argan), where one species, *S. Mermulana*, Lowe, is found on the rocky heights of the interior The order is not found in the Canary Islands, but reappears in the Cape de Verdes in a species of *Sapota*, and is well represented in the humid regions of Western Africa. It would thus appear that *Argania* and the Madeiran *Sideroxylon* are two outlying representatives of a very tropical order; and, considering the proximity of the areas they inhabit, and their position in the extreme west of the Old World, they are, in a Botanico-Geographical point of view, plants of a very high interest, as evidences of a relationship between the Floras of these areas, which must originally have been established under very different conditions from those which now prevail.

The Argan was, as stated above, introduced into England in 1811, and was long established on a south wall, but ultimately was killed in an unusually severe winter. Numerous plants were raised, from seed sent by Sir John Hay, by Mr. Grace, and from those brought by myself, and the plant may be seen in the Economic-plant House at Kew. It is of very slow growth, which has disappointed colonists and others, to whom the fruits have been largely distributed from Kew.

APPENDIX E.

On the Canarian Flora as compared with the Maroccan.

By Joseph Dalton Hooker.

In respect of their botanical relationship to neighbouring Continents, Islands or Archipelagos may be roughly classed under two divisions : namely, those which are situated within a moderate distance of continents, and whose Floras are manifestly derived from them or have had a common origin with theirs ; and those which are situated very far from any continents, and whose Floras differ so much either from that of the neighbouring continent or from that of those parts of the continent that are nearest to them, that their origin is a matter of speculation. Of the first division, the British Isles, and probably Vancouver's Island, in North-West America, are conspicuous instances, their Floras being almost identical with those of the neighbouring continents. St. Helena, the Galapagos, Mauritius, and the Sandwich Islands are instances of the opposite extreme, for their Floras differ widely from those of any continents.

Between these extreme cases there are many intermediate ones; and there are others of an exceptional character, as Iceland, which, though far removed from any part of Europe, has but one flowering plant not found on that continent (*Platanthera hyperborea*); and Ceylon, which though it is almost united to the Peninsula of Hindostan, yet in many respects differs greatly from that peninsula in its Flora.

Amongst the exceptional cases to continental proximity being accompanied by close botanical relationship is the Flora of the Canarian Archipelago, which differs so greatly from that of the northern part of its neighbouring continent, namely, from that of Marocco,[1] that it demands notice in any work treating of the vegetation of the latter country.

This diversity between the Maroccan and Canarian Floras has been pointed out in Ball's Introductory Observations to the Spicilegium Floræ Maroccanæ,'[2] where it appears that whilst Marocco, out of 1,627 species of flowering plants, contains 165 endemic plants, it has only 15 which are confined to it and to the Canaries, or to it and Madeira. And Ball goes on to remark (p. 301), in respect of these few species common to both Floras: 'I think it is safe to say that the facts rather tend to show the accidental diffusion of a few Macaronesian[3] species on the adjacent coast of Africa, than to indicate the direct connection between the continent and those islands within a geological period at all recent.'

Were this diversity due solely or chiefly to the Canaries wanting many Maroccan plants, the inquiry would not be a pressing one; but as to this deficiency is to be added the presence in the Canaries of many indigenous species, and even several genera [4]

[1] The Canary Islands are situated about 3° farther south, and 280 miles distant from Mogador. They are thus opposite a much more hot and arid part of the African coast than that north of the Atlas. The large island of Fuertaventura is only about 70 miles from the continent south of Oued Noun.

[2] *Journ. Linn. Soc.* vol. xvi. p. 297.

[3] A term first applied by Webb to the Flora of the Canarian Archipelago, but which should also include the Flora of Madeira (as Ball makes it do in the above mention of it), the Azores, and perhaps also of the Cape de Verde Islands, which together form either a distinct botanical province, or a marked subdivision of the Mediterranean province.

[4] No less than nine very distinct genera are confined to the Canaries or Madeira or both:—*Parolinia, Bencomia, Visnea, Phyllis, Plocama,*

which are absent in Marocco, and in Marocco the great rarity
of endemic genera, of which *Argania* only is arboreous, the
inquiry becomes a very important one, inviting a much closer
study than can here be given to it.

The Flora of the Canarian Archipelago, though consisting,
like the Maroccan, for the most part of Mediterranean species,
yet differs from that of Marocco, in containing many plants
that may be classed under the following categories :—

I. It contains many non-Maroccan plants, obviously introduced by man, and not from Europe only, but from various
parts of both the Old and New Worlds. This will not appear
surprising when it is remembered that Teneriffe was for several
centuries the Prime Meridian of Geographers and the resort of
all the European ocean-navigators, who took their departure
from it on their outward voyages, and made for it on their homeward ones. The *Alternanthera achyrantha*, a tropical American
plant, was no doubt imported into the Canaries, and possibly
from thence introduced into Spain (where it is now naturalised). *Argemone mexicana* is another, and there are still other
as conspicuous examples of such foreign introductions. This
maritime intercourse can, however, only partially account for
the remarkable disproportion between the number of probably
introduced plants in the Canaries and in Marocco; and we
must take into account the isolation, barbarism, and exclusive-
ness of the latter country, and the absence of any commercial
intercourse between it and the Canaries or the rest of the world.

In Webb and Berthelot's 'Phytographia Canariensis' upwards of fifty plants are enumerated as to which we have little
doubt that all have been introduced by man, and none of which
have hitherto been found in Marocco. The list includes many
weeds of the widest tropical and temperate distribution, as species of *Sida, Waltheria, Siegesbeckia, Bidens, Lippia, Physalis,
Nicandra, Euphorbia, Alternanthera, Commelyna*, and various
Cyperaceæ and Grasses.

II. The Canaries contain many apparently indigenous
plants, which, though not Maroccan, are widely distributed
elsewhere; these form a large class, and the following are
some of the most prominent of them :—

Canarina, Musschia, Bosea, and *Gesnouinia*. The only endemic genera
in Marocco are *Argania, Hemicrambe, Ceratocnemum*, and *Sclerosciadium*.

Delphinium Staphysagria	Fragaria vesca
Hypecoum procumbens	Pyrus Aria
Biscutella auriculata	Prunus lusitanica
Viola canina	Epilobium palustre
Silene Behen	Anthemis fœtida
„ nutans	„ coronopifolia
Rhus Coriaria	Cynara horrida
Spartium junceum	Lactuca sylvestris
Ulex europæus	Cressa cretica
Medicago arborea	Calamintha Nepeta
Trigonella hamosa	Atriplex glauca
Trifolium striatum	Euphorbia serrata
„ squarrosum	„ obliquata
„ suffocatum	„ Lagascæ
„ filiforme	Orchis longibracteata
Lotus angustissimus	Ophrys tabanifera
Vicia hirsuta	Iris pallida
Lathyrus odoratus	Lilium candidum
Alchemilla arvensis	

together with various *Cyperaceæ*, Grasses and water-plants, some of which, and of the above, will no doubt hereafter be found in Marocco.

III. They contain some quite peculiar plants which are more closely allied to endemic species of Marocco than to those of any other country, and may have been derived from species that originally were transported from that country. These are but few, and are almost confined to species of the genus *Monanthes*, which is limited to these countries and the Cape de Verde Islands, of Cactoid *Euphorbiæ*, of succulent *Sonchi*, and of the *Kleinia* division of *Senecio*.

IV. They contain plants not found hitherto in Marocco, and which are more closely allied to Mediterranean species than to any others; and these form a very large class. The data for a complete list would require a very careful comparison of the Maroccan species with the species described in the 'Phytographia' and discovered since, many of which are unquestionably founded on too slight or too variable characters.[1]

It will be sufficient for present purposes to contrast the results obtained from a selection of genera[2] taken for comparison from Ball's 'Spicilegium' with the same from Webb's 'Phytographia':—

[1] On the other hand, many peculiar species have been added to the Canarian Flora since the date of the publication of the *Phytographia* (1836–50).

[2] The genera, which are unduly multiplied in the *Phytographia*, are

Genera	Canary Islands		Marocco	
	Number of species in each	Species confined to Canaries	Number of species in each	Species confined to Marocco
Hypericum . .	8	7	7	0
Matthiola . . .	4	3	3	0
Cistus . . .	2	1	7	0
Helianthemum . .	6	3	14	0
Polycarpia . .	6	4	1	0
Sempervivum . .	23	23	1	1
Cytisus . . .	11	9	11	4
Lotus . . .	10	6	14	2
Dorycnium . .	3	3	1	0
Rhammus . . .	3	3	3	0
Ilex	2	2	0	0
Chrysanthemum .	12	12	11	4
Senecio . . .	9	5	11	1
Doronicum . .	5	5	0	0
Tolpis . . .	5	4	2	0
Sonchus . . .	17	12	6	0
Convolvulus sect. Rhodorhiza }	5	5	0	0
Echium . . .	12	10	9	1
Micromeria . .	17	17	1	0
Sideritis . . .	6	5	7	2
Teucrium . . .	3	1	11	4
Solanum . . .	6	2	2	0
Scrophularia . .	5	3	9	1
Digitalis . . .	2	2	2	0
Statice . . .	9	9	13	3
Plantago . . .	10	3	11	1
Beta . . .	3	2	1	0
Euphorbia . .	19	9	22	6
Ephedra . . .	3	2	2	0
Juniperus . . .	2	1	4	0
Pinus . . .	1	1	1	0
Ruscus . . .	2	2	1	0
Asparagus . .	5	4	6	1
Scilla . . .	4	4	9	0
Luzula . . .	3	3	1	0
	243	$187 = \frac{3}{4}$	204	$31 = \frac{1}{6}$

here reduced to the standard adopted in the *Spicilegium*. The species are, unfortunately, also inordinately multiplied in the former work, which seriously vitiates the table: this, however, it is impossible to set right. On the other hand, some of the Canarian genera have been largely added to by later explorers.

[1] In this, the most curious case of all, the species were elaborated by Bentham, and may, therefore, be depended upon. A second Maroccan *Micromeria*, allied to a Canarian one, has been found by M. Cosson's collectors, as I am informed by Ball, whilst this sheet was passing through the press.

The disproportion between the two Floras in the case of these selected genera is thus well shown. It is most remarkable ; the number of endemic species being in the Canaries three-fourths of the whole and in Marocco only one-sixth ; and were the peculiar genera of the Canaries added, the disproportion would of course be increased.

The total number of Canarian species enumerated by Webb and Berthelot is about 1,000, of which 367,[1] or more than one-third, are regarded as peculiar to the Archipelago (a very few only of these being also Madeiran) ; whereas out of 1,627 Maroccan species only 165, or a little over one-tenth, are peculiar. Future discoveries will probably not materially increase the Maroccan proportion of peculiar species ; whereas since the publication of Webb's ' Phytographia' many peculiar species (especially of *Statice* and *Crassulaceæ*) have been discovered in the Canaries, and but few species common to other countries ; and these additions will go far to neutralise any error introduced into the estimate, due to the great number of new species founded on insufficient data which the ' Phytographia' includes.

Under this head also should be included the peculiar Canarian genera that appear to be modifications of continental ones. They are *Bencomia*, closely allied to *Poterium*, of which there are two species, both confined to one Island (Teneriffe) ; one of these is also a native of Madeira, where only two individual trees, a male and a female, have ever been seen ! *Gesnouinia*, allied to *Parietaria*; and *Canarina*, a monotypic genus allied to *Campanula*, but having a baccate fruit. *Bosea*, also a monotypic plant, is wholly unlike any known genus, and is, in some respects, intermediate between the two very distinct natural families—*Chenopodiaceæ* and *Phytolacceæ*.

V. Many Canarian plants are representatives of Floras more distant than those of Marocco or Western Europe, and are not found in those countries. These form an exceedingly interesting group, and may be classed according to countries thus :—

a. Oriental.—These are chiefly Arabo-Egyptian, but some of them extend even into Western India, and a few are repre-

[1] This estimate is subject to the same deductions as I have referred to in note 2, p. 407. On the other hand, were the many obviously introduced species to be struck out of Webb and Berthelot's enumeration, the proportion of peculiar species would be considerably augmented.

sentatives of tropical India. Some will no doubt yet be discovered in Marocco, especially south of the Atlas; and it is not
unreasonable to suppose that such have crossed Africa in a sub-
tropical latitude, and thus reached the Canaries under conditions
now operating.

The most remarkable are the following. The genera in
capitals have not hitherto been found in Marocco :—

Polycarpon succulentum	CAMPYLANTHUS salsoloides
VISNEA Moccanera	TRAGANUM nudatum
GYMNOSPORIA cassinoides	APOLLONIAS barbusana
Trigonella hamosa	Euphorbia Forskählii
Senecio flavus	DRACÆNA Draco
CEROPEGIA dichotoma	

Of the above hardly any have been found west of the Levant,
or anywhere between Egypt and the Canaries, except, possibly,
in Southern Algeria. *Traganum* must be reckoned as an African
and Oriental desert type, and will probably be found in South
Marocco; but *Ceropegia* is mainly Indian, as is *Gymnosporia*
(*Catha cassinoides*, Webb). *Campylanthus* consists of the Canarian species, of a variety or closely allied one in the Cape de Verde
Islands, and of a third which extends from Southern Arabia to
Scinde. The nearest ally of the *Apollonias* (*Phœbe barbusana*,
Webb) is a Ceylon tree; and *Visnea* is nearly allied to the
Malayan genus *Anneslea*. *Dracæna Draco* is the most in-
teresting of all in the list; for, though the genus abounds in
tropical Africa, the Canarian form, which is also a native of
the mountains of the Cape de Verde Islands, has only one near
ally, the *D. Ombet*, which is confined to Abyssinia, Southern
Arabia, and the intervening Island of Socotra.

b. The peculiar species representing American types inhabiting the Canaries or Madeira, but not found in Marocco, are
in some respects even more remarkable than the Oriental.

They belong to the following genera :—

Bowlesia [1] (*Drusa oppositifolia*, DC.), *Clethra*, five species
of *Bystropogon*, and *Cedronella*. Of these *Bowlesia* is other-

---

[1] Whilst this sheet was passing through the press, I am informed
that M. Cosson's collectors have found *Bowlesia* in South Marocco. No
doubt this is another case of that accidental diffusion of Macaronesian
species alluded to by Ball. (See p. 405.)

wise confined to the tropical Andes of America, one species only extending as far north as Mexico; the Canarian species, which according to Webb is found on rocky shaded places in Teneriffe, from the sea-level to the wooded region, is most closely allied to a Peruvian one. *Clethra* is a genus which extends from South Brazil to the Northern United States, and is also found in Japan and the Malayan Archipelago. The Macaronesian species most resembles a North American; it is found also in Madeira. *Bystropogon* is, like *Bowlesia*, an Andean genus, extending from Peru to Columbia. All the Canarian species belong to a different section from the Andean, and there is one species of the same section in Madeira. *Cedronella* is a North American and Mexican genus, and the Canarian species differs from all its congeners in its trisect leaves; it is also Madeiran.

Of the Canarian *Laurineæ*, *Persea indica*, also a native of Madeira and the Azores, belongs to an American section of that large genus.

c. Tropical and South African types in the Canaries. Of these the most noticeable are two forest trees, belonging to the large tropical genus *Myrsine*. One of these, *M. excelsa* (*Heberdenia excelsa*, Banks) is also found in Madeira; the other, *M. canariensis*, is confined to the island whose name it bears. The tropical order *Sapotaceæ*, to which *Argania* belongs, has no representative in the Canaries, but has one in the *Sideroxylon Mermulana* of Madeira.

The only almost exclusively South African genus[1] in the Canaries is a species of *Lyperia*, of which there are numerous Cape of Good Hope species, and one doubtful one in the Somali country (North-East Africa). The widely diffused Cape shrub, *Myrsine africana*, is found in the Azores and in Abyssinia, but not in the Canaries, Cape de Verdes, Madeira, or Marocco. The two singular shrubs *Phyllis* and *Plocama*, consisting each of a single species, of which the *Phyllis* is found also in Madeira, are representatives of the *Anthospermeæ*, a very large and conspicuously South African and Australian tribe of *Rubiaceæ*, and of which the only Maroccan representative is *Putoria*, a Mediterranean genus of a single species, and which is not Canarian.

[1] The Cape of Good Hope mountain plant, *Melianthus comosus*, found at the south end of Fuertaventura, must be assumed to have been introduced by man into that island.

The *Oreodaphne fœtens* of the Canaries and Maderia is now [1] referred to the American, Madagascar, and South African genus *Ocotea*, and is most nearly allied to a species found in the latter country.

The Maroccan flowering plants are thus grouped by Ball in his ' Spicilegium Maroccanum ' [2] :—

Total number of Maroccan species	1,627
Species widely diffused, temperate or tropical	467
Of which there are common to Marocco and the Islands . .	300
Maroccan, but not Insular	167
Mediterranean species in Marocco	995
Of which there are widely spread species common to the Islands and Marocco	254
Confined to Marocco and the Islands	15
Mediterranean species in Marocco, but not in the Islands . .	726
Maroccan species exclusively	165

The proportion of *Monocotyledons* to *Dicotyledons* is in Marocco 1 to 4·6, in the Canaries 1 to 6—a very great difference.

The leading natural orders in Marocco and the Canaries respectively are :—

	Marocco Species	Canaries Species
Compositæ . . .	208	143
Leguminosæ . . .	189	104
Gramineæ	134	77
Umbelliferæ . . .	86	27
Labiatæ	81	59
Cruciferæ . . .	73	29
Caryophylleæ . . .	69	38

In each country these seven natural orders include nearly half the Dicotyledonous plants. But in the Canaries the *Crassulaceæ* with 31 species should replace the *Cruciferæ*, and the *Umbelliferæ* be excluded.

The natural orders which are indigenous to the three Archipelagos of the Canaries, Madeira, and the Azores, but which are absent in Marocco, and the reverse are :—

[1] In Bentham and Hook, f., *Gen. Plant.* (vol. iii. ined.), the *Laurineæ* are described by Bentham, who has determined, for the first time, the proper position of the Canarian Laurels.

[2] *Journ. Linn. Soc. Bot.* vol. xvi.

In the Archipelago, but not in Marocco.	In Marocco, but not in the Archipelago.
Simarubeæ (Cneorum)	Berberideæ
Pittosporeæ	Capparideæ
Ternstrœmiaceæ	Polygaleæ
Ilicineæ	Ampelideæ
Myrsineæ	Coriarieæ
Phytolacceæ (Bosea)	Saxifrageæ
Myriceæ	Apocyneæ
Commelyneæ (introduced ?)	Lentibularieæ
	Nyctagineæ
	Ulmaceæ
	Cupuliferæ
	Ceratophylleæ
	Alismaceæ
	Juncagineæ
	Melanthaceæ

In the above lists the *Commelyneæ* are most probably introduced by man into the Canaries, and the absence of *Lentibularineæ*, *Ceratophylleæ*, *Alismaceæ*, and *Juncagineæ* in the Archipelago may be due to the want of suitable localities. The total absence of *Cupuliferæ* in all the Macaronesian Archipelago is inexplicable; and of *Quercus* especially, a genus so prominently developed in number of species and individuals on both continents, and which further abounds in both the Pliocene and Miocene beds of Europe.

The apparently indigenous Macaronesian genera which are wanting in Marocco are the following. Those in capitals are confined to the Canaries, or to the Canaries and Madeira :—

Malvaceæ	Rosaceæ	PLOCAMA
Abutilon	BENCOMIA	Compositæ
Cruciferæ	Alchemilla	Chrysocoma
PAROLINIA	Fragaria	Allagopappus
Barbarea	Aquifoliaceæ	Vieræa
Simarubeæ	Ilex	Doronicum
Cneorum	Pittosporeæ	Serratula
Celastrineæ	Pittosporum	Prenanthes
Gymnosporia	Ternstrœmiaceæ	Campanulaceæ
Sapindaceæ	VISNEA	MUSSCHIA
Melianthus ?	Umbelliferæ	CANARINA
Leguminosæ	Todaroa	Wahlenbergia
Spartium	Rubiaceæ	Ericeæ
Ulex	PHYLLIS	Clethra

Asclepiadeæ	Acanthaceæ	Phytolacceæ
Ceropegia	Justicia	BOSEA
Convolvulaceæ	Oleineæ	Urticeæ
Cressa	Notelæa	GESNOUINIA
Boragineæ	Myrsineæ	Myriceæ
Tournefortia	Myrsine	Myrica
Labiatæ	Sapotaceæ	Aroideæ
Bystropogon	Sideroxylon	Dracunculus
Cedronella	Primulaceæ	Liliaceæ
Verbenaceæ	Pelletiera	Dracæna
Lippia	Chenopodieæ	Cyperaceæ
Solaneæ	Traganum	Fimbristylis
Nicandra	Laurineæ	Cladium
Scrophularineæ	Persea	Gramineæ
Campylanthus	Apollonias	Chloris
Lyperia	Ocotea	Tricholæna

There are in Marocco, out of a total of 517 genera, 202, included under 67 orders, that have no indigenous species in the Canaries or Madeira. Many of these, about a quarter, being North Maroccan, *i.e.* only found in parts of Marocco farthest from the Canaries, would not be expected to occur in those islands, were it not that the vegetation of islands near to large continents often most resembles that of a higher latitude on the continent than that in which the islands are situated.

The following is a list of the Maroccan genera which are absent in Macaronesia;—those confined to North Marocco marked * ; those which have been found in Macaronesia, but certainly introduced, marked ‖; those in italics have been discovered since our return from Marocco.[1]

Clematis	Draba	Ceratocnemum
Thalictrum	*Erophila	*Cakile
*Anemone	Malcolmia	*Hemicrambe
Aconitum	Diplotaxis	*Cleome*
Berberis	*Moricandia*	Capparis
Rœmeria	‖Lepidium	*Caylusea*
Corydalis	Thlaspi	Fumana
Cardamine	Iberis	Polygala
Morettia	Hutchinsia	Velezia
Anastatica	Isatis	Dianthus

[1] M. Cosson has published in the 22nd volume of the ' Bulletin of the Botanical Society of France ' a list including the plants received from his collectors in South Marocco up to the year 1874.

Holosteum
Buffonia
Loefflingia
Montia
Althæa
Malope
*Radiola
Peganum
Celastrus
Zizyphus
Acer
*Coriaria
Lotononis
Crotalaria
Argyrolobium
*Calycotome
Anthyllis
*Securigera
Coronilla
Colutea
*Glycyrhiza
Hedysarum
Ornithopus
Ebenus
* Pisum
Ceratonia
‖Acacia
Saxifraga
Parnassia
Ribes
*Drosophyllum
*Peplis
*Ecbalium
*Hydrocotyle
Eryngium
Deverra
Hippomarathrum
Kundmannia
*Magydaris
Sclerosciadium
Meum
Heracleum
*Peucedanum
‖*Coriandrum ?
Thapsia
Elæoselinum
Gaillonia

Putoria
Callipeltis
Asperula
Crucianella
*Valeriana
‖Centranthus
*Fedia
Nidorella
Nolletia
Micropus
Leysera
Grantia
Anvillea
*‖Xanthium
Achillea
Cladanthus
Echinops
Xeranthemum
Onopordon
Stæhelina
Crupina
*Leuzea
Carduncellus
Catananche
Hyoseris
Phœcasium
Hieracium
Scorzonera
Jasione
Trachelium
*Calluna
Armeria
Limoniastrum
Plumbago
Coris
Argania
Fraxinus
Phyllyrea
* Vinca
Nerium
Dæmia
Glossonema
Boucerosia
*Microcala
Cicendia
Trichodesma
Echinospermum

Rochelia
Nonnea
Cerinthe
Calystegia
*Mandragora
Anarrhinum
*Pinguicula
‖*Acanthus
Vitex
Lycopus
*Satureja
Hyssopus
Cleonia
Zizyphora
*Betonica
Ballota
Boerhavia
Corrigiola
Scleranthus
Sclerocephalus
Polycnemum
Telephium
*Obione
Salicornia
Caroxylon
Passerina
Osyris
Andrachne
*Ulmus ?
Celtis
Quercus
Populus
Ceratophyllum
Callitris
Cedrus
Aceras
*Serapias
*Cephalanthera
*Crocus
Leucojum
*Lapiedra
*Tapeinanthus
*Corbularia
Narcissus
Aurelia
*Alisma
Damasonium

*Triglochin
Chamærops
Gagea
*Hyacinthus
Anthericum
*Simethis
*Aphyllanthes
Colchicum
Erythrostictus
*Convallaria
*Schœnus

*Leersia
Lygeum
*Crypsis
*Alopecurus
Macrochloa
*Sporobolus
Ammophila
|| Arundo
*Ampelodesmos
Phragmites

Pappophorum
Echinaria
*Spartina
*Airopsis
Gaudinia
Glyceria
Secale
Elymus
*Lepturus
Anthistiria

These 202 genera, which are absent in the Canaries, comprise upwards of 300 Maroccan species, including *Eryngium*, with eleven species, *Coronilla* with eight, *Diplotaxis* with seven, *Narcissus, Anthyllis, Polygala, Passerina,* and *Quercus* five each, besides twenty other genera with three or four each. Not a few of them contain very common and wide-spread species, as do all the above-named, as well as *Clematis, Malcolmia, Cardamine, Dianthus, Hedysarum, Heracleum, Asperula, Achillea, Onopordon, Hyoseris, Scorzonera, Phyllyrea, Fraxinus, Calystegia, Anarrhinum, Ballota, Populus, Chamœrops.* That no species of these or of many of the other genera should exist in the Canaries is inexplicable, considering the position and extent of the Archipelago, and the means of migration which must exist between it and the mainland.

The species common to Macaronesia and Marocco exclusively, are in so far as is at present known :—

Helianthemum canariense, *Jacq.*
Polycarpia nivea, *Ait.* (also occurs in C. de Verde)
Zygophyllum Fontanesii, *Webb*
? Cytisus albidus, *DC.*
Ononis angustissima, *Lam.* (? A form of *O. Natrix*)
Astragalus Solandri, *Lowe* (Madeira only)
Astydamia canariensis, *DC.*
Bowlesia oppositifolia
Odontospermum odorum, *Schousb.*
Sonchus acidus, *Schousb.* (In Lancerotte, only a single plant, possibly introduced)
Lithospermum microspermum, *Boiss.*
Linaria sagittata, *Poir.*
Chenolea canariensis, *Moq.*
Salix canariensis, *Chr. Sm.* (rather uncertain)
Romulea grandiscapa, *Webb.* (Perhaps only a var., but Baker keeps it)
Asparagus scoparius, *Lowe.* (Not quite certain)

Although it would be out of place here to discuss all the questions raised by this slight sketch of the peculiarities of the

Canarian Flora, there are some of them which so intimately bear upon the Maroccan as to awaken attention.

The wonderful development in the Canaries of endemic species belonging for the most part to Mediterranean types, points to the very early introduction of the parent forms of these, and the long isolation both of the Archipelago and its separate islets. It is in accordance with generally accepted views, to assume that the endemic species of each genus have been derived from parent forms originally introduced into one or more of the islets; and that as the descendants of these species spread over the Archipelago they were exposed to different conditions in each islet, resulting in their varying, and in the segregation and conservation of different local varieties each in its own insular birth-place; a supposition which is in accordance with the fact that those endemic species are really very local, many being confined to a single islet. In Marocco the parent forms of its Flora would be exposed to no such diverse conditions, and the areas in which varieties occurred, not being isolated, would be exposed both to invasion on all sides by other plants, and to destruction by agencies that affected the whole surrounding country, as drought, floods, insects, and birds.

The tropical types in the Canaries, with the exception of the Egypto-Arabian and the trees mentioned under V. c., are chiefly weeds of wide distribution, which have not reached Marocco, because of its want of ports and its limited commerce.

Finally the *Dracæna*, together with the tropical trees of *Myrsineæ*, *Sapotaceæ* (in Madeira), and *Laurineæ*, and the Egypto-Arabian types, suggest the hypothesis that at a very remote period these and many other plants of warmer and damper regions flourished in the area included in North-West Africa and its adjacent islands, and that they have been expelled from the continent by altered conditions of climate, but have been preserved in the more equable climate and more protected area of the Atlantic Islands.

Ball, who has given me valuable aid on many points discussed in this article, directs my attention to the important differences that exist between the vegetation of the eastern group of the Canary Islands—Fuertaventura, Lanzarote, and the adjacent islets—the ' Purpurariæ ' of authors, and the western group, including Teneriffe, Grand Canary, &c.

In the first place, nearly all the characteristic Canarian types are absent in the eastern group. Out of fifty-four genera above enumerated as present in the Canaries but wanting in Marocco, two are in the Canaries confined to the eastern islands: one of these, *Traganum*, is an African desert type, probably to be found in South Marocco; the other, *Melianthus*, a Southern African plant, and scarcely indigenous. Of the remainder *Plocama* alone is certainly present, and three other generic types probably exist in that group; while forty-eight genera, including eight out of nine peculiar to the Canaries, are apparently absent. In the next place several characteristic desert plants, such as *Oligomeris subulata*, *Ononis vaginalis*, *Convolvulus Hystrix*, and *Traganum nudatum*, are present in the 'Purpurariæ,' but absent from the western islands.

Although the Flora of the Purpurariæ is incompletely known, and our acquaintance with that of the neighbouring African coast between the rivers Sous and Draha is extremely imperfect, these facts tend to prove that there is a closer botanical relationship between the eastern islands and the adjoining continent than there is between them and the western portion of the Canarian Archipelago. Such relationship might be brought about in three different ways.

1. The greater dryness and heat of the eastern islands may have favoured the immigration of African forms, and at the same time led to the destruction, or weeding out, of the characteristic Canarian types. In this case the cause would be of a purely local and climatic character.

2. We may believe in the trans-oceanic migration of some African species to the nearer islands, along with the transport of some Canarian species (those enumerated in p. 416, and others which may be hereafter found) to the neighbouring continent.

3. An ancient extension of the continent to the Purpurariæ, leaving the other islands separated by deep sea.

It is an objection to the latter hypothesis that a profoundly deep ocean bed lies between the lines of 100 fathom soundings that girdle the islands and the African coast respectively; and that while the 100 fathom line extends about thirty miles from the coast of the continent, it is never more than five miles, rarely more than one or two, from those of the islands.

In favour of the hypothesis of trans-oceanic transport it may be remarked that the distance between the African coast and

Fuertaventura is not more than seventy miles, and that a moderate change of level of about 600 feet would reduce that distance by one-half, while it would but slightly affect the interval that separates the Purpurariæ from the other islands.

Among the possible causes leading to an interchange of species between the Purpurariæ and the African coast the agency of man must not be omitted. The fishermen of those islands were formerly in the habit of visiting some points on the opposite coast, although intercourse of this kind has almost ceased in recent times.

It must be observed that our knowledge of the vegetation of the Canary Islands is yet incomplete. Although several additions to the Flora have been published by C. Bolle and others, no supplement to Webb's 'Phytographia' has been published. Several additional species exist in herbaria, besides those that may be hereafter found.

So little is known of the geology of Marocco, that there are no data for ascertaining whether during antecedent geological periods it contained a more tropical Flora than now; but evidence in support of such a hypothesis is forthcoming in Madeira, where fossiliferous beds which have been referred to 'some part of the Pliocene period'[1] have been discovered, containing leaves referable in part to existing species of Madeiran plants, and in part to extinct ones of tropical aspect;[2] and it is well ascertained that during preceding geological periods Western Europe was clothed with a vegetation that suggests a very much warmer climate than now prevails, and of which vegetation the *Laurus nobilis* in the south-west of the continent has been supposed to be a surviving representative.

In Grand Canary, also, Upper Miocene beds exist, containing numerous species of fossil shells, of which one is an Oregon species, and another tropical African; and in more recent deposits of the same Archipelago many shells have been found which no longer inhabit the adjacent seas, including tropical West African, Mozambique, and Mediterranean species.

We can form no conception of means of transport from the American continent that would transfer the parent species of *Bowlesia* and of the *Bystropogons* from the Andes to the

[1] Lyell's *Principles of Geology*, ed. 11, vol. ii. p. 410.
[2] Lyell's *Student's Elements of Geology*, ed. 2, pp. 538, 539.

Atlantic islands; and we can but hazard the assumption that, at some very distant date, these genera existed in more eastern parts of America, from whence seeds were transported across the ocean.   On the other hand, the transport of parent forms or existing species from the continents of Europe and Africa to the Atlantic islands may have been much facilitated by greater extensions of land in bygone ages.   Madeira, the Canaries, and the Cape de Verde Islands, are all supposed to stand on a sub-marine platform which skirts the coasts of Western Europe and North-Western Africa, and whose submerged margin imme-diately to the westward of the position of the islands descends rapidly to a profound depth.   The westward margin of this platform was possibly the coast-line in Miocene times.   An ele-vation of its surface of a few hundred feet would approximate the islands to the mainland very materially, and greatly facili-tate transport.   That they were, however, ever united to the continent is opposed to the views of most competent geologists. Lyell, speaking of this, says : ' The general abruptness of the cliffs of all the Atlantic islands, coupled with the rapid deepening of the sea outside the 100 fathom line, are characters which favour the opinion that each island was formed separately by igneous eruptions, and in a sea of great depth.'   Moreover, the Azores, whose botany in so many respects resembles that of the other Atlantic islands, as distinguished from that of the con-tinent, are enormously more distant from the mainland; and these islands stand on a platform of their own, separated from the continental one by an ocean of profound depth; so that any theory of transport which applies to the Canarian and Madeiran Archipelagos, should apply also to the Azorean.

It remains a point of some nicety to decide whether the Macaronesian islands should be regarded as a Botanical province apart from the Mediterranean, or a sub-division of the latter. The assemblage of American and Oriental genera which their Flora contains, together with the arboreous representatives of tropical *Laurineæ*, all so entirely foreign to the European Flora, would give it a title to be called a Botanical province; and to this as a further title is the prevalence of a considerable proportion of North European plants, in the Northern Archi-pelago especially.   On the other hand, fully two-thirds of the species are typical of the Mediterranean Flora, and by far the majority of the remainder are derivative species of the same origin; so that, on the whole, I am disposed to regard it as a very distinct sub-division of the Mediterranean province, which owes its peculiarities partly to the conservation of types once common to West Europe and North Africa, but which have been eliminated in those regions, and partly to the effect of isolation and climate on the progeny of species still existing in those regions.

APPENDIX F.

Comparison of the Maroccan Flora with that of the Mountains of Tropical Africa.

By JOSEPH DALTON HOOKER.

As was to have been anticipated, the Maroccan Flora contains most of the European species which have been collected on the mountains of Abyssinia and of the Bight of Biafra, which alone of the tropical African Alps have been botanically explored. Of these the former have been visited by Schimper and various collectors; whilst the mountains of the pestilential West African coast, of Fernando Po, 9,500 feet, and the Cameroons Mountains, upwards of 13,000 feet, have been ascended for botanical purposes only by Gustav Mann, when employed for the Royal Gardens of Kew.

The results of the latter were published by myself in the 'Journal of the Linnæan Society of London' (vol. vii. p. 171), from whence the following observations are for the most part extracted. They included 26 European species, gathered at elevations between 5,000 and 10,000 feet. Most of them are also natives of the Abyssinian Alps, and two-thirds of them are also Maroccan, whilst others will probably yet be found in the latter country.

The following is a catalogue of all the European plants found in the upper regions of the Cameroons Mountains and Fernando Po :—

	Height	Where found
	feet	
Cardamine hirsuta	7,000–10,000	Marocco and Abyssinia
Cerastium vulgatum (viscosum, Fr.)	8,000	,, ,,
Radiola Millegrana	7,000	Marocco

	Height	Where found
	feet	
Oxalis corniculata . .	7,000–8,500	Marocco and Abyssinia
Umbilicus pendulinus . .	7,000–10,000	„ „
Sanicula europæa . . .	4,000–7,500	Abyssinia
Galium rotundifolium . .	7,000–12,000	Marocco and Abyssinia
Galium Aparine . . .	7,000–10,000	Marocco and Abyssinia
Scabiosa succisa . . .	10,500 ?	
Myosotis stricta . . .	8,000–10,000	Marocco and Abyssinia
Limosella aquatica . .	9,000–10,000	Abyssinia
Sibthorpia europæa . .	7,000–7,500	„
Solanum nigrum . . .	7,000–11,000	Marocco and Abyssinia
Rumex obtusifolius . .	7,000	Abyssinia
Parietaria mauritanica . .	7,000–8,000	Marocco and Abyssinia
Trichonema Bulbocodium .	7,000–9,000	„ „
Juncus capitatus . . .	7,000	Marocco
Luzula campestris . .	8,000–10,000	Abyssinia
Deschampsia cæspitosa . .	9,000–12,000	„
Aira caryophyllea. . .	7,000–8,000	Marocco and Abyssinia
Poa nemoralis . . .	7,000–10,000	Abyssinia
Kœleria cristata . . .	8,000–12,000	„
Vulpia bromoides . . .	7,000–10,000	Marocco and Abyssinia
Festuca gigantea . . .	8,500	Marocco
Brachypodium sylvaticum .	7,000	Abyssinia
Andropogon distachyus .	7,000	Marocco and Abyssinia

The most remarkable features of the Temperate vegetation
of these West African tropical mountains are :—

1. Its poverty.

2. The preponderance of Abyssinian genera and species.

3. The considerable proportion of European plants.

4. The paucity of South African genera and species.

5. The great rarity of new genera.

6. The absence of St. Helena and Canarian types.

Upon each of these propositions I have a few general remarks to offer.

In the poverty of their Flora the Cameroons range and
Peak of Fernando Po seem to partake of the characteristics of
the Abyssinian Alps. We know far too little of the physical
geography of either of these districts to hazard many conjectures upon this point, which must to a certain extent be dependent on the arid volcanic nature of the soil and the limited
area of the Temperate region. Mr. Mann spent many weeks,
and at various seasons, in his explorations, and yet 237 flowering plants were all that rewarded his toil. Geological causes
have probably had, in the case of the Cameroons Mountains,

much to do with the dearth of species, some parts of the range even now presenting evidence of subterranean heat.

The preponderance of Abyssinian forms is proved by almost all of the genera and half the species being natives of Abyssinia, and by many other species being very closely related to, or obvious representatives of, plants of that country. There are, further, several of the genera and many of the species peculiar to Abyssinia and the peaks of Biafra.

The number of European genera amounts to 43, and species to 26, the greater part of which are British. Very few of them extend into South Africa. The greater part are Abyssinian; the remarkable exceptions being *Radiola, Scabiosa succisa, Luzula campestris,* and *Festuca gigantea,* all of which, however, may have been hitherto overlooked in Abyssinia.

I find no other evidence of relationship between the Biafran mountain Flora and that of Marocco than what is afforded by the European species common to both. In most other respects the Floras differ totally, the other mountain plants of Biafra being Abyssinian or Cape types, or more nearly related to tropical African ones.

APPENDIX G

On the Mountain Flora of Two Valleys in the Great Atlas of Marocco.

By JOHN BALL.

ALTHOUGH an attempt to discuss the character and relations of the Flora of a region so wide and so little known as the mountain region of the Great Atlas would as yet be quite premature, it appears that the materials at our disposal suffice for an examination of the vegetation of the valleys lying south and south-west of the city of Marocco, which may be an acceptable contribution to botanical geography. For this purpose it seems best to limit the discussion to the two valleys where our collections were sufficiently extensive to give a tolerably complete representation of the vegetation, as far as this was developed at the season of our visit, and to exclude altogether the plants found along the skirts of the great range below the level of about 1,200 metres above the sea. The Flora of the zone below that level is largely mixed up with extraneous elements, represented by plants of the low country that reach the base of the mountains, but do not penetrate the interior valleys, and if these had been admitted the special features of the mountain Flora would have become less apparent. The plants collected in the course of a somewhat hurried excursion from Seksaoua, when we reached a height of about 1,600 metres, have been designedly omitted. On such occasions attention is to a great extent monopolised by the new and rare species not hitherto seen in the same region, while comparatively familiar forms are less carefully noted. Collections made under such circumstances rarely give a moderately complete report of the vegetable population.

In ascending to the higher region of mountains that approach to the limit of vegetation the absolute number of species is so much smaller that this source of error is far less apparent; and it is not likely that in the two ascents which we made to the dividing ridge of the Atlas many species that came within our range of vision were overlooked. The following tabular arrangement shows that in the Aït Mesan valley, where we spent the greater part of six days, we collected 375 species of phænogamous plants, to which have been added three subsequently found there by MM. Rein and Fritsch; while in the Amsmiz valley only 223 species—or less than three-fifths of the above number—were collected. Of these 146 species are common to both valleys; so that our list does not in all exceed 455 species of flowering plants, to which I have added 10 vascular cryptogams, of which two only were found in the Amsmiz valley.

In the following list I distinguish a middle zone, extending from 1,200 to 2,000 metres above the sea, and a superior zone including all above that limit; the latter corresponding pretty nearly with the sub-alpine and alpine zones of the higher mountains of Europe. With reference to their distribution I have divided the species into four categories : 1, Mid-European, those extending to Central Europe, of which more than three-fourths belong to the British Flora : 2, wide-spread Mediterranean, extending beyond the bounds of the three adjoining regions, Algeria, the Spanish peninsula, and the Canary Islands : 3, confined to adjoining regions; that is, to one or more of those just enumerated : 4, endemic, known only in the Great Atlas, or the neighbouring provinces of Marocco.

Tabular View of the Mountain Flora of the Great Atlas, showing the distribution of the Species found in the Valleys of Aït Mesan and Amsmiz. S indicates the superior zone from 2,000 m. to 3,500 m. above the sea; M the middle (or mountain) zone, from 1,200 m. to 2,000 m.

Name of Species	Aït Mesan	Amsmiz	Mid-European	Widespread Mediterranean	Confined to adjoining regions	Endemic
Ranunculus spicatus, L. var.	–	S	–	–	*	–
" atlanticus, Ball	M	M	–	–	–	*
" bulbosus, L. var.	M	M	*	–	–	–
" arvensis, L.	M S	–	*	–	–	–
" muricatus, L.	–	M	–	*	–	–
" Reinii, nov. sp.	S	–	–	–	–	*
Aquilegia vulgaris, L. var.	–	M	*	–	–	–
Delphinium Balansæ, B. et R. var. ?	S	S	–	–	*	–
Berberis cretica, L. var.	–	S	–	*	–	–
Papaver tenue, Ball	M S	M S	–	–	–	*
" rupifragum, B. et R. var.	S	S	–	–	*	–
Rœmeria hybrida, D. C.	–	M	–	*	–	–
Hypecoum pendulum, L.	–	M	*	–	–	–
Corydalis heterocarpa (Dur.)	M	–	–	–	*	–
Fumaria officinalis, L.	S	–	*	–	–	–
" parviflora, Lam.	M	M	*	–	–	–
" media, Lois var.	M	–	*	–	–.	–
" agraria, Lag. var.	M	–	–	*	–	–
" tenuisecta, Ball	M	–	–	–	–	*
Nasturtium officinale, R. Br.	S	–	*	–	–	–
" atlanticum, Ball	S	–	–	–	–	*
Arabis albida, Stev.	S	–	–	*	–	–
" erubescens, Ball	S	–	–	–	–	*
" auriculata, Lam.	S	S	*	–	–	–
" decumbens, Ball	S	S	–	–	–	*
" conringioides, Ball	S	S	–	–	–	*
Cardamine hirsuta, L. var.	–	S	*	–	–	–
Alyssum alpestre, L. vars.	M	M	*	–	–	–
" montanum, L. var.	–	S	*	–	–	–
" campestre, L.	M S	–	*	–	–	–
" calycinum, L.	M S	–	*	–	–	–
" spinosum, L.	S	–	–	*	–	–
Draba hispanica, Boiss.	–	S	–	–	*	–
Sisymbrium Thalianum	–	S	*	–	–	–
" runcinatum, Lag. var.	–	M	–	–	*	–
Erysimum australe, Gay, var.	M S	M S	*	–	–	–
Brassica rerayensis, Ball	S	–	–	–	–	*
Capsella bursa-pastoris, L.	S	–	*	–	–	–

TABULAR VIEW OF THE MOUNTAIN FLORA—*continued.*

Name of Species	Aït Mesan	Amsmiz	Mid-Euro-pean	Wide-spread Mediter-ranean	Con-fined to adjoin-ing regions	En-demic
Lepidium nebrodense, Raf. var.	S	—	—	*	—	—
Biscutella lyrata, L. var. . .	S	—	—	*	—	—
Thlaspi perfoliatum, L. et var.	S	S	*	—	—	—
Hutchinsia petræa, R. Br. .	S	S	*	—	—	—
Isatis tinctoria, L. var. . .	M	M	*	—	—	—
Crambe hispanica, L. . .	—	M	—	—	*	—
Capparis spinosa, L. . . .	M	—	—	*	—	—
Reseda attenuata, Ball . .	S	—	—	—	—	*
„ phyteuma, L. . .	—	M	*	—	—	—
„ lanceolata, Lag. .	M	—	—	—	*	—
Cistus polymorphus, Willd. .	M	—	—	*	—	—
Helianthemum niloticum, L. var. ?	S	—	—	*	—	—
„ rubellum, Presl. .	—	M	—	*	—	—
„ glaucum, Cav. .	—	M	—	*	—	—
„ virgatum, Desf. et var.	M	M	—	—	*	—
Fumana glutinosa, L. . .	M	M	—	*	—	—
„ calycina, Claus. .	M	—	—	—	*	—
Viola tezensis, Ball . . .	—	S	—	—	—	*
Polygala rupestris, Pourr. .	M	M	—	*	—	—
„ Balansæ, Coss. .	M	M	—	—	—	*
Dianthus attenuatus, Sm. .	M S	—	—	—	*	—
„ virgineus, L. . .	M	M	*	—	—	—
Tunica compressa, Desf. .	—	M	—	—	*	—
„ prolifera, L. . .	M	M S	*	—	—	—
Silene inflata, Sm. var. . .	M	—	*	—	—	—
„ nocturna, L. . . .	M	—	—	*	—	—
„ corrugata, Ball . .	M	—	—	—	—	*
„ muscipula, L. . . .	M	—	—	*	—	—
„ italica, L. . . .	M	M	—	*	—	—
Holosteum umbellatum, L. .	S	S	*	—	—	—
Cerastium glomeratum, Thuill.	S	S	*	—	—	—
„ brachypetalum, Desf.	S	S	*	—	—	—
„ arvense, L. . .	M S	M S	*	—	—	—
Stellaria media, L. . . .	M	—	*	—	—	—
„ uliginosa, Murr. .	S	—	*	—	—	—
Arenaria pungens, Clem. et var.	S	S	—	—	*	—
„ serpyllifolia, L. . .	S	S	*	—	—	—
„ procumbens, Vahl. .	M	—	—	*	—	—
„ fasciculata, Gouan .	S	—	*	—	—	—
„ setacea, Thuill. var. .	S	S	*	—	—	—
„ verna, L. var. .	S	S	*	—	—	—
Buffonia tenuifolia, L. . .	M	—	—	*	—	—
Sagina procumbens, L. var. .	S	S	*	—	—	—
„ Linnæi, Presc. .	S	—	*	—	—	—
Polycarpon tetraphyllum, L. .	—	M	*	—	—	—
„ Bivouæ, J. Gay .	S	S	—	*	—	—

TABULAR VIEW OF THE MOUNTAIN FLORA—*continued.*

Name of Species	Aït Mesan	Amsmiz	Mid-European	Widespread Mediterranean	Confined to adjoining regions	Endemic
Montia fontana, L.	S	–	*	–	–	–
Hypericum perforatum, L.	M	–	*	–	–	–
„ coadunatum, Chr. Sm. var.	M	–	–	–	*	–
Malva sylvestris, L.	M	–	*	–	–	–
„ rotundifolia, L.	M S	–	*	–	–	–
Linum corymbiferum, Desf.	M	–	–	–	*	–
Fagonia cretica, L.	M	–	–	*	–	–
Geranium malvæflorum, B. et R.	S	S	–	–	*	–
„ pyrenaicum, L.	M	M	*	–	–	–
„ molle, L.	S	–	*	–	–	–
„ rotundifolium, L.	M	M	*	–	–	–
„ lucidum, L.	S	M S	*	–	–	–
„ robertianum, L. var.	M	–	*	–	–	–
Erodium Jacquinianum, F. et M.	M S	–	–	*	–	–
„ malacoides, L.	M	–	–	*	–	–
„ guttatum, W.	M	–	–	–	*	–
Oxalis corniculata, L.	M	–	*	–	–	–
Ruta chalepensis, L.	M S	–	–	*	–	–
Rhammus Alaternus, L.	M	M	–	*	–	–
„ lycioides, L.	–	M	–	–	*	–
Acer monspessulanum, L.	M	–	*	–	–	–
Pistacia Lentiscus, L.	M	M	–	*	–	–
Lotononis maroccana, Ball	M	M	–	–	–	*
Argyrolobium Linnæanum, Walp. var.	M S	–	–	*	–	–
„ stipulaceum, Ball	–	M	–	–	–	*
Adenocarpus anagyrifolius, Coss.	M	M	–	–	–	*
Genista dasycarpa (Coss.)	M	–	–	–	–	*
„ myriantha, Ball	–	M	–	–	–	*
„ florida, L. var.	M	–	–	–	*	–
Cytisus Balansæ, Boiss. var.	S	S	–	–	*	–
„ albidus, D.C.	–	M	–	–	–	*
„ Fontanesii, Spach	M	–	–	–	*	–
Ononis atlantica, Ball	–	M	–	–	–	*
„ antiquorum, L.	–	M	–	*	–	–
Trigonella monspeliaca, L.	M	–	–	*	–	–
„ polycerata, L. et var.	S	M	–	–	*	–
Medicago lupulina, L.	M	–	*	–	–	–
„ suffruticosa, Ram.	S	S	–	–	*	–
„ turbinata, W. vars.	M	M	–	*	–	–
„ denticulata, W.	–	M	*	–	–	–
„ minima, Lam.	M	–	*	–	–	–
Melilotus indica, All.	M	–	–	*	–	–
Trifolium atlanticum, Ball	S	–	–	–	–	*
„ glomeratum, L. var.	S	–	*	–	–	–
„ repens, L.	M	–	*	–	–	–
„ humile, Ball	S	–	–	–	–	*
„ tomentosum, L.	M	–	–	*	–	–

TABULAR VIEW OF THE MOUNTAIN FLORA—*continued.*

Name of Species	Aït Mesan	Amsmiz	Mid-European	Widespread Mediterranean	Confined to adjoining regions	Endemic
Anthyllis Vulneraria, L. et var.	M S	–	*	–	–	–
„ tetraphylla, L.	M	–	–	*	–	–
Lotus cytisoides, D. C.	M	–	–	*	–	–
Coronilla pentaphylla, Desf.	M	–	–	–	*	–
„ ramosissima, Ball	M	–	–	–	–	*
„ minima, L.	M	–	*	–	–	–
„ scorpioides, L.	M	–	–	*	–	–
Hippocrepis atlantica, Ball	S	–	–	–	–	*
„ multisiliquosa, L. var.	M	–	–	*	–	–
Psoralea bituminosa, L.	M	–	–	*	–	–
Colutea arborescens, L.	M	–	*	–	–	–
Astragalus sesameus, L.	M	–	–	*	–	–
„ Reinii, Ball	S	–	–	–	–	*
„ Glaux, L. var.	M	–	–	–	*	–
„ atlanticus, Ball	M	–	–	–	–	*
„ ochroleucus, Coss.	S	–	–	–	–	*
„ incurvus, Desf.	M	–	–	–	*	–
Vicia onobrychoides, L.	S	–	–	*	–	–
„ glauca, Presl. var.	S	–	–	*	–	–
„ sativa, L. vars.	M S	–	*	–	–	–
Lathyrus aphaca, L.	M	–	*	–	–	–
„ sphæricus, Retz.	M	–	–	*	–	–
Ceratonia siliqua, L.	M	M	–	*	–	–
Prunus prostrata (Labill.)	S	S	–	*	–	–
Poterium sanguisorba, L.	M	–	*	–	–	–
„ verrucosum, Ehrnb. var.	M	–	–	*	–	–
„ anceps, Ball	S	S	–	–	–	*
„ ancistroides, Desf.	–	M	–	–	*	–
Rosa canina, L. var.	M	M	*	–	–	–
„ Seraphini, Viv.	S	S	–	*	–	–
Saxifraga globulifera, Desf.	M S	M S	–	–	*	–
„ tridactylites, L.	S	–	*	–	–	–
„ granulata, L.	S	S	*	–	–	–
Ribes Grossularia, L.	M S	S	*	–	–	–
Cotyledon umbilicus, L.	M	M	*	–	–	–
Sedum modestum, Ball	M	M	–	–	–	*
„ dasyphyllum, L. var.	M	M	*	–	–	–
„ acre, L.	M S	M S	*	–	–	–
Sempervivum atlanticum, Ball	M	–	–	–	–	*
Monanthes atlantica, Ball	–	S	–	–	–	*
Bryonia dioïca, Jacq.	M	–	*	–	–	–
Eryngium Bourgati, Gouan, var.	–	S	–	–	*	–
„ variifolium, Coss.	M	–	–	–	–	*
Bupleurum spinosum, L.	S	S	–	*	–	–
„ acutifolium, Coss.	M	M	–	–	*	–
„ oblongifolium, Ball	M	–	–	–	–	*
„ lateriflorum, Coss.	M	–	–	–	–	*

TABULAR VIEW OF THE MOUNTAIN FLORA—*continued.*

Name of Species	Aït Mesan	Amsmiz	Mid-European	Wide-spread Mediterranean	Confined to adjoining regions	Endemic
Deverra scoparia, Coss. et Dur.	–	M	–	–	*	–
Carum mauritanicum, B. et R.	M	M	–	–	*	–
Pimpinella Tragium, Vill.	M	–	–	*	–	–
Tinguarra sicula, L.	–	M	–	*	–	–
Scandix pecten Veneris, L.	M	–	*	–	–	–
Kundmannia sicula, L.	M	–	–	*	–	–
Meum atlanticum, Coss.	–	s	–	–	–	*
Heracleum Sphondylium, L.	s	–	*	–	–	–
Bifora testiculata, L.	M	–	–	*	–	–
Caucalis latifolia, L.	M	–	*	–	–	–
„ daucoides, L.	–	M	*	–	–	–
„ leptophylla, L.	M s	M	–	*	–	–
„ cœrulescens, Boiss.	M	–	–	–	*	–
Elœoselinum meoides, Desf.	M	–	–	*	–	–
Hedera Helix, L.	M	–	*	–	–	–
Sambucus nigra, L.	M	M	*	–	–	–
Viburnum Tinus, L.	–	M	–	*	–	–
Lonicera etrusca, Santi	M	–	–	*	–	–
Putoria calabrica, L.	M	–	–	*	–	–
Callipeltis cucullaria, L.	–	M	–	*	–	–
Rubia tinctorum, L.	s	–	–	*	–	–
„ peregrina, L. et var.	M	–	*	–	–	–
Galium Poiretianum, Ball	M	–	–	–	*	–
„ corrudæfolium, Vill.	–	s	–	*	–	–
„ sylvestre, Poll. var.	s	–	*	–	–	–
„ acuminatum, Ball	M s	–	–	–	–	*
„ noli-tangere, Ball	–	M	–	–	–	*
„ tunetanum, Lam.	M	–	–	–	*	–
„ parisiense, L. var.	M s	M s	*	–	–	–
„ tricorne, With.	s	M	*	–	–	–
„ spurium, L.	M s	s	*	–	–	–
„ murale, L.	–	M	–	*	–	–
Asperula aristata, L. var.	–	M	–	*	–	–
„ hirsuta, Desf.	M	–	–	–	*	–
Crucianella angustifolia, L.	M	–	–	*	–	–
Sherardia arvensis, L.	M	–	*	–	–	–
Centranthus angustifolius, D. C.	M s	–	–	*	–	–
„ calcitrapa, L.	s	s	–	*	–	–
Valerianella discoidea, W.	M	–	–	*	–	–
„ auricula, D. C.	M s	M	*	–	–	–
„ carinata, Loisel.	–	M	*	–	–	–
Scabiosa stellata, L.	M	–	–	*	–	–
Pterocephalus depressus, Coss.	M s	M	–	–	–	*
Bellis annua, L. var.	–	M	–	*	–	–
„ cœrulescens, Coss.	M s	M s	–	–	–	*
Evax Heldreichii, Parl.	s	–	–	*	–	–
Micropus bombycinus, Lag.	M	M	–	*	–	–
Filago germanica, L. var.	–	M	*	–	–	–
„ heranthâ, Rafin.	s	s	–	*	–	–

TABULAR VIEW OF THE MOUNTAIN FLORA—*continued*.

Name of Species	Aït Mesan	Amsmiz	Mid-Euro-pean	Wide-spread Mediter-ranean	Con-fined to adjoin-ing regions	En-demic
Filago gallica, L. . . .	M	M	*	–	–	–
Phagnalon saxatile, L. . .	M	M	–	*	–	–
„ atlanticum, Ball .	M	–	–	–	–	*
Gnaphalium luteo-album, L. .	M	M	*	–	–	–
„ helichrysoides, Ball	S	–	–	–	–	*
Inula montana, L. . . .	M	–	–	*	–	–
Pulicaria mauritanica, Coss. .	M	–	–	–	–	*
Odontospermum aquaticum, L.	M	–	–	*	–	–
Anacyclus depressus, Ball .	S	–	–	–	–	*
„ valentinus, L. . .	–	M	–	*	–	–
Achillea ligustica, All. et var. .	M	M	–	*	–	–
Anthemis tuberculata, Boiss. .	–	S	–	–	*	–
„ heterophylla (Coss.).	M	M	–	–	–	*
Chrysanthemum Gayanum (Coss.) et var. . . }	M S	M S	–	–	*	–
„ atlanticum, Ball	S	S	–	–	–	*
„ Catananche, Ball	S	S	–	–	–	*
Senecio lividus, L. var. . .	S	–	–	*	–	–
„ giganteus, Desf. .	M	M	–	–	*	–
Calendula maroccana, Ball .	M	M	–	–	–	*
Echinops spinosus, L. . .	M	–	–	*	–	–
Xeranthemum modestum, Ball	M	M	–	–	–	*
Atractylis cancellata, L. . .	M	M	–	*	–	–
„ macrophylla, Desf. .	M	–	–	–	*	–
Carduus macrocephalus, Desf. .	M	–	–	*	–	–
„ Ballii, H. fil. .	M S	M	–	–	–	*
Cnicus echinatus Desf. . .	M	–	–	*	–	–
„ ornatus, Ball .	–	M	–	–	–	*
„ chrysacanthus, Ball .	M	–	–	–	–	*
„ Casabonæ, L. . .	S	–	–	*	–	–
Stæhelina dubia, L. var. . .	–	M	–	*	–	–
Centaurea incana, Desf. var. .	M	–	–	*	–	–
„ Salmantica, L. var. .	M	M	–	*	–	–
Carthamus cœruleus, L. var. .	M S	–	–	*	–	–
Carduncellus lucens, Ball .	M S	–	–	–	–	*
Catananche cœrulea, L. et var.	M	M	–	*	–	–
„ cæspitosa, Desf. .	–	S	–	–	*	–
Tolpis umbellata, Bert. . .	M	–	–	*	–	–
Rhagadiolus stellatus, L. .	M	M	–	*	–	–
Crepis taraxacifolia, Thuil. var.	M	–	*	–	–	–
„ Hookeriana, Ball .	–	S	–	–	–	*
Phæcasium pulchrum, L. . .	S	S	*	–	–	–
Hieracium Pilosella, L. . .	S	S	*	–	–	–
Hypochæris glabra, L. var. .	M	–	*	–	–	–
„ leontodontoides, Ball . . }	S	–	–	–	–	*
Leontodon autumnalis, L. var..	S	–	*	–	–	–
„ Rothii, Ball . .	M	M	–	*	–	–
„ helminthioides, Coss.	–	M	–	–	–	*

TABULAR VIEW OF THE MOUNTAIN FLORA—*continued.*

Name of Species	Alt Mesan	Amsmiz	Mid-European	Widespread Mediterranean	Confined to adjoining regions	Endemic
Taraxacum officinale, Wigg. var.	S	S	–	*	–	–
Lactuca viminea, L.	M	–	–	–	*	–
„ tenerrima, Pourr.	M	M	*	–	–	–
„ saligna, L.	M	–	*	–	–	–
Sonchus oleraceus, L.	–	M	–	*	–	–
„ asper, Vill.	M	–	*	–	–	–
Microrhynchus nudicaulis, L.	M	–	*	–	–	–
„ spinosus (Forsk.)	M	–	*	–	–	–
Scorzonera undulata, Vahl.	M	–	–	*	–	–
„ pygmæa, S. et S.	S	–	–	*	–	–
Jasione atlantica, Ball	S	–	–	–	*	–
Campanula maroccana, Ball	M	–	–	*	–	–
„ rapunculus, L.	M	–	–	–	–	*
„ Loefflingii, Brot.	M	–	–	–	–	*
Specularia falcata (Ten.)	–	M	*	–	–	–
Trachelium angustifolium, Schousb.	M	–	–	–	*	–
Arbutus Unedo, L.	M	–	–	*	–	–
Armeria plantaginea (All.)?	–	S	–	–	–	*
Asterolinum linum-stellatum, L.	–	M	*	–	–	–
Anagallis linifolia, L. et var.	M S	–	*	–	–	–
Jasminium fruticans, L.	M	–	–	*	–	–
Fraxinus oxyphylla, M. B.	M	M	–	*	–	–
„ dimorpha, Coss. et Dur.	M	–	–	*	–	–
Phillyrea media, L.	M	M	–	–	*	–
Olea europæa, L.	M	M	–	–	*	–
Nerium Oleander, L.	M	M	–	*	–	–
Convolvulus Cantabrica, L.	M	M	–	*	–	–
„ undulatus, Cav.	M	–	–	*	–	–
„ sabatius, Viv. var.	M S	–	–	*	–	–
„ siculus, L.	–	M	–	*	–	–
„ althæoides, L.	M	–	–	*	–	–
Hyoscyamus albus, L.	M	–	–	*	–	–
Anchusa atlantica, Ball	M	–	–	*	–	–
Lithospermum arvense, L.	S	–	–	*	–	–
„ incrassatum, Guss. var.	S	M S	–	–	–	*
„ apulum, Vahl.	M	–	*	–	–	–
Myosotis sylvatica, Hoffm. var.	S	S	–	*	–	–
„ hispida, Schlecht. var.	S	–	–	*	–	–
„ stricta, Link	S	S	*	–	–	–
Cynoglossum Dioscoridis, Vill. et var.	M S	–	*	–	–	–
Rochelia stellulata	M	M	*	–	–	–
Verbascum calnycium, Ball	M	–	–	*	–	–
Celsia maroccana, Ball	M	–	–	–	–	*
Linaria ventricosa, Coss.	M	M	–	–	–	*

TABULAR VIEW OF THE MOUNTAIN FLORA—*continued*.

Name of Species	Aït Mesan	Amsmiz	Mid-Euro-pean	W de-spread Mediter-ranean	Con-fined to adjoin-ing regions	En-demic
Linaria heterophylla, Desf. .	M	M	–	*	–	–
„ galioides, Ball et var. .	S	–	–	–	–	*
„ arvensis, L. var. . .	S	M S	–	*	–	–
„ marginata, Desf. . .	S	–	–	–	*	–
„ lurida, Ball . . .	S	–	–	–	–	*
„ Munbyana, Boiss. et } Reut. . . .	M	–	–	–	*	–
„ Tournefortii (Poir.) .	S	S	–	–	*	–
„ rubrifolia, Rob. et Cast.	M	–	–	*	–	–
Anarrhinum pedatum, Desf. .	M	–	–	–	*	–
„ fruticosum, Desf. .	–	M	–	*	–	–
Scrophularia canina, L. var. ? .	–	M	*	–	–	–
Digitalis lutea, L. vár. . .	M	–	*	–	–	–
Veronica Beccabunga, L. . .	S	–	*	–	–	–
„ cuneifolia, Don. var. .	S	S	–	*	–	–
„ arvensis L. et var. .	S	S	*	–	–	–
„ triphyllos, L. . .	–	M	*	–	–	–
„ agrestis, L. . .	S	–	*	–	–	–
„ hederifolia, L. et var.	M S	M S	*	–	–	–
Phelipæa cœrulea, Vill. . .	M	–	*	–	–	–
Orobanche Hookeriana, Ball .	M	–	–	–	–	*
„ barbata, Poir. .	M	–	–	–	*	–
Lavandula dentata, L. et var. .	M	M	–	*	–	–
„ tenuisecta, Coss. .	M S	M S	–	–	–	*
Mentha rotundifolia, L. . .	S	–	*	–	–	–
Thymus saturejoides, Coss. et } var.	M	M	–	–	–	*
Thymus Serpyllum, L. var. .	S	–	*	–	–	–
„ lanceolatus, Desf. var.	M	M	–	*	–	–
„ maroccanus, Ball .	M	–	–	–	–	*
Micromeria microphylla, Benth	M	–	–	*	–	–
Calamintha graveolens, M. B. .	M	–	–	*	–	–
„ alpina, L. var. .	S	S	*	–	–	–
„ atlantica, Ball .	M S	M S	–	–	*	–
Hyssopus officinalis, L. . .	S	S	*	–	–	–
Salvia Maurorum, Ball . .	–	M	–	–	–	*
„ clandestina, L. var. .	M S	–	*	–	–	–
Nepeta multibracteata, Desf. .	M	–	–	–	*	–
„ atlantica, Ball .	–	M	–	–	–	*
Sideritis villosa, Coss. . .	M S	M S	–	–	–	*
„ scordioides, L. var. .	–	S	*	–	–	–
Lamium amplexicaule, L. .	M S	M S	*	–	–	–
„ album, L. var. . .	M	–	*	–	–	–
Teucrium granatense, B. et R. } var. . .	–	M	–	–	*	–
„ polium, L. vars. .	M	M	–	*	–	–
Ajuga Iva, L. . . .	M S	M S	–	*	–	–
Globularia Alypum, L. . .	–	M	–	*	–	–
Plantago albicans, L. var. .	–	M	–	*	–	–

TABULAR VIEW OF THE MOUNTAIN FLORA—*continued.*

Name of Species	Aït Mesan	Amsmiz	Mid-European	Wide-spread Mediterranean	Confined to adjoining regions	Endemic
Plantago coronopus, L. var.	M S	M	*	–	–	–
„ mauritanica, B. et R.	–	M	–	–	*	–
Paronychia argentea, Lam.	M S	M S	–	*	–	–
„ capitata, Lam. var.	M S	–	*	–	–	–
„ macrosepala, Boiss. et var.	M	–	–	*	–	–
Scleranthus annuus, L. var.	–	S	*	–	–	–
Polycnemum Fontanesii, Dur. et Moq.	M	–	–	–	*	–
Rumex scutatus, L. var.	M S	–	*	–	–	–
„ Papilio, Coss.	M	–	–	–	–	*
Polygonum aviculare, L.	–		*	–	–	–
Daphne Gnidium, L.	M	M	–	*	–	–
„ Laureola, L.	M	–	*	–	–	–
Thymelæa virgata, Endl. var.	M	M	–	–	*	–
Osyris alba, L.	–	M	–	*	–	–
Aristolochia Pistolochia, L.	M	–	–	*	–	–
Euphorbia rimarum, Coss.	M	M	–	–	–	*
Quercus Ilex, L. et var.	M	M S	–	*	–	–
Salix purpurea, L. var.	M	–	*	–	–	–
Populus alba, L. var.	M	M	*	–	–	–
„ nigra, L.	M	–	*	–	–	–
Ephedra altissima, Desf.	M	–	–	*	–	–
„ procera, F. et M.	M	–	–	*	–	–
Pinus halepensis, Mill.	–	M	–	*	–	–
Callitris quadrivalvis, Vent.	M	M	–	–	*	–
Juniperus oxycedrus, L.	M	M	–	*	–	–
„ phœnicea, L. et var.	M	–	–	*	–	–
„ thurifera, L.	S	–	–	–	*	–
Orchis pyramidalis, L.	M	–	*	–	–	–
„ latifolia, L.	S	–	*	–	–	–
Ophrys apifera, Huds.	M	–	*	–	–	–
Iris germanica, L.	M	–	–	*	–	–
Chamærops humilis, L.	M	M	–	*	–	–
Gagea foliosa, Schult.	–	S	–	*	–	–
Muscari comosum, L.	M	M	–	*	–	–
Scilla hispanica, Mill.	–	S	–	–	*	–
Ornithogalum comosum, L.	S	–	*	–	–	–
„ tenuifolium Guss.	–	S	–	*	–	–
„ orthophyllum, Ten.	–	M	–	*	–	–
„ pyrenaicum, L. var.	–	M	*	–	–	–
Allium paniculatum, L. var.	–	M	–	*	–	–
Asphodelus microcarpus, Viv.	M	–	–	*	–	–
Anthericum Liliago, L. var.	M	–	*	–	–	–
Colchicum Civonæ, Guss.	–	S	–	*	–	–
„ arenarium W. K. var.?	S	–	–	* ?	–	–

Tabular View of the Mountain Flora—*continued.*

Name of Species	Aït Mesan	Amsmiz	Mid-European	Widespread Mediterranean	Confined to adjoining regions	Endemic
Smilax mauritanica, Desf.	M	–	–	*	–	–
Asparagus acutifolius, L.	M	–	–	*	–	–
„ scoparius, Lowe?	M	–	–	–	–	*
Juncus bufonius, L.	M	M	*	–	–	–
Scirpus Savii, S. et M.	M	–	*	–	–	–
Carex Halleriana, Asso.	–	M	*	–	–	–
„ ambigua, Link.	M	–	–	–	*	–
„ fissirostris, Ball	S	M S	–	–	–	*
Phalaris nodosa, L.	M	–	–	*	–	–
Piptatherum cœrulescens, Desf.	M S	M	–	*	–	–
Stipa parviflora, Desf.	M	–	–	*	–	–
„ gigantea, Lag.?	M	–	–	*	–	–
„ nitens, Ball	S	–	–	–	–	*
Agrostis verticillata, Vill.	M	–	–	*	–	–
Phragmites communis, Trin.	M	–	*	–	–	–
Echinaria capitata, Desf.	M S	M S	–	*	–	–
Aira caryophyllea, L.	–	M S	*	–	–	–
Trisetum flavescens, L.	M	–	*	–	–	–
Avena bromoides, Gouan. var.	M	–	–	*	–	–
Arrhenatherum elatius. L.	–	M	*	–	–	–
Poa annua, L.	M S	M	*	–	–	–
„ bulbosa, L.	M S	M S	*	–	–	–
„ pratensis, L.	M S	M S	*	–	–	–
„ trivialis, L.	M S	–	*	–	–	–
Melica ciliata, L. var.	M	–	*	–	–	–
„ Cupani, Guss. var.	M S	M	–	*	–	–
Dactylis glomerata, L. vars.	M S	M S	*	–	–	–
Cynosurus elegans, Desf.	–	M S	–	*	–	–
Festuca rigida, L.	M	–	*	–	–	–
„ unilateralis, Schrad. var.	M	–	–	*	–	–
„ geniculata, L., et var.	M	M	–	*	–	–
„ duriuscula, var.	S	S	*	–	–	–
„ arundinacea, Schreb.	M	M	*	–	–	–
Brachypodium pinnatum, L. var.	M	–	*	–	–	–
„ distachyum, L.	M	M	–	*	–	–
Bromus tectorum, L.	M S	–	*	–	–	–
„ madritensis, L.	M	M	*	–	–	–
„ mollis, L. vars.	M S	M	*	–	–	–
„ macrostachys, Desf. var.	M	–	–	*	–	–
Lolium perenne, L.	M	–	*	–	–	–
Triticum hordeaceum, Coss. et Dur.	M	–	–	–	*	–
Secale montanum, Guss.	M	–	–	*	–	–
Elymus Caput-medusæ, L. var.	M	–	–	*	–	–
Hordeum murinum, L.	M	M	*	–	–	–
Ægilops ovata, L.	M	M	–	*	–	–
„ ventricosa, Tausch.	M	–	–	*	–	–

TABULAR VIEW OF THE MOUNTAIN FLORA—*continued.*

Name of Species	Aït Mesan	Amsmiz	Mid-Euro-pean	Wide-spread Mediter-ranean	Con-fined to adjoin-ing regions	En-demic
Andropogon hirtus, L. var. .	M	–	–	*	–	–
Cistopteris fragilis, Bernh. .	S	S	*	–	–	–
Cheilanthes fragrans, L. . .	M	–	–	*	–	–
Pteris aquilina, L. . . .	S	–	*	–	–	–
Asplenium trichomanes, L. .	M S	–	*	–	–	–
„ viride, L. . .	S	–	*	–	–	–
„ Adiantum-nigrum, L.	S	–	*	–	–	–
Notochlæna vellea, Desv. .	M	–	–	*	–	–
Ceterach officinarum, Willd. .	M S	M S	*	–	–	–
Equisetum ramosissimum, Desf.	M	–	*	–	–	–
Selaginella, rupestris, (Spreng.)	M	–	–	*	–	–
Total number of species . 465	388	225	161	168	61	75

Before discussing the inferences to be derived from this list, it may be well to notice some sources of error that, to a slight extent, affect the results. Although the season of our visit— the second half of May—was probably the best as regards the middle zone, it was too early to find the vegetation fully developed in the superior zone, especially on the highest ridges. It is probable that on this account the proportion of *Umbelliferæ* and *Gramineæ* found in the higher region is smaller than it would have been at a later season. At first sight it would appear that the shorter time that we were able to devote to an examination of the upper region, and the snow-storm which we encountered in the ascent to the Tagherot Pass, make the proportion of species found there, as shown by our lists, unduly small. There can be no doubt that we must have lost several species owing to these causes, but not enough to vitiate the results to a serious extent. In confirmation of this opinion it may be mentioned that although a native employed by M. Cosson has since made a large collection in the same part of the Great Atlas, and two German naturalists—MM. Rein and Fritsch—have visited the head of the Aït Mesan valley, very few species have been added to the Flora of the higher mountain region.

The first conclusion that strikes a botanist on examining the foregoing list is that the general type of the vegetation clearly marks this as belonging to the great Mediterranean Flora, which extends, with local peculiarities, from Persia and Belutschistan to the Atlantic Islands. Out of 248 genera represented in the Flora of these valleys there is not one which is not common to other portions of the Mediterranean region, and one only (*Monanthes*) is confined to the Great Atlas and the Canary and Cape de Verde Islands, all the others being types more or less widely spread. Further than this, the proportion borne by each of the prevailing natural orders to the whole vegetable population is pretty nearly the same that we are accustomed to find in the mountain regions of the Mediterranean region.

The materials for a comparison are unfortunately yet incomplete as regards many of the mountain districts which are best fitted for the purpose. The Flora of the Lesser Atlas of Algeria, as well as that of the rest of the French possessions in Africa, will be fully known only on the appearance of the important work promised by M. Cosson. The Flora of Spain by MM. Willkomm and Lange is yet unfinished, and there is the further difficulty that those authors have admitted a large number of plants to the rank of species which many botanists reckon only as varieties. M. Boissier's great work, the 'Flora Orientalis,' is also unfinished, and no adequate materials exist for compiling lists of the plants of the Greek mountains, of those of Asia Minor, or of the Lebanon chain, all of which would afford interesting materials for comparison. In the following table I have taken for comparison the Flora of the Sierra Nevada, with the neighbouring mountains of the ancient kingdom of Granada above the level of about 800 metres, compiled from Boissier's 'Voyage botanique dans le Midi de l'Espagne;' that of the Bulgardagh (the principal group of the Cilician Taurus), from a list published by M. Pierre de Tchihatcheff in the 'Bulletin of the French Botanical Society;' that of Dalmatia, from Visiani's excellent 'Flora Dalmatica;' and that of the southern slopes of the chain of the Alps from Nice to the Karst, formed by myself from all available sources.

In the same table I have introduced, for the purpose of further comparison, separate columns for the middle and superior regions of the Great Atlas valleys, and in connection with the last I have added in a separate column the results for the higher zone of the Sierra Nevada. Under each heading I have stated

TABLE. I.—Showing the number of species of each of the principal groups and natural orders of plants in two valleys of the Great Atlas compared with other mountain districts in the Mediterranean region.

	Gt. Atlas valleys. 455 sp.		Middle Zone, Gt. Atlas. 341 sp.		Superior Zone, Gt. Atlas. 176 sp.		Sierra Nevada, &c., above 800 m. 890 sp.		Superior* Zone, Sierra Nevada. 486 sp.		Bulgardagh. 882 sp.		Dalmatia. 2,002 sp.		Southern side of the Alps. 2,545 sp.	
Dicotyledones	391	86·0	286	83·9	154	87·5	762	85·6	419	86·2	808	91·6	1594	79·6	2035	80·0
Monocotyledones	64	14·0	55	16·1	22	12·5	128	14·4	67	13·8	74	8·4	408	20·4	510	20·0
Compositæ	63	13·8	46	13·5	22	12·5	119	13·4	63	13·0	97	11·0	235	11·7	343	13·5
Leguminosæ	48	10·5	38	11·1	14	8·0	67	7·5	32	6·6	93	10·5	222	11·1	172	6·8
Gramineæ	39	8·6	37	10·8	14	8·0	59	6·6	37	7·6	38	4·3	173	8·6	176	6·9
Caryophylleæ	26	5·7	14	4·1	15	8·5	40	4·5	29	6·0	81	9·2	74	3·7	121	4·8
Cruciferæ	25	5·5	7	2·1	21	11·9	49	5·5	37	7·6	84	9·5	98	4·9	139	5·5
Labiatæ	23	5·0	18	5·3	11	6·3	54	6·1	28	5·8	67	7·0	100	5·0	89	3·5
Scrophularineæ	21	4·6	13	3·8	10	5·7	37	4·2	28	5·8	39	4·4	66	3·3	109	4·3
Umbelliferæ	20	4·4	16	4·7	5	2·8	50	5·6	23	4·7	33	3·7	113	5·6	113	4·4
Rubiaceæ	18	4·0	15	4·4	7	4·0	20	2·2	12	2·5	19	2·2	26	1·3	34	1·3
Papaveraceæ	10	2·2	8	2·3	3	1·7	10	1·1	6	1·2	12	1·4	14	0·7	14	0·6
Geraniaceæ	10	2·2	8	2·3	4	2·3	10	1·1	6	1·2	6	0·7	17	0·8	23	0·9
Liliaceæ	10	2·2	6	1·8	4	2·3	15	1·7	9	1·9	23	2·6	61	3·0	52	2·0
Boragineæ	9	2·0	5	1·5	6	3·4	18	2·0	11	2·3	23	2·6	40	2·0	39	1·5
Ranunculaceæ	8	1·8	5	1·5	4	2·3	28	3·1	15	3·1	11	1·2	53	2·6	87	3·4
Cistineæ	7	1·5	6	1·8	1	0·6	23	2·6	10	2·1	1	0·1	11	0·5	10	0·4
Rosaceæ	7	1·5	4	1·2	3	1·7	26	2·6	20	4·1	21	2·4	57	2·8	93	3·7
Campanulaceæ	6	1·3	5	1·5	1	0·6	9	1·0	6	1·2	21	2·4	26	1·3	46	1·8
Convolvulaceæ	5	1·1	5	1·5	1	0·6	3	0·3	2	0·1	4	0·5	10	0·5	9	0·1
Coniferæ	5	1·1	4	1·2	1	0·6	10	1·1	8	1·6	15	1·7	15	0·7	11	0·4
Saxifrageæ (inclusive of Grossulariæ)	4	0·9	4	1·2	4	2·3	12	1·3	9	1·9	2	0·2	9	0·4	52	2·0
Cyperaceæ	4	0·9	4	1·2	1	0·6	17	1·9	12	2·5	7	0·8	43	2·2	119	4·7
Gentianeæ	0	0	0	0	0	0	7	0·8	5	1·0	4	0·5	14	·07	31	1·2
Primulaceæ	2	0·4	2	0·6	1	0·6	7	0·8	6	1·2	8	0·9	12	0·6	60	2·4
Juncæe	1	0·2	1	0·3	0	0	11	1·2	9	1·9	0	0	8	0·4	31	1·2

* In the Superior Zone of the Sierra Nevada I include all species found above the level of about 1,600 m., considering this to correspond with the level of 2,000 m. in the Great Atlas.

the whole number of phanerogamous species included in the
Flora of the region, and opposite the name of each natural order
I have entered the number of species found in each region, and
the percentage proportion which this number bears to the entire flora.   Besides the orders which bear the largest proportion
in the Great Atlas Flora I have enumerated those that usually
characterise the vegetation of high mountains in this part of
the world, though several of these are little, or not at all, represented in the Flora of the Great Atlas.

Confining the comparison in the first instance to the figures
given for the Atlas Flora as a whole in the first column, and those
given in the fourth, sixth, seventh, and eighth columns respectively, for the Sierra Nevada, the Bulgardagh, Dalmatia, and
the southern side of the Alps, we remark in the first place that
Monocotyledons bear about the same proportion to *Dicotyledons*
in the Great Atlas that they do in the Sierra Nevada, the percentage here being much larger than it is in the Bulgardagh, and
considerably less than in Dalmatia or the Southern Alps.   In
this part of the world this percentage in the Flora of a given
region mainly depends upon the number of *Gramineæ* and
Cyperaceæ.   The abundance of the latter group in the Alps
doubtless arises from the fact that at a former period physical
conditions favoured the migration of a large number of
northern species that have been unable to extend to the more
southern mountain regions of the Mediterranean area.

In all the regions under consideration we find, with a single
exception, that the same eight natural orders take precedence
of all others as regards the number of species that they exhibit,
the aggregate in every case exceeding one-half of the whole
phanerogamous Flora.   These natural orders are *Compositæ*,
Leguminosæ, *Gramineæ*, *Caryophylleæ*, *Cruciferæ*, *Labiatæ*,
Scrophularineæ, and Umbelliferæ.   The exception arises from
the prevalence, already noticed, of *Cyperaceæ* in the Flora of
the Southern Alps.   In comparing the figures in the Great
Atlas column with those for the other areas above enumerated,
it is well to recollect that our materials are taken from a
district much more limited in extent than the others, and are
necessarily imperfect, because obtained from a single short visit
to each valley at a season when many species are yet undeveloped.   It is probable, for instance, that the proportion of

Umbelliferæ would be increased if the whole Flora were better known. Subject to this remark, it will be seen, as might be expected, that the constituents of the Great Atlas Flora show more analogy with those of the Sierra Nevada and Bulgardagh Floras than with those of Dalmatia and the Southern Alps ; but the proportion of *Compositæ* is larger than in any of them (nearly 14 per cent). In comparing the vegetation of a small district with that of a large one it must be recollected that a small natural group containing a few widely spread species, such as *Geraniaceæ*, is likely to show a larger percentage proportion to the whole Flora in the small district than in the larger one. It may happen that the same species are spread through both regions ; but in one case the number is to be compared with a small total, in the other with a much larger one. This remark has a bearing on the fact that in the Great Atlas Flora the natural orders that bear an unusually large proportion to the total number of the Flora are *Leguminosæ, Caryophylleæ, Rubiaceæ, Papaveraceæ, Geraniaceæ,* and *Convolvulaceæ*. On the other hand, there is a remarkable deficiency in the natural orders that especially characterise the Flora of the Alps, and in a less degree, the high mountains of Southern Europe. These are *Ranunculaceæ, Rosaceæ, Saxifrageæ, Primulaceæ, Junceæ,* and *Cyperaceæ* ; not to speak of *Gentianeæ*, which are here altogether absent.

If, instead of regarding the Atlas Flora as a whole, we examine separately the figures given in the several columns for the middle and superior zones respectively, we find very different proportions for the chief natural orders, except for *Compositæ* and *Leguminosæ* which are in both very numerous. In the middle region of the Atlas these two orders represent very nearly one-fourth of the phænogamous Flora. After these *Gramineæ, Rubiaceæ, Papaveraceæ, Geraniaceæ, Cistineæ,* and *Convolvulaceæ* are, in the middle region, unusually frequent, while *Cruciferæ, Rosaceæ, Boragineæ,* and *Liliaceæ* are remarkably deficient. In the superior zone, on the other hand, the proportion of *Compositæ* and *Leguminosæ* is less excessive, making jointly a little over one-fifth of the whole Flora of the upper region. The most marked characteristic here is the very large proportion of *Cruciferæ*, being less by one species only than the number of *Compositæ*. Taking into account the number of individuals as well as that of species, this must be regarded as the dominant element in the Flora of the higher region of the Great Atlas, affording as it does 12 per cent of the whole Flora. The only region in which this characteristic is approached is the Bulgardagh in Cilicia, where *Cruciferæ* supply near one-tenth of the whole list. *Caryophylleæ* also form an unusually large element in the Flora of the upper zone of the Atlas; but, unlike *Cruciferæ*, this order exhibits no endemic species, and four-fifths of the whole number are common plants of Central and Northern Europe. *Rubiaceæ* and *Boragineæ* have more representative species than is usual in mountain Floras; while there are but three species of *Rosaceæ* in our list; and *Campanulaceæ*, *Primulaceæ*, *Coniferæ*, and *Cyperaceæ* are each represented by a single species, and *Gentianeæ* and *Junceæ* are altogether absent from the higher zone.

Although statistical results, such as those given above, are not without interest, as throwing light upon the general characteristics of the Flora of a given region, any rational grounds for speculation as to the real affinities and past history of the vegetation must be derived from a closer examination of the individual species of which it is constituted. It is at least conceivable that two Floras should exhibit similar proportions of species belonging to the several natural groups, with no identical species, and with little or no indication of community of origin. The particulars given in our general list will have already led the reader to infer that the results of an examination into the distribution of the individual species that go to make up the Great Atlas exhibit some very peculiar features. Taking the totals at the foot of our list, and excluding cryptogams, it is seen that more than one-third of the species are plants of Middle and Northern Europe, while about one-sixth is made up of endemic species peculiar to Marocco, and, with few exceptions, not known out of the Great Atlas, more than half of the whole list belonging to one or other of these categories. The results, as shown in the following table, are still more remarkable when we separately examine the zones into which mountain vegetation is naturally divided. As in the former table the figures first entered in each column represent the number of species belonging to each category, those next given showing the percentage proportion borne by that number to the total proportion of each region.

TABLE II.—*Showing the distribution of the species of flowering plants included in the Flora of the Great Atlas, and of the Sierra Nevada of Granada,*[1] *and the Bulgardagh in Cilicia.*[2]

	Mid-Euro-pean		Wide-spread Mediter-ranean		Confined to adjoining regions		Endemic	
Great Atlas, including all species found above 1,200 m. 455 sp.	154	33·8	165	36·2	61	13·4	75	16·6
Middle Zone of Atlas, from 1,200 m. to 2,000 m. 341 sp.	106	31·1	141	41·3	46	13·5	48	14·1
Superior Zone of Atlas, from 2,000 m. to 3,500 m. 176 sp.	78	44·3	43	24·4	20	11·4	35	19·9
Superior Zone of Sierra Nevada, above 1,600 m. 486 sp.	209	43·0	74	15·2	104	21·4	99	20·4
Bulgardagh in Cilicia. 882 sp.	159	18·0	359	40·7	157	17·8	207	23·5

From this table we see that while over one-third of the whole Atlas Flora consists of plants of Central and Northern Europe, the proportion reaches nearly to one-half in the higher region (above 2,000 metres); and also that the proportion of endemic species, which in the aggregate is one-sixth of the whole, rises to one-fifth in the upper zone. On the other hand, the proportion of purely Mediterranean species, which amounts to 55 per cent. in the Flora of the middle zone, falls below 36

[1] The name Sierra Nevada is here used in a wide sense, and is intended to include the Serrania de Ronda, and the other mountains of Andalusia. Under this head, the plants classed as 'confined to adjoining regions' are either common to the Sierra Nevada and the mountains of Northern Spain, including the Pyrenees, both Spanish and French, or else are common to the Sierra Nevada and the mountains of Northern Africa.

[2] The Bulgardagh has been introduced into this table rather for the sake of contrast than as showing similarity to the conditions in the Great Atlas. The species classed as 'confined to adjoining regions' are all found in the other mountain districts of Asia Minor, and it has been necessary to include under the heading 'Wide-spread Mediter-ranean' a large number of Oriental species, whose western limit is in Greece or Crete. As compared with the Great Atlas, the number of species common to the western and south-western parts of Europe is here quite insignificant.

per cent. in the upper region. Of these Mediterranean species the large majority (more than two-thirds) are widely distributed plants, several of them extending to the mountains of Asia Minor, and twenty species only are exclusively confined to the Great Atlas and to the mountains of Southern Spain, the Lesser Atlas, or the Pyrenees. There is nothing in the distribution of these latter plants to indicate any special connection between the Atlas and any one of the mountain regions above mentioned. Six Atlas species are common to Southern Spain and the Algerian Atlas, six more are known only on the mountains of Southern Spain, five have been hitherto supposed to be peculiar to the Lesser Atlas, and three are elsewhere confined to the Pyrenees.

Some further light may be thrown on the origin of the Great Atlas Flora by considering the affinities of the plants which are reckoned in our list as endemic in Marocco, nearly all being confined, so far as we know, to the chain of the Great Atlas. Although all of these, along with some that we have classed as mere varieties, would be counted as distinct species by many botanists, a considerable number, amounting to more than a quarter of the whole, are, according to the views expressed elsewhere by the writer,[1] to be ranked as sub-species. But here again we fail to discover indications of special relations between the Great Atlas Flora and that of neighbouring mountain regions. Ranking as sub-species twenty-one out of the seventy-five endemic forms enumerated in our list, we find that ten of these are allied to widely spread Mediterranean species, three are related to plants of Central Europe, three to species common to Algeria and Southern Spain, three more to species confined to the Spanish peninsula, and two to endemic Algerian forms.

If we scrutinise in the same manner the endemic forms of the higher region of the Great Atlas, we find that out of the thirty-five enumerated eight, or less than one-fourth, are to be ranked as sub-species. Of these, three are nearly allied to widespread Mediterranean species, one to a plant common to Spain and Algeria, two to endemic Spanish species, one to an Algerian

---

[1] See 'Spicilegium Floræ Maroccanæ,' in *Proceedings of the Linnæan Society,* 'Botany,' vol. xvi. parts 93 to 97 inclusive.

endemic form, and one is related to a species indigenous in the Alps and other high mountains of Central Europe.

While recognising the fact that the relations between the vegetable population of the Great Atlas and that of the south of Spain are less close than might have been expected on theoretical grounds, we must yet admit that, on the whole, the Great Atlas is more nearly connected in a botanical sense with this than with any other mountain region that is known to us; and it becomes a matter of some interest to compare closely the list of species obtained by us in the Atlas, with the comparatively well known Flora of Southern Spain. The results of this comparison are given for the Great Atlas generally, and for the superior zone separately, in the following table, in which the Atlas species are distinguished under five heads : 1, those found in the higher region of the Sierra Nevada; 2, in the mountain region of Andalusia; 3, in the lower warm region below the level of about 2,000 feet; 4, absent from Southern Spain, but found in the central or northern provinces; and 5, those not included in the Spanish Flora.

TABLE III.

	Superior region, Sierra Nevada	Mountain region of Andalusia	Lower region of Southern Spain	Central, or Northern Spain, exclusively	Absent from Spain
Great Atlas Valleys. 455 sp.	103	82	100	44	126
Superior region of the Great Atlas. 176 sp.	61	19	20	21	55

The figures given in this table are of much interest, proving, as they do, the wide differences that exist between the Floras of two mountain regions not widely separated from each other, and exposed to climatal conditions not altogether dissimilar. We see that three-sevenths of the plants found in the higher region of the Great Atlas are absent from the South of Spain, and that the same remark applies to considerably more than one-third of all the plants found in the portion of the Great Atlas visited by us, although a notable proportion (in both cases) is to be found in Central and Northern Spain. Especially noteworthy is the fact that many of the species thus absent in Southern Spain are plants of Central Europe, most of which extend to the northern part of the Spanish Peninsula, although some of them are altogether wanting in the Floras of Spain and Portugal.

A simple inspection of our list suffices to show that it discloses no trace of affinity between the Great Atlas Flora and that of the Canary Islands, or, to use a term of wider geographical import, that of Macaronesia. The few species belonging exclusively to the latter region and to Marocco are nearly all confined to the coast region.[1] Almost all the species common to the Atlas and to Macaronesia are widely spread Mediterranean plants that ascend from the low country into the valleys. The solitary mountain plant belonging to this category is *Arabis albida*, the southern form of *A. alpina*, common in the East, and in the Apennines of Central and Southern Italy, but which, strange to say, has not been found in Spain. In Teneriffe, as in the Atlas, it ascends to about the level of 2,700 metres above the sea. The only fact suggesting a remote affinity between the Great Atlas and Macaronesian Floras is the presence in the former of a species of *Monanthes*, a generic group hitherto found only in the Canary and Cape de Verde Islands. But the absence of any closer connection clearly shows that the separation between the Macaronesian group and the main land of Africa must date from a period, even geologically speaking, remote.

When we come to sum up the results of the foregoing discussion, bearing always in mind the fact that we possess a mere fragment of the Flora of the Great Atlas, and that future exploration may largely modify our conclusions, we find as its most striking characteristic the presence of a large proportion of plants of Central and Northern Europe, along with a considerable number of peculiar species not hitherto known elsewhere; and we observe that these two constituents, which

[1] The only possible exception to this statement among the plants enumerated in our list is that entered as *Asparagus scoparius*, Lowe (?) From the differences between the foliage and that of other known species it was at first entered as a new species peculiar to the Atlas. Subsequent comparison with a Madeira specimen from the late Mr. Lowe suggested their possible identity. Should this be hereafter verified, the number of endemic species in the tables given above must be reduced from 75 to 74.

together form about one-half of the Flora of the region here discussed, amount to very nearly two-thirds of the species found in the higher zone. We remark that of these northern plants none are of Alpine or Arctic type, that nearly all belong to what has been called the Germanic Flora, and all are plants of the plain, not in Europe characteristic of mountain vegetation.[1]

Of the species belonging to the Mediterranean region, which constitute more than one-half of the vegetation of the middle zone, and about one-third of that of the higher zone of the Atlas, the large majority are widely diffused species. The remaining number, for the most part mountain plants, may be divided into three nearly equal sections, some being common both to Southern Spain and Algeria, others to the Atlas and Southern Spain exclusively, and others to the Great Atlas and the Lesser Atlas of Algeria. Nothing indicates any special connection with the Floras of either of those regions.

The absence of any distinct generic types from the Great Atlas Flora has already been remarked. It is not less important to note the absence of any of the southern types, characteristic of the sub-tropical zone, some representatives of which are found in the same or even in higher latitudes, in Arabia, Syria, Persia, and Northern India, and which also appear in the Canary Islands. We finally are led to regard the mountain Flora of Marocco as a southern extension of the European temperate Flora, with little or no admixture of extraneous elements, but so long isolated from the neighbouring regions, that a considerable number of new specific types have here been developed. The physical causes which have operated to bring about these conditions are doubtless numerous and complicated, but the most important of them are easily indicated. The influence of the Atlantic climate, and the prevailing direction of the aërial and oceanic currents, have fitted this region for the habitation of such northern species as do not require a long period of winter repose. In the present condition of the African continent, the Great Desert, extending for a distance of 700 or 800 miles between the Atlas and the river region of

---

[1] The only apparent exception is *Sagina Linnæi*. This is habitually a mountain plant; but in Germany it is often seen in the moorland region, at a level of about 2,500 feet above the sea.

tropical Africa, effectually prevents the northward extension of most forms of animal and vegetable life ; while in a period geologically recent, it is most probable tha the same area was occupied by a wide gulf, which served the same purpose of barring the migration of southern forms.

It may be premature to attempt to trace in further detail the origin of the Great Atlas Flora ; but the facts already ascertained certainly authorise some negative inferences. The absence of plants of Arctic type proves that if some mountains of Southern Europe received contributions to their vegetation during the glacial period by means of floating ice-rafts, that mode of diffusion did not extend to the Great Atlas. If we suppose that during the glacial period the temperature of the region north of the Atlas had fallen so low as to permit the migration of northern species across the intervening low country, we find it difficult to understand why so many species which, according to this theory, must have retreated to the Atlas on the subsequent rise of temperature, should have failed also to find a refuge in the mountains of Southern Spain.

It is a further difficulty that if the constituents of the Great Atlas Flora had, to a large extent, travelled by the route here indicated, other species, now inhabiting the mountains of Southern Spain, could scarcely fail to take the same road, and a much nearer connection than is now apparent would have been established between the Floras of these two mountain regions.

It is, at least, possible that the wide diffusion of many of the species constituting the so-called Germanic Flora may date from a period much more remote than is ordinarily supposed ; and it is a circumstance not without significance that so many species of this type prove themselves capable of tolerating wide variations in conditions of soil and climate.

APPENDIX H.

Notes on the Geology of the Plain of Marocco and the Great Atlas.

By GEORGE MAW, F.G.S., F.L.S., &c.

OF the Geology of Barbary little information has hitherto been put on record. The only publications with which I am acquainted are some notes on the geological features of the district between Tangier and Marocco in Lieut. Washington's 'Geographical Notice of the Empire of Marocco,' published in the first volume of the 'Journal of the Royal Geographical Society;' a few cursory remarks on the Marocco Plain by Dr. Hodgkin, in his account of Sir Moses Montefiore's 'Mission to Morocco in 1864;' a short paper, by Mr. G. B. Stacey, on the subsidence of the coast near Benghazi, published in the twenty-third volume of the 'Quarterly Journal of the Geological Society;' a report by M. Mourlon on some rocks and fossils in the Museum of Brussels, collected in the north-west of Marocco by M. Desquin, a Belgian engineer, published in Vol. XXX. of the 'Bulletin de l'Académie Royale de Belgique,' for 1870, to which I shall have further occasion to refer;. a geological memoir, by M. Coquand ('Bull. de la Soc. Géolog. de France,' vol. iv. p. 1188), on the environs of Tangier and northern part of Marocco; and finally, a paper I read before the Geological Society of London in 1872.

Barbary, with the exception of the immediate neighbourhood of a few of the ports, has been almost inaccessible to Europeans; and the extreme jealousy of the Moorish Government with reference to the mineral riches of the country has hitherto prevented any geological investigation. In the year 1869 I visited the northern portion of Marocco, including the Tangier and Tetuan promontory, and during the spring of 1871 accompanied Dr. Hooker and Mr. Ball to Mogador, the city of Marocco and the Great Atlas, permission for our visit having been obtained from the late Sultan through representations made to the Moorish Government by Lord Granville through Sir John D. Hay, our Minister Plenipotentiary at Tangier.

The object of the second journey was mainly botanical; and as an engagement was given by Dr. Hooker that we should not collect minerals, the opportunities for geological investigation were very limited.

The observations I was able to make on the structure of the great chain, which had not been previously ascended by a European, and of the plain of Marocco, are embodied in the accompanying section. Stopping for about a fortnight at Tangier, we made several excursions in the neighbourhood. The western part of the northern promontory of Marocco, facing the Straits of Gibraltar, consists of highly-contorted beds of hard courses interstratified with brindled yellowish sand-stones and variegated puce and grey marls, having a general dip to the south-east, but so twisted about that the dip and strike are often reversed within a few feet. The country has a general undulating contour, here and there rising up into ridges of from 2,000 to 3,000 feet, in which the hard bands weathered out from the softer strata are strikingly prominent from a great distance.

We observed no palæontological evidence of their age; but, judging from their resemblance to the cliff-sections near Saffi, where fossils occur, they are presumably Neocomian or Cretaceous.

Fucoids were collected by M. Coquand in the vicinity of Tangier, in beds considered by him to be representatives of the Upper Chalk; but M. Mourlon, referring to the works of Pareto and Studer on the nummulitic rocks of the Northern Apennines and Switzerland, inclines to place the Tangier fucoid beds above the nummulitic horizon, and as part of the Upper Eocene. But near the villages of Souani and Meharain, a little to the south of Tangier, undoubted Cretaceous fossils were met with by M. Desquin, including

Inoceramus,
Ostrea Nicaisei,
O. syphax,
Globiconcha ponderosa (?),
Trigonia (casts), and
Echinodermata (undeterminable);

and M. Mourlon concludes that the Tangier promontory consists of Eocene beds resting on Cretaceous.

The eastern half of the northern promontory, including Tetuan and Apes' Hill facing Gibraltar, consists of beds of a different character, for the most part of a hard metamorphic limestone, in which dip and strike are very obscure: these may be a southern extension of the Gibraltar limestone; but I had no opportunity of tracing the connection to Tetuan.

The late James Smith, of Jordan Hill (in 'Journal of Geological Society,' vol. ii. p. 41), mentions the occurrence of casts of Terebratula fimbriata and T. concinna, belonging to the Lower Oolite, in the Gibraltar limestone. M. Coquand also assigns to the Jurassic period the beds in the neighbourhood of Tetuan, and divides them into four stages, characterised respectively by marls, dolomites, a calcareous sandstone with the odour of petroleum, and a lithographic limestone containing siliceous concretions. I am of opinion that the Tetuan series, ranging with the Gibraltar limestone, and probably extending far to the south, is separated from the more recent Cretaceous series to the west and north-west by a great north and south fault, which divides nearly equally the Tangier promontory.   M. Mourlon, referring to some specimens of shelly limestone in the Brussels Museum, collected near the river Mhellah in the district of Ouled Eissa, between Fez and Tetuan, resembling the Muschelkalk in aspect, and associated with beds resembling those at Tetuan, considers that they may also be of Jurassic age.

The Tetuan limestone has given rise to enormous beds of brecciated tufa, on terraces of which the city is built.   The flow seems to have taken place from the hills to the north-west of the city, and has produced beds of a collective thickness of 60 or 70 feet.   This is evidently true tufa, due to aqueous deposition, and is of a different character from the great calcareous sheet, to which I shall have occasion further to refer, which shrouds over the entire plain of Marocco.

Respecting the Mediterranean coast-line of Barbary, I will not add much to a paper read before the British Association at Liverpool, in which I remarked on the singular absence of coast-cliffs of any height.   The undulating contour of the land-surface extends down to the water's edge, a continuation of the form of the bottom of the straits without the intervention of cliff-escarpments, from which I surmised that the present sea-level and coast-line of the straits had not been of long duration.

Of frequent changes of level on the Barbary coast there is abundant evidence.   The more recent seem to be, first, an elevation of from 60 to 70 feet along the entire coast, implied by the existence of concrete sand-cliffs with recent shells exactly similar to the raised beaches of Devon and Cornwall.   These occur in Tangier Bay to a height of 40 feet, resting on the up-turned edges of nearly vertical mesozoic beds ; to the south of Cape Spartel, as a long cliff nearly 50 feet high ; as low shoals near Casa Blanca ; as a compact cliff about 50 feet high at Saffi, and as a coast-cliff and islands at Mogador, where the concrete sand-beds attain a height of 60 or 70 feet above the sea-level. It seems probable that this elevation of coast-line was coincident with a similar rise, implied by the existence of concrete sand-cliffs, all along the Spanish and Portuguese coasts, viz. on the eastern face of Gibraltar, where stratified raised beaches are seen cropping up at a considerable height from under the great mass of drift-sand in Catalan Bay; at Cadiz, as low cliffs 40 to 50 feet high, forming a hard coarse freestone of which the city is built; and also at the Rock of Lisbon, where, at a height of from 150 to 180 feet, isolated fragments of stratified concrete sandstone are seen clinging to the sea-escarpment of the older rocks.

The great range of latitude included in this simultaneous coast-rise suggests the probability that the elevation of similar coast-beds in Devon and Cornwall may pertain to the same movement.

Judging from the evidence afforded by the coast near Mogador, a subsequent submergence appears to be taking place. The island is probably diminishing in bulk; and, from observations made by M. Beaumier, the French Consul, it appears to have been reduced about one-fourth in area in twenty years; but whether from denudation or subsidence is not clear. The sea is, however, sensibly encroaching, as an old Portuguese fort and some Moorish buildings are now environed with sand and salt-marsh close to the sea, in a position where they would not have been built. This submergence of the coast at Mogador may perhaps be contemporaneous with the subsidence at Benghazi, Barbary, described by Mr. G. B. Stacey in the twenty-third volume of the 'Quarterly Journal of the Geological Society.' The general absence of cliffs characterises nearly the whole of the Barbary coast. A few low cliffs occur at scattered intervals west of Tangier; but from Cape Spartel to Cape Cantin a low monotonous coast shelves under the waters of the Atlantic, and not a cliff is to be seen, save an occasional raised beach. After rounding Cape Cantin the coast trends nearly north and south; and here the first good coast-section presents itself as a vertical cliff nearly 200 feet high (fig. 1), consisting of nearly level stratified alternations of grey and reddish marl, and fine-grained sandstone with beds of argillaceous carbonate of iron resembling the cement-stone of the Kimmeridge clay.

At a distance the cliff has a massive rocky aspect due to the vertical infiltration of tufaceous seams, which support the softer beds and stand out in prominent masses. The cliffs continue southwards to Saffi, where I obtained a small series of fossils from the section represented in fig. 1, amongst which Mr. Etheridge has determined *Exogyra conica, Ostrea Leymerii,* and *O. Boussingaulti.* He considers the beds to be of Neocomian age. The hard band c is almost entirely made up of *Exogyra conica.*

I am indebted to the late Mr. Carstensen, H.B.M. Viceconsul at Mogador, for a specimen of *Ostrea Leymerii,* brought

FIG. 1.

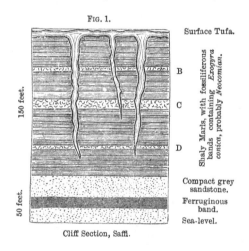

Cliff Section, Saffi.

to him by a Moor from Agadir, and obtained, at a height of 1,500 feet, on the flanks of the maritime termination of the Great Atlas range, 160 miles south of the Saffi section.

Two or three miles south of Saffi another section occurs, known as the ' Jew's Cliff;' and from this Dr. Hooker, who landed on his homeward voyage, obtained a few fossils, viz. several undeterminable species of *Pecten* ; an *Ostrea* allied to *O. Virleti,* and a scutelliform *Echinus* of an unknown type, which Mr. Etheridge proposes to place under a new genus, and names *Rotuloidea fimbriata.* All these Mr. Etheridge supposes to be of Miocene age; and the ' Jew's Cliff' section may probably give the key to the age of the beds of the Marocco plain in which we found no fossils. In connection with the occurrence of these Tertiary beds at Saffi, I must refer to MM. Desquin and Mourlon's observations in the neighbourhood of Mazagan to the north-west, near which, at a place called Sidi Moussa, calcareous tufas associated with flints occur, containing

FIG. 2.

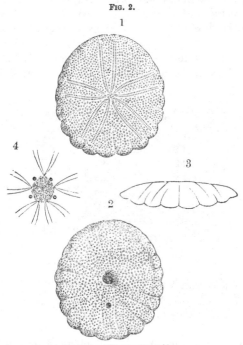

Rotuloidea fimbriata, Etheridge.

1. Dorsal aspect, showing the twelve fimbriations, subpetaloid ambulacra and central madreporic tubercle.
2. Ventral aspect, showing mouth, position of vent, and ramifying furrows.
3. Posterior border and height of test.
4. Apical disk, with the madreporic tubercle, the four genital pores, and place of the five oculars.

Solen, Venus, Modiola, Cardium, &c.; the deposit in its main characters resembling the description given by M. Coquand of the fluvio-marine travertines of the north of Marocco, and also the Sahara beds described by M. Ville; with the difference that the Sahara deposits are characterised by the presence of little *Paludinas,* whilst those of Sidi Moussa are full of vermiculiform

perforations.   The depressions are occupied by a very porous
conglomerate, passing into a calcareous sandstone used for
building.   This conglomerate contains an abundance of *Helix
vermiculata*, a species living in the country, and also found in
the calcareous sands which are supposed to be of post-Pliocene
age.   The plain of Doukala (*Ducaila* of Washington), at a level
of about 140 feet above the sea, is covered with these sands.
At Sidi Ammer an escarpment was observed, the base of which
consisted of clay and red ferruginous marls, containing a stratum
formed for the most part of oysters, in which also *Teredina
personata* occurred, supposed by M. Nyst to belong to the
Eocene formation; succeeded by another fossiliferous bed containing

> *Balanus sulcatus,*
> *Pecten Beudanti,*
> *Arca,*
> *Buccinum prismaticum,* and
> *Conus,*

supposed by M. Nyst to be Miocene, the upper part of the
escarpment resembling the beds of later age before described.

An examination of the higher points of the western coast
near Saffi, and at Azfi in the province of Abda, near Mazagan,
tended to establish the fact of the occurrence of Pliocene beds
in the district.

At Cape Saffi, 180 metres in altitude, a reddish calcareous
sand was met with abounding in *Cyclostoma, Cylindrellas,* and
a species of *Helix* differing from that at Mazagan; and at
other points, including the hill of Aher and at Sidi Bousid,
white marls and sands associated with calcareous sandstones
were met with analogous to the supposed quaternary beds in
the neighbourhood of Mazagan.

The only other point in the geology of the coast-line I have
to refer to is the great mass of blown sand surrounding Mogador,
presenting a weird expanse of sea-like waves of sand, on a scale
vastly greater than anything of the kind on our own coast,
mimetic of mountain-chains and bold escarpments in miniature,
differing only from true hill-and-valley structure in the absence
of continuous valley-lines, the hollows being completely surrounded by higher ground.   Many of the ranges of sand are
from 80 to 100 feet in height, and their perfectly straight

scarped faces are produced by the violent westerly gales blowing the sand *up* the angle of repose, and accumulating it in fountain-like showers over the rounded backs of the sand-hill ranges.

It is worthy of note that the sub-aërial ripple-markings superimposed on the greater undulations, occupy a reversed position with reference to the prevalent winds, their long side facing the wind, with the more vertical straight scarps on the lee side. The moving sand in this case is drifted up the long side, and falls over the scarp at the angle of repose.

The Plain of Marocco.—We now turn inland; and before referring to the details of the structure of the Great Atlas range, it will save repetition if I briefly describe the general contour of the district under consideration. Leaving the sand-hills, which die out inland, and travelling westward, we gradually ascend over an undulating country, in aspect somewhat like the Weald of Sussex, covered for 30 miles with Argan Forest, till we reach, at 60 miles inland, the average level of the plain, about 1,700 feet above the sea.

The fundamental rock is here rarely to be seen; for the entire face of the country is shrouded over by a sheet-like covering of tufaceous crust (fig. 3), rising over hill and valley, and following all the undulations of the ground. Only in river-beds and here and there by the side of a hill were the fundamental beds visible, and seen to consist of alternations of hard and soft cream-coloured calcareous strata, dipping and undulating in various directions at low angles, and so closely resembling the surface crust that it was difficult to distinguish the one from the other, unless the surface crust happened to lap unconformably over the scarped exposures of the stratified beds. This singular deposit varies in thickness from a few inches to two or three feet, and is taken advantage of by the Moors for the excavation of cellars in the soft ground, over which the crust forms a strong roof. These are termed *matamoras*, and are used for the storage of grain, and as receptacles for burying the refuse from the villages. The calcareous crust in the neighbourhood of Marocco is extensively burned for lime. In section it presents a banded agatescent structure, often much brecciated. It is impossible it can have been deposited by any waterflow, as completely isolated hills are shrouded over by it as thickly as the valley bottoms; and the only satisfactory explanation of its origin I can suggest is, that it results from the intense heat of the sun rapidly drawing up water charged with soluble carbonate of lime from the calcareous strata, and drying it layer by layer on the surface, till an accumulation several feet thick has been produced. The rapid alternations of heavy rains and scorching heat which take place in the Marocco plain are conditions favourable to this phenomenon, which is unknown in northern temperate climates.

FIG. 3.

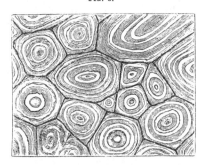

Surface.

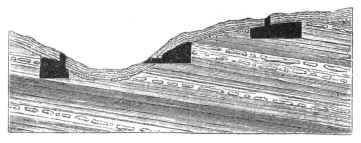

Section.

A familiar illustration of the same kind of action is seen in what brickmakers term 'limewash.' A brick formed of marl containing soluble carbonate of lime, if rapidly dried or placed in the clamp in a wet state, will have on its upper surface, after burning, an unsightly white scum or crust, by the accretion of soluble matter driven upwards and outwards by the quick evaporation. Before we left Mogador on our journey inland, we were told of great beds of shingle covering the plain, and fully anticipated some interesting drift phenomena; but these shingle-beds were found to be nothing more than the broken débris of the surface tufa, covering the plain for hundreds of square miles with stony fragments. Of marine drift there is not a vestige, the few isolated patches of waterworn stones and alluvial shingle being always connected with river valleys, excepting only the huge boulder deposits of the Atlas hereafter to be referred to.

About midway between Mogador and the city of Marocco, the monotony of the plain is broken by a curious group of flat-topped hills, which rise two or three hundred feet above its

FIG. 4.

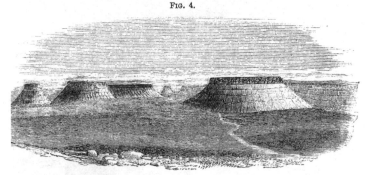

'Camel's Back,' flat-topped hills in the Plain of Marocco.

general surface. They present straight scarped sides, on which are exposed cream-coloured calcareous strata capped with a flat tabular layer of chalcedony, which seems, in arresting denudation, to have determined their peculiar and symmetrical form. In these we found no fossils; and I am doubtful whether they are an inland extension of the Miocene beds observed by Dr. Hooker at the 'Jew's Cliff,' near Saffi, or are some members of the Cretaceous series, of which there are sections on the coast north of Saffi and on the flanks of the Atlas.

At this point the main boundaries of the plain come into full view,—on the north a rugged range of mountains trending east and west, which we estimated at from 2,000 to 3,000 feet in height; and on our right the great chain of the Atlas, rising 11,000 feet above us and between 12,000 and 13,000 feet above the sea, bounds the view to the south, framing-in the great plain, here some 50 miles broad, which is lost as a level horizon in the eastern distance.

The Atlas Range.—Commencing at Cape Guer, on the Atlantic sea-board, the range, which at a little distance has the aspect of a single ridge, averages at its western extremity from 4,000 to 5,000 feet in height, from which it slightly falls off in height for a few miles, and then gradually increases in height as it recedes from the coast. In the eastern part of the province of Haha the summits probably attain to a height of about 10,000 feet. At a point about 60 miles from the sea there is a comparatively deep breach in the range, through which runs the main road to Tarudant. Eastward of that pass the projecting summits appear to lie between 11,000 and 11,500 feet above the sea to a distance from the coast of about 100 miles, and about SW. of the city of Marocco, where a second depression occurs, affording a pass to the south, at an altitude of about 7,000 feet. Immediately east of this, and due south of the city of Marocco, the range for 30 miles in length presents a long unbroken ridge, 12,000 feet in height, on which are deposited a few isolated crags and peaks rising from 500 to 800 feet above the general level; and it is doubtful whether this part of the chain attains an extreme height of 13,000 feet. Still farther east the ridge-like character is lost, the range becoming broken up into a series of less continuous peaks (including Miltsin, estimated by Lieut. Washington to be 11,400 feet in altitude, and supposed by him to be the highest point in the chain) of diminished height: beyond this, eastward, little or nothing is known either of the altitude or character of the range, excepting that it trends NE. by E. towards the southern borders of Algeria on the Sahara.

Rohlfs, in his journal of his overland journey from Marocco to Tripoli, speaks of mountains to the east of Marocco being covered with *perpetual* snow; but this is a character which has been erroneously attributed to the Maroccan section of the Atlas range. When we arrived at Marocco in the first week of May, the snow was limited to steep gullies and drifts—all the exposed parts, including the very summit, being entirely bare. There were, however, frequent storms, which intermittently covered the range down to 7,000 or 8,000 feet; but it is certain that these occasional falls would be rapidly cleared off by the summer heat; and we came to the conclusion that there was nothing like perpetual snow on any portion of the chain we visited, included in the section (apparently the highest part) lying due south of the city of Marocco.

As seen from the city, the great ridge appears to rise abruptly from the plain some 25 miles off; and so deceptive is the distance, that it looks as though it were a direct ascent from the plain to the snow-capped summit, even too steep to scale; but in reality this wall-like ridge represents a horizontal distance of 15 miles or more from the foot to the summit. As we approached it, an irregular plateau four or five miles wide was seen to form a sort of foreground to the great mass of the chain, from 2,000 to 3,000 feet above the plain, and 4,000 to 5,000 feet above the sea-level. This is intersected by occasional narrow ravines, which wind up to the crest of the ridge; and its face, fronting the plain, is for the most part exposed as an escarpment of red sandstone and limestone beds dipping away from the plain, and again rising from a synclinal against the crystalline porphyrites of the centre of the ridge, and unconformably overlying nearly vertical grey shaly beds with a strike ranging with the general trend of the Atlas range. Against the plateau escarpment rest enormous mounds of boulders spreading down to the level plain.

These, then, are the general features of the chain of the Atlas and plain of Marocco, the further details of which it will be convenient to consider under the following heads :—

(a) Surface Deposits and Boulder Beds.
(b) Moraines of the higher valleys.
(c) Stratified Red Sandstone and Limestone Series.
(d) Grey Shales.
(e) Metamorphic Rocks.
(f) Porphyrites.
(g) Eruptive Basalts.

(a) *Surface Deposits and Boulder-beds.*—Next to the Tufa crust already described, which extends over almost the entire plain of Marocco, perhaps the most remarkable feature in the physical geology of the country is the enormous deposit of boulders that occurs in the lateral valleys, and flanks the great chain on its confines with the plain. Of marine drift there is not a trace; and alluvial drift and valley gravels are very limited in their distribution, being confined to the borders of a few insignificant rivers that intersect the plain and the localities of occasional waterflows; but as soon as the flanks of the Atlas are reached, new and distinct drift phenomena present themselves. It was on our second day's journey from Marocco to the Atlas that the great boulder-beds came under our notice, first in a valley leading up from Mesfioua to Tasseremout, as scattered blocks of red sandstone, remarkable for their large average size, many of them of from ten to twenty cubic yards; but here the method of their disposition scarcely enabled us to decide that they were other than stream-borne masses from the higher ground. From Tasseremout we turned west, and at the

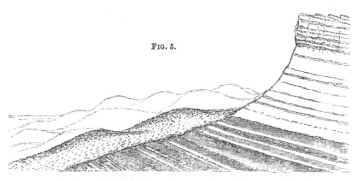

FIG. 5.

Boulder-mounds, skirting Atlas Plateau Escarpment. (Section.)

mouth of a second valley, two miles from the village, suddenly came upon a huge development of these Red Sandstone boulder-beds as great ridge-like and very symmetrical masses with terminal faces three or four hundred feet high, and, like the more scattered blocks NW. of Tasseremout, intermixed with but a very small proportion of fine matter. From this valley we turned out northwards, skirting the escarpment facing the plain; and for more than ten miles no lateral valley breaks into the cliff-like face; but below it the great boulder-beds (figs. 5, 6) still occur in huge masses not resting directly against the escarpment, but as isolated mounds two or three hundred feet in advance, sloping down towards the escarpment in one direction, and in the other rolling away in great wave-like ridges and undulating sheets,

which terminate at a well-marked line of demarcation, just
where the level portion of the plain commences. I measured
by aneroid the height of these mounds; and at one point their
summit was 3,950 feet above the sea-level, from which they
spread down uninterruptedly to the edge of the plain nearly
2,000 feet below. They bear a striking resemblance to the
glacial ridges or escars between Edinburgh and Perth; their
mound-like structure is distinctly visible from the city of Marocco, twenty-five miles off, appearing like a row of pyramidal
tali resting against the face of the escarpment as though they
had been cast down from its edge on to the plain. The internal
structure of the mounds also suggests such a deviation from the

FIG. 6.

Boulder-mounds, skirting Atlas Plateau Escarpment.

disposition of the boulders in layers sloping away from the escarpment towards the plain; and on a nearer approach it is seen
that the individual mounds are not connected with channels or
valleys breaking through the escarpment.

The depression between the escarpment and the drift-
mounds is a remarkable feature, and suggests an entire change
of conditions since the boulder-beds were deposited. If they
are a mere sub-aërial talus, they should rest directly against the
cliff face, and the depression separating them must have been
formed after the accumulation had ceased; and yet no satisfactory reason can be assigned for such cessation, if rain and
river action were the only operating causes. The form of the

mounds in the valley west of Tasseremout at once conveyed to me the impression that they were of glacial origin; and the discovery of undoubted moraines in the higher valleys strengthened my conviction that the boulder-mounds and ridges flanking the Atlas plateau can only be satisfactorily explained as the result of glaciers covering the escarpment, leaving on their recession the intermediate depression.

(b) *Moraines of the Higher Atlas.*—Kindred phenomena occur higher up in the Atlas valleys, most notable in the case of unquestionable moraines, commencing at the village of Adjersiman, in the province of Reraya, at an altitude of 6,000 feet. Here we met with a gigantic ridge of porphyry blocks, having a terminal angle of repose of between 800 and 900 feet in vertical height, and grouped with several other mounds and ridges of similar scale, all composed of great masses of rock with little or no admixture of small fragments, and completely damming up the steep ravine and retaining behind it a small alluvial plain 6,700 feet above the sea-level.

We failed to detect any scratched blocks or striæ; but that these ridges are true glacial moraines no one who has seen them and compared them with other glacial phenomena would for a moment doubt; and their interrupted occurrence at various heights is strictly in accordance with the distribution of moraines in many of the Swiss and Scotch valleys.

Lieut. Washington, in referring to the pointed mountainous hills NW. of the city of Marocco, crossed on his homeward journey, describes one of them as being 'covered with masses of gneiss and coarse-grained granite (? diorite), many of the blocks being several tons in weight,' and asks, 'how got they there?' 'If granite, the nearest granite mountains are at a distance of twenty-five to thirty miles : can they be boulders?' As far as my own observations go, there was no rock *in situ* in the part of this range I visited near Marocco resembling granite or diorite; and in connection with the boulder-mounds of the Atlas, the occurrence of foreign blocks north of the plain of Marocco so far from the parent source, is a circumstance of great interest.

(c) *Stratified Red Sandstone and Limestone Series.*—A long line of comparatively low and flattish hills, forming a plateau, with an average height of about 4,500 feet above the sea, and 2,800 feet above the plain of Marocco, intervenes between it and the main ridge of the Atlas. The edge of this plateau facing the plain is for some distance an escarpment, exposing stratified beds of limestone containing bands of chalcedonic concretions, underlain by grey and puce-coloured marls. As this plateau is crossed from north to south towards the Atlas ridge, its central line would represent a synclinal, from which the beds rise northwards towards the plain and southwards towards the Atlas; but it is locally broken and contorted, and near Tasseremout the limestone beds stand up nearly on end. South of the synclinal, *i.e.* between the centre of the irregular plateau and the Atlas, great deposits of red sandstone and dark-red conglomerate, interstratified with cream-coloured shelly limestone, occur, which appear to be inferior members of the series of limestones and marls exposed in the escarpment facing the plain. Lieut. Washington, who ascended Miltsin to a height of 6,400 feet, describes hard red sandstone with an east and west strike dipping 10° south, as occurring at this elevation, which is nearly 2,000 feet higher than we observed the Red Sandstone series in the province of Reraya farther west, and also both in his approach and descent from Miltsin of ranges of limestone running NE. and SW. dipping 70° SE. with abrupt sterile sandstone mountains rising above them. From the few obscure fossils, including an *Ostrea*, I was able to collect from the limestone bands, Mr. Etheridge considers that they are of Cretaceous age. They are, like the beds of the plain, remarkable for containing great deposits of chalcedonic concretions; but the latter may possibly be of more recent age. They rest unconformably on the upturned edges of grey shaly beds, and extend also over the porphyries that form the great mass of the Atlas chain. They appear to have been deposited subsequently to the porphyry ridge assuming its present hill-and-valley contour, as little isolated fragments are seen clinging to the sides of a narrow ravine leading out of the valley we ascended through the province of Reraya to the Atlas. Their relation to the few exposures of stratified beds in the plain is somewhat uncertain, as no fossils were obtained in the latter, and there are no direct connecting links; but, judging from petrological similarity, and from the fact that Neocomian fossils occur in exposed beds on the coast cliffs, and Cretaceous fossils in the beds forming the crest of the plateau, it seems possible that an unbroken series

occurs from the cliff north of Saffi to the plateau skirting
the Atlas, representing the whole of the Cretaceous epoch ; but
it is also open to question whether the level beds of the plain
may not be an inland extension of the strata of Miocene age
from which Dr. Hooker obtained fossils at the Jew's Cliff south
of Saffi.

(d) *Grey Shales.*—At several points on entering the lateral
valleys of the Atlas, almost vertical shaly beds are crossed,
having a strike nearly east and west, corresponding with the
trend of the chain.   They clearly underlie, and are unconform-
able to, the Red Sandstone and Limestone series ; and their
almost vertical position appears connected with one of the
several upheavals that have affected the chain.   Of their
geological age there is no evidence, except that they are pre-
Cretaceous.   In places, as at Assghin, they abound in nodules
of carbonate of iron.   Pale shales, containing quartz veins, crop
up near the village of Frouga, in the plain south-west of
Marocco, which may possibly belong to this series ; and if the
porphyries forming the mass of the Atlas are contemporaneous,
they are probably interbedded with these grey shaly beds.
Lieut. Washington speaks of the occurrence of clay-slate
dipping 45° east between El Mansoria and Fidallah, and again
of a hilly country of clay-slate near the plain of Smira, and at
Peira, farther south ; but it is impossible to say whether these
beds are related to the grey shales of the Atlas.

(e) *Metamorphic Rocks.*—The most important development
of metamorphic rocks in the neighbourhood of Marocco is on
the north side of the city.   In its immediate neighbourhood,
three miles to the north-west, a low rugged hill occurs, composed
of a very hard and compact dark-grey rock, containing knotted
white concretions elongated in the line of stratification, which
dips from 50° to 80° south-west, the strike being north-west and
south-east.   The whole of the north side of the plain is
bounded by ranges of rugged hills of similar form, and
apparently rising from 2,000 to 3,000 feet above the plain.   We
had not an opportunity of visiting them ; but, judging from
their outline, they are identical in formation with the hill close
to Marocco.   We observed nothing in the Atlas resembling it.
Lieut. Washington, who crossed these hills on his journey to
Marocco at about the point I visited, and again forty miles to

the east, near the source of the river Tensift, on his homeward journey, speaks of them as from 500 to 1,200 feet in height, consisting of micaceous schist and a schistose rock with veins of quartz dipping 75°, with a strike north by east and south by west. The strike may vary a little at different points, and, taking Lieut. Washington's and my own observations together, would average about north and south; and it is worthy of note that these apparently ancient rocks are nearly at right angles to the strike of the rocks of the Atlas chain a few miles to the south.

The only other metamorphic rocks that came under our notice were:—first, white marble or metamorphic limestone, intercalated with the porphyrites at the summit of the ridge of the Atlas south of Arround; secondly, mica-schists, pierced by red porphyry dykes, forming the mass of Djebel Tezah, a peak 11,000 feet in height, and fifteen miles farther west, ascended by Dr. Hooker and Mr. Ball after my return. It is possible that the mica-schists may be a portion of the grey-shale series, metamorphosed by the intrusion of the porphyry dykes. Lieut. Washington, on his first day's journey south of Tangier, refers to the occurrence of rounded schistose hills about 300 feet high, strike north-west and south-east, dip 75° south-west, containing mica-slate with veins of foliated quartz; but I have no recollection of observing any such metamorphic rocks between Tangier and Tetuan.

(f) *Porphyrites.*—Of the eruptive rocks of the Atlas, porphyrites and porphyritic tuffs occupy by far the most prominent position, forming the great mass of its ridge.

On entering the lateral valleys, after crossing the vertical shaly beds, great masses of red porphyrites and tuffs are met with, associated with specular iron and occasional green porphyries. The harder portions of the latter are seen as *Verde antique* pebbles in the river-beds; but we failed to detect this *in situ*. From the large proportion of tuffs that occurs the porphyrites appear to be interbedded, and are possibly contemporaneous with the vertical grey shales to which they are adjacent. They are overlapped unconformably by the Red Sandstone and Limestone series of Cretaceous age. The late Mr. D. Forbes informed me that they bear a strong likeness to the porphyrites of the Andes, of Oolitic age; but beyond the

fact that they were in existence and had undergone denudation into hill-and-valley contour before the Cretaceous beds
were deposited over them, there is no certain evidence as to
their age.

There may have been at least one or two subsequent intru-
sions of red porphyrites, viz. of the dykes of Djebel Tezah,
metamorphosing grey shales into mica-schists, and of the dykes
that break up through the stratified beds of the plain east of
Sheshaoua—which may probably be more recent than the porphyrites of the Atlas, as they appear to penetrate strata which
extend over the denuded surface of the Atlas mass ; but I cannot speak with certainty as to the relative age of the stratified
beds and the porphyritic bosses which rise up out of the plain.

(g) *Eruptive Basalts.*—Of these we met with three distinct
species :—

(1) Black vesicular basalt (porous and compact pyroxenic
lava with olivine) on the coast near Mogador, and imbedded in
the base of the post-Tertiary concrete sandstone cliffs : but it
was nowhere seen *in situ* ; and I think it possible that the
fragments may have been derived from the Canary Islands,
which are only 70 or 80 miles distant, or possibly from some
point of eruption nearer the land.

(2) Amygdaloid green Basalt, which rises up in dykes, in
many places penetrating the Red Sandstone and Limestone
series on the flanks of the Atlas, and also piercing the diorite
of the Arround valley. We observed numerous dykes at Tasseremout, Tassgirt, and Asni, south-east and south of Marocco
city. Beyond the fact that they are probably post-Cretaceous,
there is no evidence as to their age. From what we could
see of their distribution, the whole range of the Atlas seems
abundantly intersected by these dykes.

(3) Diorite rises up in considerable masses among the porphyrites in the valley of the Arround, due south of Marocco,
but forms no great proportion of the bulk of the ridge. Its intrusion may have been contemporaneous with the dislocation
and upturning of the Red Sandstone and Limestone series overlying the porphyrites.

General Summary.—It now only remains briefly to reca-
pitulate the order of sequence of the geological phenomena
observed in the plain of Marocco and the Atlas.

The oldest rocks that have been noticed are :—

(1) The ranges of rugged metamorphic rocks north of the city of Marocco, and forming the northern boundary of the plain, respecting the age of which, and the period of their upheaval and metamorphism, there is no evidence.

(2) The interbedded porphyrites and porphyritic tuffs of the Atlas, forming the backbone of the ridge, the age of which, and of the grey shales with which they seem to be interbedded, is also uncertain.

(3) Mica-schists of Djebel Tezah, in the Atlas, south-west of Marocco, pierced with eruptive porphyritic dykes, which may be an altered condition of the vertical grey shales adjacent to the interbedded porphyrites.

These rocks are our starting point, respecting which there is no evidence of their age, or even relative age.

(4) We now come to a long period of denudation of the Atlas ridge, and its sculpturing into hill-and-valley contour, before the deposition of the Red Sandstone and Limestone series.

(5) The deposition over what is now the Marocco plain, of the Cretaceous Red Sandstone and Limestone series (and beds possibly of Miocene age), which also occupies pre-existing valleys in the older porphyrites of the Atlas.

(6) The intrusion of diorite into the porphyrites and porphyritic tuffs, probably accompanied by a further elevation of the Atlas range, disturbing the stratified Red Sandstone and Limestone series, throwing them into a synclinal trough, from which the beds rise northwards towards the plain, and southwards towards the Atlas.

(7) A further long period of denudation of the Red Sandstone and Limestone series, rescooping out the lateral valleys of the Atlas, in continuation of the valleys that existed in the porphyrite ridge prior to their deposition, and also denuding the beds in the Marocco plain to the extent of at least 300 feet, leaving isolated remnants as flat tabular hills rising above the present general level of the plain.

(8) A further possible emission of red porphyrites through the stratified beds of the plain, which may have been contemporaneous with the eruption of the red porphyry dykes of Djebel Tezah, in the High Atlas; but I could not clearly

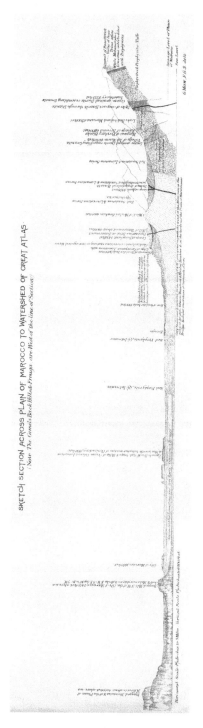

The material originally positioned here is too large for reproduction in this reissue. A PDF can be downloaded from the web address given on page iv of this book, by clicking on 'Resources Available'.

ascertain whether these bosses really pierced the stratified beds, or were existing before their deposition.

(9) A post-Cretaceous eruption through the Red Sandstone and Limestone series of a multitude of dykes of amygdaloid basalt, the age of which is uncertain.

The more recent changes commence with :—

(10) The formation of gigantic boulder-beds flanking the northern escarpment of the Atlas plateau, and spreading down in great mounds and undulating ridges from a height of 3,900 feet to the borders of the plain, 1,900 feet above the sea, with a range in vertical height of about 2,000 feet, and extending up the entrances of several of the lateral valleys, as well-defined and symmetrical moraines.

(11) The formation of moraines at the heads of the Atlas valleys, commencing at a height of 5,800 feet, and spreading up to the cliffs of the Atlas ridge, to a height of between 7,000 and 8,000 feet, with a terminal angle of repose 850 feet in vertical height.

(12) The formation of a plain of shingle behind the moraines, at a height of about 6,700 feet, which seems to be the bed of a small lake.

(13) The recession and extinction of glaciers in the Atlas range, on which there is now not even perpetual snow.

(14) An elevation of the coast-line of at least 70 feet, represented by the height of the raised beaches of concrete sand at Mogador and other parts of the coast, which may possibly be contemporaneous with the elevation of similar raised beaches on the coasts of Spain and Portugal, and with the raised beaches of our south-western coast.

(15) A slight subsidence of the coast-line, now going on, with an accumulation of extensive deposits of blown sand at Mogador.

(16) The formation of a tufaceous surface-crust over almost the entire plain of Marocco, due to the drawing up to the surface, by rapid evaporation, of water from the subjacent calcareous strata, depositing, layer by layer, laminated carbonate of lime.

APPENDIX I.

MOORISH STORIES AND FABLES.

FROM much information that has been kindly furnished to us by Mr. Freeman Rogers, a gentleman who was several years resident in Marocco, and had become familiar with the people and their language and manners, the following extracts have been taken for the sake of the light which they throw on the condition of the country. It being the main object of this volume to relate our personal experiences, we have not been able to avail ourselves of much information supplied to us by Mr. Rogers, and other competent witnesses; but it has appeared to us that the extracts here given form a useful supplement to the facts which came to our knowledge during our short stay in Marocco, and will help the reader to form a truer conception of its present condition.

The stories, which may be said to have a political character, furnished to us by Mr. F. Rogers, all refer to events that have occurred during the last twenty-five years, and are precisely similar in character to others which were passing at the time of our visit. They are accepted as substantially accurate by our informant, and we see no reason to refuse them credence. They certainly tally with the universal belief of the natives as to the conduct of their rulers. Any one who is familiar with the chronicles of the Middle Ages, who has marvelled at the deeds of ferocious cruelty recorded of German petty rulers, or the more refined atrocities of Italian princes, must sometimes have felt a wish to know what manner of men they were who committed these deeds. To satisfy such a curiosity, he cannot do better than pay a visit to the interior of Marocco. If duly commended to their good offices, he will be received by men of stately and courteous manners, prompt to display a lavish hospitality, who will inevitably send him away with a favourable impression; but before he has been many weeks in the country, he will become aware that these amiable hosts are habitually guilty of deeds of combined ferocity and treachery that equal, if they do not surpass, those of the dark periods of European history.

The popular fables, which were taken down from the mouth of an old Moorish story-teller, and literally translated by Mr. Rogers, complete the impression derived from the fragments of contemporary history. They all turn upon the success of fraud and force in the affairs of life. The moral, so to speak, of all is 'woe to the weak and the confiding;' but admiration is mainly given to those who supply the place of strength by successful perfidy.

Abd el Saddock, Kaïd of Mogador, Duquallah, Abda, and Sous.[1]

On one occasion this Kaïd was sent to Sous by the Sultan to reduce some provinces to submission. When arrived there, a grand entertainment was given to him by the refractory Sheiks, and immense quantities of provisions sent in to supply the guests, among which was a large quantity of a particular dish of which the Kaïd was known to be very fond, and this was all poisoned. The Kaïd, suspecting from the Sheiks' importunity for him to eat of it that it was poisoned, ordered his soldiers to guard the doors and let no one escape, and then called upon the Sheiks one by one to partake of the dish. Most of the Sheiks refused to eat, and some few came cheerfully forward at the Kaïd's call; those who refused were compelled to eat, and those who came cheerfully forward were not allowed to eat; and so the Kaïd in one day not only got rid of his enemies, but saved his friends, whom he rewarded by putting them in the place of those who fell by their own treachery.

Kaïd Boh Djemma.

Some short time after the news of the foregoing had spread over the country, a revolt took place at Shedma, and many of

---

[1] Abd el Saddock was the father of Hadj Hamara, the Kaïd of Mogador, by whom we were hospitably entertained soon after our arrival. The father appears to have cumulated important offices to an extent now rare, if not unknown, in Marocco. As the Sultan's hold over the province of Sous is very feeble, and limited to the occasional receipt of tribute, there is no resident Kaïd, but the title is given to any official sent, *pro hâc vice*, to represent the Sultan. But the provinces of Duquallah and Abda, like the rest of the settled country, are ordinarily administered by resident governors.

the Sheiks made themselves conspicuous by their opposition to
the Kaïd, who determined to get rid of all his enemies at one
blow; he therefore made peace with them, and all seemed well
and tranquil for some time. At last came the holiday *l'ashora*,
or the day of the Sultan's tenthing, when an invitation was
issued by the Kaïd to all his Sheiks to appear at his entertainment; none dare refuse, and so all went. The Kaïd had, in
the meantime, prepared a large room, into which he sent the
Sheiks known to be his enemies, and another into which he
sent those known to be his friends. When all had feasted until
they could eat no more, the Kaïd quietly ordered the windows
and doors to be closed, the men to be bound, burning charcoal
to be placed in the room; and the doors then to be built up, and
all left to their fate. Nine days afterwards, when the room
was opened, nothing remained of all those men, some twenty-
two or twenty-three, but bones, attesting the fatal effects of
burning charcoal and the daring ferocity of the rats; except
one man whom the Kaïd pardoned, believing him to be inno-
cent, as his life seemed to be so miraculously preserved.

What the Sultan means when he bestows a Wife.

The Kaïd of Shedmah, Boh Djemma, had distinguished himself against some rebels who had risen against the Sultan, and
the praise bestowed upon him openly by his enemies in the
hearing of the Sultan, excited the suspicious sovereign's anger
and jealousy to such an extent that he was determined to get
rid of such a dangerous enemy; in order to which he called for
the Kaïd and praised his exploits in the presence of all his
great men, ordered him a suit of his own royal clothing and
a favourite horse, and promised him a wife out of his own
seraglio. The Kaïd rejoiced, and his enemies too: the Kaïd,
because he regarded himself as a favourite; and his enemies, who
were older and knew better, because that he was doomed. In
a few days the Kaïd was sent home and his new wife along with
him in great state, and in ten days more the Kaïd was carried
to the grave, he having died suddenly (poisoned by the Sultan's
female executioner) in the night.

A similar occurrence took place with the Kaïd of Haha;
but he had a watchful and wise mother, who watched the new

wife until she saw her prepare a dish for her son, when she
presented herself before him, charged the new wife with her
intended crime, and dared her to eat of her own dish. The
Kaïd's eyes were opened, and he compelled the Sultan's lady,
his new wife, to eat of the dish which she had prepared for him,
when she immediately died from the effects of her own poison.
This Kaïd ever after kept away from the Sultan until, a few
years ago, his evil genius prevailed on him to obey the Sultan's
call, when he died within an hour after taking supper with
the Sultan.

Abd el Saddock, Kaïd of Mogador, Duquallah, Abda, and Sous, and his False Friend.

Some years since, the Kaïd of Mogador[1] (father of the Kaïd
Hadj Amara who entertained you when there) ruled over the
provinces of Duquallah, Sous, and Abda, and made a great deal
of money during his administration, and secured the love of all
good Moors by his making the Jews acknowledge their inferiority to the Moors. But suspecting that his time to be
squeezed by the Sultan had nearly arrived, he determined to prepare for it, and so outwit a false friend, who was an enemy of
his, and the Sultan at the same time; in order to which, he
called upon his false friend, and invited him to dine with him
that evening in private as he had something to tell him. After
dinner the Kaïd told his guest that he was getting afraid of the
Sultan seizing him in order to get his money. 'Now,' said he,
'I have a favour to ask of you, which is that you will carefully
preserve the treasure which I will show you, and when I am
seized upon take the keys of my house, but do not live in it,
and when my son Hamara knows how to use my money, then
tell him of the box and give him the keys; and further, I want
you to swear that you will never tell where I have hid my
treasure, and that you will not tell any one of what has passed
this night.' The false friend took the oath with mental reser-
vations, as would appear from the sequel. The Kaïd then
ordered four slaves to attend upon him, and all descended to
the cellar, where the money was concealed in a large strong
wooden box, buried in the ground. The box was then opened

---

[1] The same of whom the first story is related above.

and was seen to be full of silver and gold, &c. The Kaïd then
had the box covered up, and the false friend took his departure.
After he went away, the Kaïd returned with his slaves and had
the money, but not the box, removed to a really secure place,
and had the box filled with bits of stones and broken pottery
and *recovered* over in the same manner as it was before, when
seen by the Kaïd's false confidant; he then had his slaves carried off to prison and put to death on some pretext or other.
The next day when the · Kaïd's confidant heard of the slaves
being dead, he knew it was to prevent their telling, and concluded that it was the secret which he possessed which the Kaïd
wanted to guard, and that he alone knew of the secret of the
Kaïd's wealth and its hiding place. In some short time afterwards, an order came from the Sultan ordering Abd el Saddock
up to Marocco; upon which the Kaïd told his confidant of his
trouble and begged him to be true to his oath, blessed him,
kissed him, and then went to wait upon the Sultan.   The
Sultan upon seeing him ordered his arrest and torture, accusing
him of robbing him and his people, &c.; upon which the Kaïd
was carried off to the torture, when he kept denying having
any money, and being guilty of the charges brought against
him.   At last the Sultan, losing all patience, sent him word
that he had received information, so the Kaïd had better speak
the truth at once, for such a one (the false friend) had declared
that he had a large box full of treasure, but was sworn not to
tell of its whereabouts. The Kaïd, therefore, must either tell
or suffer death by torture. At this the Kaïd pretended to be
much alarmed, and declared that nothing could be concealed
from Seedna, so he would confess the whole truth, and that
what such a one said was true and that it was concealed in such
a place, and put there in presence of such a one (the informer,
the Kaïd's confidant), and that if the Sultan sent for it he
would have it all. The Kaïd was then put in prison, and
notaries and soldiers sent for the money under the guidance of
the informer who was in great glee, thinking that now his
fortune was made and his favour with the Sultan secure; but
upon arriving at the cellar and the box being opened, nothing
but stones and broken pottery was found where there had been
gold and silver.   Imagine the wretch's horror as the notaries
said he himself must inform the Sultan, as they dare not

do so ; however, as there was no use in lamenting, they returned
to Marocco, and the informer had to tell Seedna that there
was nothing in the box but rubbish ; upon which the Sultan
ordered the Kaïd to be brought before him and demanded the
meaning of such a thing.   The Kaïd answered, 'True, our lord,
it is that I did not oppress your people, and the money hidden
in that box was made by lawful means, and I reposed confidence
in my friend here, and left the money for my son ; and so I
told your majesty truly that I had nothing, because it then
became by my gift my son's money, and this, my false friend,
has broken his trust, robbed my son and Seedna, and then to
cover his knavery, sought my life by trying to turn our lord
against me.   I therefore beg that our lord will make him confess what he has done with Seedna's money.'   The Sultan
thought the informer simply wanted to make him a fool to
cover his knavery, and at once, in a passion, ordered him to be
flogged until he confessed.   But as he could not confess that he
had taken the money and had none of his own to replace it, the
lash was continued until the wretch died under it.   The Kaïd
was set free and restored to Mogador, and the informer's son is
now assistant weigher at the Custom House, Mogador.

SIX MOORISH FABLES.

1. *Fable of a Hedgehog and a Fox.*

Once upon a time a fox accidentally meeting a hedgehog
addressed him as follows, ' I am much oppressed with thirst ; '
to which the hedgehog replied, ' So am I, and I know a well
where we can drink.'   The fox then said, ' Come along.'   They
travelled on till they reached the well where they found two
buckets worked by a pulley, one ascending whilst the other descended.   ' Now,' said the hedgehog, ' I will go in first, and
when I tell you, jump into the other bucket.'   The hedgehog
went down and had his drink, and then shouted to the fox,
' Now you jump in.'   He did as he was told, and as he went
down met the hedgehog coming up in the ascending bucket ;
upon seeing which he said, ' What does this mean ?'   The
hedgehog answered thus, ' It is the world goes round :' and
when he was safely at the top, and the fox had reached the

bottom, he called down to him and said, ' Those who want to
kill me I catch them in a trap, and to those who do me a good
turn I do the same to them.'

2. *The Camel, the Hedgehog, and the Lion.*

Once upon a time there was a camel who met with a hedgehog, and the camel tried to trample on him. The hedgehog said
to the camel, ' Wait till I call my brother, he is able to kill both
you and me.' ' No,' said the camel, ' if he comes he will perhaps kill me.' ' No, no,' said the hedgehog, ' if you wish to see
him, lie down on your belly or on your back, open your mouth
and let the flies come in, and appear as if you were dead.' The
camel said, ' All right.' ' Well, well,' said the hedgehog, ' I
will go call my brother.' The hedgehog went away to look for
a lion, and meeting with one, said, ' Your servant, my lord ;
there is a wild beast which wishes to eat me.' The lion replied,
' Will he eat *me* ? ' The hedgehog said if he were there he
certainly would ; but he has gone away to get food, ' but my
lord, if you would like to see what he has procured for his
breakfast, come along with me.' The lion said, ' You go first.'
' Very well I will do so,' said the hedgehog, ' and when you follow and get near, roar with all your might.' The lion said, ' All
right.' The hedgehog said, ' I shall go first.' So away went the
hedgehog, and said to the camel, ' Now, he is coming you lie
still ; don't stir or he will eat you.' The camel said, ' All
right ; ' and whenever they heard the roarings of the lion, the
camel said to the hedgehog, ' Listen to the noise he makes
while talking.' The lion then drew nearer and roared again ;
when the camel exclaimed, ' In the name of the most mer-
ciful God, is he going to eat me ? ' The hedgehog said,
' Don't stir, don't fear.' The camel said, ' All right.' They
waited till the lion came, when the hedgehog addressed him,
and said, ' This is a morsel of the breakfast the monster is
going to eat.' The lion and the hedgehog now bade each other
adieu ; and when the lion had departed the hedgehog said to the
camel, ' Now you may get up, but tell me which of the two is
master.' The camel replied, ' It is you, it is you ; good
morning.' The hedgehog said to the camel, ' Are you going
away ? ' The camel replied, ' Yes, my lord, I am ; ' and from
that day to this they have never spoken.

3. *The Snake, the Hedgehog, the Man, and the Hunters.*

The Hedgehog personates the Kadi.
The Hunters   ,,   the Soldiers.
The Snake   ,,   the People.
The Man   ,,   the Sanctuary.

Once upon a time there were some hunters, who went out to hunt a fairy embodied as a snake. The snake being pursued meeting with a man passing by, said to him, 'Will you afford me protection, for there are hunters following, who want to catch me?' The man answered, 'Very well,' and allowed the snake to be concealed in his clothes; presently the hunters came up to the man and asked him whether he had seen a snake, to which he answered no, and the hunters passed on in pursuit. After the hunters had left, the man asked the snake to go down. The snake said, 'No, and if you attempt to force me down, I will kill you.' The man said to the snake, 'Very well, let us go to the Kadi and hear what the law says.' The snake said, 'Very well, come along.' So they went on till they came to the hedgehog, who was Kadi; and the man said, 'Your servant, my lord; here is a snake that I have saved from the hunters, and I have told him to get down, but he would not.' The hedgehog addressed the snake, and said, 'I will decide the law for you, but first get down.' The snake at once got down, and then demanded of the hedgehog what the law said. The Kadi then addressed the man as follows : 'The snake is on the ground and a stick is in your hand.' The man, taking the hint, struck the snake on the head and killed it.

4. *The Sheep, the Fox, the Lion, and the Shepherd.*

Once upon a time a fox met a lion, and the lion, addressing the fox, said, 'Will you be my servant to catch sheep for me? The fox said, 'I will, if you will give me my share.' The lion said, 'No, and if you eat a single bit, I will kill you.' 'Very well,' said the fox, 'if that is the bargain, I will agree to go hunt for you.' So away went the fox and hunted about till he found some sheep, one of which he killed and ate. He then went off to the owner of the sheep and said, 'The lion sent me to hunt your sheep for him, but I would not do so, and he came

himself and ate one, and I have run to tell you.' ' Very well,'
said the shepherd, ' you shall be the guard over my sheep, and
let me know when the lion comes;' and the fox said, ' All
right.' So he waited till the evening, and then went to guard
the sheep; and whilst on guard he killed and devoured two
more, and afterwards, making a little wound in his own leg,
he ran off to the shepherd, and said, ' The lion has come and
eaten two sheep, and wanted to eat me also; see the wound he
made in my leg.' The shepherd said, ' I see it is true; I will
put two men to assist you to guard.' The fox said, ' All right;
I will go hunt for something to eat, and then return.' So he
went off in search for the lion, and meeting him, said, ' I know
where there are lots of sheep ; they are in such a place, you come
in the evening to eat them.' The lion said, ' All right.' The fox
then ran back to the shepherd, and said, ' The lion is coming to-
night,' and directed the men who were to assist in guarding to
conceal themselves, but before doing so to bring a big sack into
which the fox put a great stone, and waited till the lion came.
When the lion came he said to the fox, ' Why have not you killed
me a sheep?' The fox said, ' Because I was afraid of a great monster that none but you can master, and there he is in that sack; go
in and kill him.' The lion said, ' All right,' and went in, when the
fox tied securely the mouth of the bag, so that the lion could
not get out. The fox then said to the lion, ' Have you found
him?' The lion replied, ' No, no, I have not, and it is funny
I cannot get out.' The fox said, ' Push away, try and get out.'
The lion said, ' I cannot.' The fox said, ' Probably the monster
holds you there.' The fox said to the lion three times, ' Can't
you get out?' and three times the lion replied, ' I cannot.' The
fox said, ' He who wishes to kill me I catch him in a trap, and
to him who does good to me I will do good in return.' So he
went away and called the guard, and said, ' There he is, beat
him;' so they beat him, and beat him till they were tired ; and
at last broke his leg. The fox said to the guard, ' Now let him
out, he has broken his leg and cannot escape;' and when the
lion got out the fox, addressing him, said, ' Now, which is
master?' The lion replied, ' You are my master;' and the fox
said, ' Whoever wishes to eat me at one time will try again,'
and turned to the men and ordered them to kill him. After
which the fox said, ' Now that we have killed the lion, good-

bye;' and the men replied, 'Good-bye.' He went away, and
waiting till evening, and returning at supper time, he wounded
the leg of each remaining sheep, and ran off to the owner and
said, 'A monster bigger than the last has come and wounded
all your sheep.' Previous to this the fox went to the market,
and bought a suit of clothes, and sprinkled them with the
blood of the sheep, and made it appear as if the big monster of
whom he spoke had also killed a man. He then said to the
owner, 'All your sheep are spoiled; we had better kill them
and make a feast.' So the shepherd killed the rest of the sheep,
and ate them with the fox; and when they had finished, the
fox filled a bowl with dirty water. The owner after he had
finished his breakfast, said, 'What am I to do now? The
sheep are killed.' The fox said, 'Now I will tell you how it all
happened.' The man said, 'How?' The fox said, 'Not till you
open that door so that the light is let in;' and then said
quickly, 'It is I who killed your sheep;' he then threw the
dirty water in his face, and made off.

5. *The Pigeon and the Monkey.*

Once upon a time as a pigeon was passing by he met with a
monkey. The monkey said to the pigeon, 'Come, let us play;'
and the pigeon said, 'Very well, what shall we play at?' The
monkey put up a stick and proposed they should get up it.
The pigeon agreed, and the monkey said, 'Which shall go up
first?' The pigeon said, 'You.' The monkey said, 'All right,'
and tried but could not manage it; upon which the pigeon immediately flew to the top. The pigeon said, 'Now it is my
turn to say, 'What shall we next try?' The monkey said, 'All
right.' So the pigeon challenged the monkey to tie his tail to
his leg, and when he had accomplished it, and untied it, he said,
'Come, let us see whether you can tie your wing to your leg.'
The pigeon said 'All right,' and fastened his wing to his leg,
but could not undo it; and the monkey devoured him.

6. *The Hyena and the Hedgehog.*

Once upon a time there was a hedgehog travelling in quest
of something to eat, and saw a hyena coming towards him with
intent to devour him. As soon as the hyena had reached him,

he said, 'My lord, I observe that you are dirty and stand in need of a bath, and if you require one I have a bath at your service in my house.' The hyena replied, 'Yes, it is true, I am much troubled with fleas, please come along and give it me; but first come to my house and breakfast, and then we will go to the bath.' The hedgehog said, 'That is just what I want, for I am out now looking for food.' So the hedgehog went to the hyena's house and had his breakfast. The hedgehog then said, 'Now come along and take your bath.' The hyena said, 'All right.' So they went to the hedgehog's house and the bath was heated to boiling. The hedgehog said to the hyena, 'Now jump into the bath, and scratch yourself.' So the hyena jumped in; whereupon the hedgehog closed down the lid, and tied it with a string. The hyena said, 'This is too hot for me, I want to get out.' 'No, no,' said the hedgehog, 'it is far better for you to be there than for me to be in your belly; bawl away till you are dead.'

APPENDIX K.

On the Shelluh Language.

By JOHN BALL.

JACKSON in his 'Account of Marocco' refers to the opinion of Marmol, that the Shelluhs of Marocco and the Berebers (Kabyles) of Algeria speak the same language, as altogether incorrect, and positively affirms, on the contrary, that these languages are quite distinct. In proof of this assertion, he gives a short list of Shelluh words or short phrases, with the Bereber equivalents of most of them, and concludes, from the differences between these, that the languages are profoundly, if not radically, different. A comparison of this kind is so notoriously misleading that no importance would have been attached to the conclusion derived from it, were it not for the fact that Jackson was well acquainted with the Shelluh language, probably better than any other European has since been; and that although not versed in comparative philology, a science not yet come into existence in his time, he was a man of good general intelligence who seems to have had frequent occasion to compare the two languages.

The first person who was able to speak on the subject with any authority was Venture de Paradis, a man of remarkable linguistic attainments, who died prematurely while accompanying the French Syrian Expedition in 1799. His grammar and vocabulary of the Bereber language were not published until 1844, and his conclusions were not until then made known to the world. It appears that in the year 1788 two Shelluhs, one a native of Haha, the other from Sous, went to Paris. Notwithstanding the difficulty of communicating with men who possessed no written language, Venture de Paradis contrived to obtain from them a list of Shelluh words and short phrases. He was very soon after attached to a mission sent to Algiers, where he was detained for more than a year. He made acquaintance with two Kabyles, theological students, at Algiers, and, finding that his list of Shelluh words corresponded very nearly with the Kabyle equivalents, he devoted himself to the study of the Kabyle dialect of the Bereber tongue, and prepared the grammar and dictionary which remained for more than half a century unpublished. It might be sufficient to refer the reader to the judgment of so competent an authority; but a slight examination of the subject has afforded such confirmation to the conclusions of Venture de Paradis as seems to place them beyond the reach of controversy.

It must be remarked in the first place that, from the want of sacred books or other written records among the races of the Bereber stock, there is no one of the many dialects spoken by them that can be taken as the classical standard to which others may be compared. French writers in treating of what they style 'la langue Berbère' usually mean the Kabyle, spoken by most of the mountain tribes of Algeria. The same language, with dialectic differences, is used by many tribes of the Sahara; but throughout the larger part of the vast region lying between the southern borders of Algeria and Marocco and the Soudan, the prevailing tongue, though unquestionably belonging to the Bereber family, deserves to rank as a distinct language from the Kabyle. A slight examination of the latter shows that it has been largely adulterated by contact with the Arab population, who from an early period have ruled the open country and carried on all commercial intercourse ; while the characteristic grammatical features have been in many respects obscured or effaced. On the other hand, it appears from a recent publication by General Faidberbe[1] that the dialect spoken at the south-western limit of the Bereber races, adjoining the river Senegal, while preserving the chief Bereber grammatical characteristics, has undergone much etymological alteration, whether from contact with the Negro tribes, or from inherent causes. As far as the available materials enable us to form a judgment, it seems clear that the best living representative of the Bereber language is that spoken by the Touarecks of the Great Desert, and especially by the great tribes, the Azguer and Ahaggar, who occupy between them a territory measuring at least half a million of square miles. Of this, which is properly called Tamashek', a grammar was published by General Hanoteau in 1860, and another by Mr. Stanhope Freeman in 1862. The Tamashek' is distinguished from the other languages of the same family by the greater regularity and completeness of its grammatical system, by the comparative absence of Arab words, of which the Kabyle shows a large infusion ; but especially by the possession of a system of writing, rude, indeed, and imperfect, but not known to any other branch of the Bereber stock. This privilege has not led to the growth of a national literature ; the written characters are used only for rock inscriptions, for mottoes on shields, and occasionally for verses on festive occasions ; but their use is widely spread among men of the higher class, and still more among the women, and, however restricted, has doubtless tended to give comparative fixity to the language.

Of the Shelluh tongue the materials available are, indeed, very scanty. The most considerable document is contained in the ninth volume of the ' Transactions of the Royal Asiatic Society,' where Mr. Francis Newman has given a literal Latin version of a story written in Arabic characters by a native of South Marocco. It would require far more knowledge of the Shelluh language and familiarity with Arabic writing than I possess to enter on any examination of that document ; and there is the

---

[1] *Le Zénaga des Tribus Sénégalaises.* Paris, 1877.

further difficulty that the natives who learn to write their own
language in Arabic characters are usually those who also acquire
the Arabic language, and in so doing learn to adopt Arabic
phrases and forms of speech. In the following table I have introduced all the Shelluh words given by Jackson and Washington, of which I have been able to find equivalents in Kabyle or
Tamashek', and have endeavoured to adopt a uniform mode
of orthography. The vowels are intended to have the sounds
to which they correspond in most European languages, and not
those peculiar to England. *Th* and *sh* have nearly the same
sounds as in English; *gh* before *e* or *i* has the hard sound; and
r' indicates the peculiar sound intermediate between the
guttural and the ordinary *r*, which European travellers indicate
sometimes by *r*, and sometimes by *gh*. In several instances
synonyms are given in brackets.

English	Shelluh	Kabyle	Tamashek'
Man . .	argaz . .	ergaz .	ales
Woman .	{ tamraut } { tamtout }	{ themthout } { themgart } .	{ tameth { tamethout
Boy . .	ayel . .	ashish . .	{ abaradh { amaradh
Girl . .	tayelt . .	tehayalt . .	tamarat
Slave . .	issemgh .	ismigh . .	akli
Horse . .	ayiss . .	{ eïss { aghmar } .	ayiss
Camel . .	{ aroum } { algrom } .	{ aram } { elgroum } .	{ amnis { amagour
Sheep . .	{ izimer } { djellib } .	{ thiksi } { thili } .	{ izimer { ekraz
Mule . .	tasardount .	aserdoun .	—
Boar . .	amouran .	mourran . .	azibara
Cow . .	tafounest .	tefonest . .	tes
Green lizard .	tasamoumiat .	tesermoumit .	—
Water . .	amen . .	{ eman } { aman } .	aman
Bread . .	{ tagora } { aghroum } .	aghroum .	tagella
Milk . .	akfaï .	{ aifki } { aghfaï } .	akh
Meat . .	ouksoum .	aksoum .	—
Eggs . .	tikellin .	{ tighliim } { thimillim } .	—
Barley .	toumzïn . .	toumsin . .	timzin
Dates . .	{ tena } { tinie } .	tini . . .	teini
Green figs .	akermous .	tibaksisin .	—
Honey . .	tamint .	thament . .	—

English	Shelluh	Kabyle	Tamashek'
Sun . .	atfoukt . .	tefoukt . .	tafoukt
Mountain .	{ adrar (*plur.*) idrarn)	{ edrar (*plur.*) ouderan)	{ adrar (*plur.* idrarn)
Palm tree .	taghinast .	jat faroukt .	—
Year . .	aksougaz .	ezoughaz .	aouétai
Morning .	zir , . .	ighilwas . .	ifaout
To-morrow .	azgah . .	ezikka . .	toufat
Village . .	thedderth .	tedert . .	—
House . .	{ tikimie tigameen }	{ tighimi akham }	—
Wood . .	asr'oer . .	esghar . .	asr'er
Dinner . .	imkelli . .	elles . .	amekchi
Head . .	akfie . .	{ ikf akfai } .	ir'ef
Eyes . .	alen . .	ellin . .	{ tiththaouin (*sing.* tith)
Nose . .	tinzah . .	inzer . .	—
Feet . .	idarn . .	{ idaren (*sing.*) adar . }	—
Go (*imper.*) .	aftou . .	eddou . .	eg'al
Come . .	ashi . .	{ as eshkad }	{ as (come, or go)
Give . .	fikihie . .	efki . .	ekf
Eat . .	aïnish . .	itch . .	eksh
Call . .	irkerah . .	kera . .	—
Sit down .	gaouze . .	{ ghaouer aguim } .	{ r'im ekk'im
Good . .	egan ras . .	delâli . .	elkir r'as

It will be seen that, as regards thirty out of thirty-five
Shelluh substantives here enumerated, the Kabyle equivalents
are distinguished only by dialectic differences, and the same holds
as to at least four out of six verbs. It thus appears, as far as the
evidence goes, that there is as much verbal resemblance between
these tongues as between Italian and Spanish, or other allied
languages belonging to the same stock. The comparison with
the Tamashek' shows a less close etymological relationship. Out
of twenty-four substantives for which Tamashek' equivalents
have been found, twelve only, and two only out of five verbs,
show identity of origin. But it is interesting to find indications that the Shelluh retains a closer conformity to the rules
of Tamashek' grammar than does the better known Kabyle
language. In the very few cases where a comparison is possible
we find, indeed, absolute identity. Thus the Shelluh word for
boy (*ayel*), is apparently not found either in Kabyle or Tamashek;' but the feminine form (*tayelt*), for girl, precisely follows

the rule of Tamashek' inflexion for gender, and a slight modifi-
cation of this (*tehayalt*) is found in the Kabyle. A somewhat
similar example is the word *tasardount* for mule, this being the
regular feminine form of the Kabyle name, *aserdoun*. The
word *adrar* (mountain) forms its plural *idrarn* exactly accord
ing to rule, and both singular and plural are identical with the
Tamashek' forms; while the Kabyle shows dialectic differences,
especially in the plural where the final *r* of the singular is lost.
The last word in the list affords an illustration of the liability
to error incurred by a traveller attempting to form a vocabulary
of a language with which he has but a slight acquaintance.
Good is here used in the sense of a satisfactory answer to
inquiries, pretty much as *all right* is adopted in colloquial
English. Jackson was doubtless familiar with the expression
egan ras, which he gives as the Shelluh equivalent, and which
we also often heard from the natives; but the *ras* of the Shelluh
is obviously the same as the Tamashek' adverb *r'as*, meaning
only, or *exclusively*, which invariably follows the word *elkir* in
the corresponding Tamashek' reply, *elkir r'as*.

It has not appeared necessary to add to the table given above
a column for the corresponding words in the Zénaga language
from the vocabulary given by General Faidherbe. The amount
of verbal similarity between this and the Shelluh is very
trifling, and the distinguished author referred to was doubtless
misinformed when led to express a belief in their close connection.

The time is perhaps not yet come for forming a definitive
judgment as to the origin of the Bereber languages, and the
precise nature of the relations between them and the ancient
language of Egypt on the one hand, and those of the Semitic
family on the other. The present writer feels his own incom-
petence to grapple with questions of such difficulty, and will
merely refer the reader to the conclusions recently announced
by M. de Rochemonteix as those which appear to carry with
them the greatest weight.

In his essay, published in 1876,[1] the learned writer finds
that the ancient Egyptian and the Bereber possessed the same
pronominal roots, and employed the same methods for forming

---

[1] *Essai sur les rapports grammaticaux entre l'Egyptien et le Berbère*,
par le Marquis de Rochemonteix. Paris, 1876.

their inflexions and derivatives; and he arrives at the same
opinion with reference to the inflexions of the substantives. He
further asserts that the modifications which time and external
conditions have effected are of a superficial character, and in no
way conceal the close grammatical affinity of these languages.
Whether this affinity be due to direct inheritance, or to common
descent from a more remote ancestral stock, is a question not
touched by the writer, who bases his conclusions on a study of
two only of the Bereber dialects, the Kabyle and the Tamashek.'[1]

With reference to the relation indicated by the conjugation
of the Bereber verb, in which the grammatical processes show a
considerable affinity with those of the Semitic languages, while
the comparison of the verbal elements shows no token of common origin, M. de Rochemonteix expresses the opinion that at

[1] It is of some interest to remark that the latest conclusions of
philologists on the affinity of the North African dialects, substantially
agree with the testimony of the earliest writer who came in contact
with them. The following passage is taken from the original version
of the description of Africa by Leo Africanus, published by Ramusio
in his famous work ' Delle Navigationi et Viaggi :' Venetia, 1563, vol. i.
p. 2 f. The Moorish writer divides the indigenous white population of
Northern Africa into five races, enumerated by him, and then con-
tinues: ' Tutti i cinque popoli—i quali sono divisi in centinaja di
legnaggi, et in migliaja di migliaja d' habitationi, insieme si con-
formano in una lingua la quale comunemento è da loro detta Aquel
Amarig, che vuol dir lingua nobile. Et gli Arabi di Africa la
chiamano lingua barbaresca, che è la lingua africana nathia. Et
questa lingua è diversa et differente dalle altre lingue : tuttavia in
essa pur trovano alcuni vocaboli della lingua araba, di maniera che
alcuni gli tengono et usangli per testimonianza, che gli Africani siano
discesi dall' origine d' i Sabei, popolo, come s' è detto nell' Arabia
felice. Ma la parte contraria afferma, che quelle voci arabe che si
trovano nella detta lingua, furono recate in lei dapoi che gli Arabi
entrarono nell' Africa, et la possederono. Ma questi popoli furono di
grosso inteletto et ignoranti, intanto che niun libro lasciarono, che si
possa addurre in favore nè dell' una nè dell' altra parte. Hanno ancora
qualche differenza tra loro non solo nella pronontia, ma etiandio nella
signification di molti et molti vocaboli. Et quelli che sono più vicini
a gli Arabi, et più usano la domestichezza loro, più similmente tengono
de loro vocaboli arabi nella lingua. Et quasi tutto il popolo di
Gumera ' (the Rif Country) ' usa la favella araba, ma corrotta. Et
molti della stirpe della gente di Haoara parlano pure arabo, et tuttavia
corrotta. Et ciò aviene per haver lunghi tempi havuta conversazione con
gli Arabi.'

an early period of their development, the Bereber people must
have been brought into contact with the Semitic stock, and may
well have been struck by the advantage of precision obtained by
systematic conjugation of the verb, and thus gradually moulded
their own rude tongue on the model supplied to them.

APPENDIX L.

*Notes on the Roman Remains known to the Moors as the Castle
of Pharaoh, near Mouley Edris el Kebir.*

Communicated by Messrs. W. H. RICHARDSON and
H. B. BRADY, F.R.S.

LEARNING that a party of English travellers had visited these
ruins in the spring of 1878, and believing that they had not
been seen by any European traveller since Jackson visited the
place early in the present century, we were anxious for information respecting them ; and in reply to our request we
received an account of their visit kindly drawn up by Messrs.
W. H. Richardson and H. B. Brady, F.R.S. We have also
been favoured with the loan of a sketch executed by Mr. G. T.
Biddulph, who formed one of the same party, from which the
vignette given p. 487 is taken.

After the notes were in the hands of the printer the appearance in the 'Academy,' No. 32, p. 581, of a very full account
of the ruins by Dr. Leared, already well known as a successful
Marocco traveller, informed us that the ruins had been visited
by him in 1877, in company with the members of the Portuguese mission to the Sultan, and about the same time by some
members of the German Diplomatic Mission. Dr. Leared has
fully succeeded in establishing the identity of the so-called
Castle, or Palace, of Pharaoh with the Roman town of Volubilis,
and has left little to be said on that point. Nevertheless the
ruins are interesting enough to make the additional notes of
other travellers useful and valuable ; and we have therefore
availed ourselves of the greater part of the paper kindly sent
to us by Messrs. Richardson and Brady

'One of the points we had determined to visit on our tour was the ruin known by the Moors as "Pharaoh's Palace," or "Pharaoh's Tomb." The time of our journey was in some respects unfortunate for visiting places held in veneration by the natives; we were, in fact, staying in Fez at the time of Mohammed's birthday, when religious fanaticism exhibits itself, not merely in holidays and powder-play, alternating with devotional exercises, but in processions to the shrines of saints, and in sundry manifestations of ill-will to unbelievers. We had considerable difficulty in obtaining intelligible information as to the exact site of the ruins. Our idea had been that they ought to have been accessible from the road between Alcazar and Fez, striking off near Sidi Guiddar. The interpreter and the mounted soldiers who were with us, overruled this when it was proposed, and we therefore continued our journey. They were probably right; but in our various conversations with them on the subject they managed to convey the impression that either they did not themselves know the precise locality, or that they did not intend that we should visit the place.

'During our stay in Fez we were joined by two Englishmen, Messrs. G. T. Biddulph, and F. A. O'Brien, whose acquaintance we had made in Tangier, and we proceeded to Mekinez in company. Mekinez is a sort of Mecca to the Aissowies—the most fanatical of all the sects of western Mohammedans—and the road was thronged with devotees returning from their annual pilgrimage to the city of Mohammed-ben-Aissa, their prophet. We were kindly received by the Lieutenant Bashaw (Kaïd Hamo), who seemed desirous to forward our views in every way in his power. He thought it necessary on our departure to provide us with a soldier who knew the district thoroughly, so that altogether we had a guard of four regular soldiers. Thus furnished, the tone of our interpreter changed, and we had no more obstacles thrown in our way.

'We proposed to make the Roman station the first stage on our road from Mekinez to Rabat. Whether it would have been better to have taken it, as we had originally intended, on the way to Fez, or subsequently, between Fez and Mequinez, it is needless now to inquire; it certainly is a good deal out of the direct route between Mekinez and Rabat, if maps are to be trusted. However, we got on the way on March 23 a

little after 10 A.M. The site of the ruin is some fifteen miles
north-west of Mekinez, at no great distance from Mouley Edris
el Kebir; both are on the southern slopes of one of the ranges
that constitute the Lesser Atlas. There was little of interest
by the way. Part of the road was on the horizon of a bed of
white, friable, microzoic, tertiary limestone, which forms a conspicuous feature in the mountain strata of this district. This
is traceable for a great distance, and its exposure at one or two
points in the heights to the far east, we had at first mistaken

Roman ruins of Volubilis.

for snow. After about four hours' riding we had to diverge
from the main road; and here we learned that it would be
necessary to encamp at some distance to the south-west, in the
last *douar* within the government of Mekinez. The hill-
country, it was said, was so infested by a lawless set of Berebers that we should not be safe out of the jurisdiction of the
Bashaw. Before the evening was out we had reason to know
that these fears were not entirely groundless. Leaving our servants with the luggage, therefore, we took two soldiers and rode
across country to the object of our journey.

'The ruins stand on a little hill, a mile or more from the road. At the base of the hill runs a bright little mountain stream. The ground for many acres is strewed and heaped with squared stones, the débris of ancient buildings; lines of wall-foundations appear in every direction, and pillar bases in rows or squares, arranged as though for the support of colonnades surrounding courts or *patios*. The demolition is no doubt largely due to spoliation, but it is also partly the result of the unstable character of the mortar, which has to a great extent weathered out from between the stones. In some of the walls still standing the stones appear to retain their places by their own weight rather than by the help of any cement that is left to hold them together. Two perfect Roman arches still remain, and one or two nearly complete, but even these look as though they might not long withstand the mountain winds.'[1]

'In the present condition of the place it is impossible even to guess what was the original ground-plan of the buildings. The principal frontage appears to have had the west aspect, and there is still the remnant of a sort of façade. Amongst the fallen stones of this front is part of an entablature which has borne an inscription in four lines. We could only find one stone of it; and this bore the following letters, about eight inches in height. There were the shafts of many marble columns amongst the fallen stones, and not a few capitals, some simple, others more or less carved with volutes, Ionic fashion, and one at least with the remnant of acanthus leaves, as though derived from a Corinthian building. Some of the mouldings had the common egg and arrow ornament, and there was a portion of a narrow frieze on the western side with one of the common frets of classic architecture formed of a double series of interlacing

> A X C
> P I A E
> B S I
> T I

curved lines; but beyond these there was but little decorative sculpture. There is clearly a basement storey of very large

[1] The vignette is taken from a sketch by G. T. Biddulph, Esq.

stones lying underneath the present ground level. Here and there the subsidence of the ground, or the falling in of the masonry, reveals passages and what appear to be small rooms or vaults, with solid, well constructed walls. In the short hour that circumstances permitted us to linger, it was impossible to do more than observe things as they stood. The mere removal of the loose stones would do a good deal, and a very little excavation would do much more, to indicate the history of the original structures, and we have little doubt that many inscribed stones might still be found that would help materially to the same end. On the north side of the western arch and façade is a sort of enclosure formed of loose stones piled together as a rude wall, and whitewashed. This wall has been reared by pilgrims, each of whom has carried and placed a stone, according to their custom, at what they regard as a " saint's place." '

Up to the year 1877 no traveller appears to have visited the ruins since Jackson, who twice refers to them in his 'Account of the Empire of Marocco.' In a note to p. 21 (3rd Edition, 1814) he says : 'The father of the Sultan Sulieman built a magnificent palace on the banks of the river of Tafilelt, which bounds his dominions on the eastward ; the pillars are of marble, and many of them were transported across the Atlas, having been collected from the (Ukser Farawan) Ruins of Pharaoh near the sanctuary of Muly Dris Zerone, west of Atlas.' In another place (p. 146) he says : ' When I visited these ruins in my journey from the sanctuary of Muly Dris Zerone, near to which they are situated, in the plain below, the jealousy of the (Stata) protecting guide sent by the Fakeers to see me safe to the confines of their district was excited, and he endeavoured to deter me from making any observations by insinuating that the place was the haunt of large and venomous serpents, scorpions, &c. A good number of cauldrons and kettles filled with gold and silver coins have been excavated from these ruins.'

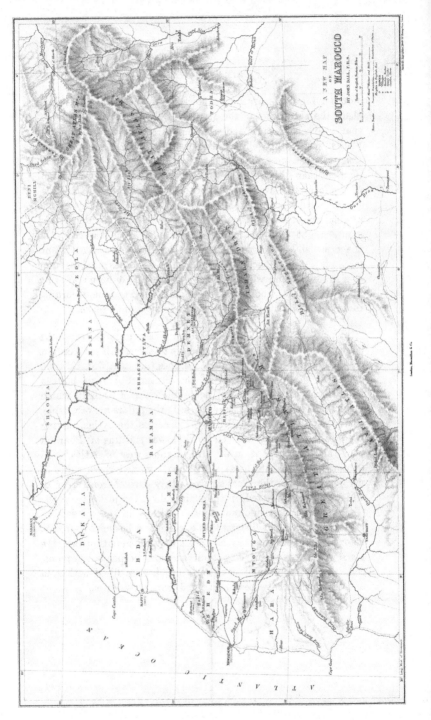

The material originally positioned here is too large for reproduction in this reissue. A PDF can be downloaded from the web address given on page iv of this book, by clicking on 'Resources Available'.

INDEX.

LONDON : PRINTED BY
SPOTTISWOODE AND CO., NEW-STREET SQUARE
AND PARLIAMENT STREET

Printed in the United States
By Bookmasters